ICE manual of
health and safety
in construction

Institution of Civil Engineers

ICE manual of health and safety in construction

Edited by

Ciaran McAleenan
Expert Ease International, Northern Ireland
David Oloke
University of Wolverhampton, UK

ice | **manuals**

Published by Thomas Telford Limited, 40 Marsh Wall, London E14 9TP, UK.
www.thomastelford.com

Distributors for Thomas Telford books are
USA: Publishers, Storage and Shipping Corp., 46 Development Road, Fitchburg, MA 01420
Australia: DA Books and Journals, 648 Whitehorse Road, Mitcham 3132, Victoria

First published 2010

ISBN: 978-0-7277-4056-4

Future titles in the ICE Manuals series from Thomas Telford Limited

ICE manual of geotechnical engineering
ICE manual of construction law
ICE manual of highway design and management

Currently available in the ICE Manual series from Thomas Telford Limited

ICE manual of bridge engineering – second edition. 978-0-7277-3452-5
ICE manual of construction materials – two volume set. 978-0-7277-3597-3

www.icemanuals.com

A catalogue record for this book is available from the British Library

This book is published on the understanding that the authors are solely responsible for the
statements made and opinions expressed in it and that its publication does not necessarily imply
that such statements and/or opinions are or reflect the views or opinions of the publishers. While
every effort has been made to ensure that the statements made and the opinions expressed in
this publication provide a safe and accurate guide, no liability or responsibility can be accepted in
this respect by the authors or publishers.

The authors and the publisher have made every reasonable effort to locate, contact and
acknowledge copyright owners. The publisher wishes to be informed by copyright owners who
are not properly identified and acknowledged in this publication so that we may make necessary
corrections.

Typeset by Newgen Imaging Sytems Pvt. Ltd., Chennai, India
Index created by Indexing Specialists (UK) Ltd, Hove, East Sussex
Printed and bound in Great Britain by Latimer Trend & Company Ltd, Plymouth

Contents

Contents

Foreword

We readily admire the achievements of the great Victorian civil engineers: their vision, their innovation, their flair, and the way they shaped our society and the world. But these achievements often had a terrible human cost, with many killed and injured during construction.

Thankfully, the last four decades have seen a quantum leap in the way that we care for the health, safety and welfare of our workforce, and anyone who may be affected by our works. This is partly because of legislation – the Health and Safety at Work Act, the Construction (Design and Management) Regulations, the Corporate Manslaughter Act and so on.

But should we take care of health, safety and welfare simply because legislation forces us to? Or should we do so because it's the right thing to do in the first place?

As professionals, we have our core values and our code of conduct. Accordingly, the Institution of Civil Engineers (ICE) requires all its members to have a sound knowledge of health and safety, and a high regard for the consequences of professional activities on the safety of workers and others. This means that we must: work with integrity; tackle only the work we are competent to do; act always in the public interest; and develop our knowledge. If we follow these principles we will ensure that health, safety and welfare will be at the heart of everything we do.

And while there have been many health and safety text books published over recent years there has never been such a comprehensive book written by, and for, practicing construction professionals. Much care has gone into the selection of topic experts. So I commend the ICE for producing this practical handbook. It is not an academic textbook. Rather, it demonstrates how working safely can become second nature, improve the way we do things, and provide the key to successful construction.

David Orr CBE FREng CEnv FICE FIAE FIHT
ICE Past President 2007–2008

Preface

'Construction Sites can be dangerous'

That sign was displayed recently outside a construction site. The interesting point to note is that it suggests construction can be dangerous. Not that construction is dangerous. It is true that there are many hazards associated with working in the construction industry but there is no reason why the existence of any of these hazards should cause workers to be harmed.

In the field of construction design and management you must, with good designs and good operational safety controls, endeavour to ensure that construction sites are safe and healthy places to work. As professionals (mostly concerned with design and/or project management) we all need to remember that:

- A hazard does not equal danger.
- It is hazards without appropriate control that equal danger.

Always remember there can be no good reason for accepting fatalities in our industry. We have procedures, guidance and good, committed professional people. But clearly there is still a way to go and the fact that you are reading this manual shows that you agree and are ready to do something about it.

Towards the end of 2008 the Secretary of State at the Department for Work and Pensions in Great Britain announced an inquiry, chaired by Rita Donaghy, into the underlying causes of construction fatalities, stating;

> The construction industry is one of the most dangerous sectors in the country – over 2800 people have died from injuries they received as a result of construction work in the past 25 years.

The subsequent 'Donaghy Report' commented that it is important that we 'ensure that sufficient consideration is given to [design, leadership and planning] in our university courses'. Donaghy went on to highlight comments from Egan, that are still pertinent today, regarding the call for designers to match their high standards of professional competence 'by a more practical understanding of the needs of clients and of the industry more generally'. Egan, it is reported, also stated that 'They [designers] need to develop greater understanding of how they can contribute value in the project process and the supply chain'.

The professional has a vital role in promoting health, safety and welfare in construction projects, and educators and mentors (including supervising civil engineers) have a role to play in professional development. Accordingly, as design professionals (and key Duty Holders under the Construction (Design and Management) (CDM) Regulations) our approach should reflect that we value workers and accept that protecting all construction personnel and those who use and maintain our creations is the right reason for safety, health and welfare. Ultimately, we must be assured that what we design can be built, used, maintained and eventually demolished safely. Our increasing involvement as CDM Coordinators and Principal Contractors should also amplify these values.

Through a robust construction design and management planning process, professional engineers need to demonstrate commitment to examining the means to design out as many of the hazards as possible and to consider the impact of the hazards that cannot be removed at the design stage. This involves considering:

- the buildability of your projects
- the time and the resources needed to achieve success
- how contractors will safely realise your designs
- the needs of maintenance workers, and
- the needs of end users.

Clients, contractors, designers and workers all have a professional responsibility to work towards eliminating accidents and ill-health. That is our challenge and our opportunity, and through the pages of this book eminent professionals share their thoughts on how this can be achieved.

This manual is aimed at providing you with practical guidance on understanding and achieving the various requirements of sound construction health and safety practices. Clearly it is not to be treated as a treasured item, to be carefully locked away for future generations to discover. This is a clear and concise working document filled with practical guidance for you to act upon. However, it is appreciated that knowledge and experience evolve continuously and if, in the process of making use of the principles within the manual, you find better or different ways to improve standards then these should also inform your courses of action.

It is not the intention of this manual to repeat advice that is already available in abundance. Rather the authors have appropriately referenced their sources and provided suggestions for further reading.

Part 1 addresses legal principles and introduces the reader to the relationship between construction processes and health hazards.

Part 2 deals with workforce issues and introduces the reader to key duty holders as well as the various conventions and interventions currently or recently initiated within the industry.

Part 3 addresses management of the construction process and includes:

- an examination of the different phases of construction
- establishing control processes
- management and planning
- designers' responsibilities for assessing both health and safety issues, and
- procurement.

Part 4 covers health hazards, including:

- chemical hazards (such as asbestos, respirable crystalline silica dust, lead and solvents) and how to control exposure to them
- biological hazards (such as leptospirosis, legionella, tetanus and hepatitis) and how to control exposure to them
- physical hazards (such as compressed air, noise, vibration, radiation and temperature extremes) and how to control exposure to them.

Part 5 addresses safety issues in construction and covers topics such as:

- working at height, in excavations, in confined spaces and near water
- demolition
- falsework design, and
- transportation, vehicle movements, work and lifting equipment.

Competent and professional engineers will have a sound knowledge of scientific, engineering and technical principles, experience of construction processes and knowledge that extends to future use, maintenance and demolition. Readers with the appropriate position and influence need to ensure that their actions reflect a commitment to continuously improving the safety and health of construction workers.

So while reading through the manual, it is suggested that you consider how your contribution to the construction industry can make a positive impact and feel free to share those ideas with colleagues and with the community of experts within the Institution of Civil Engineers. Remember that good ideas can spread rapidly if there are the resources and the commitment to disseminate them. Donaghy identified the need for resources to bring together examples of good and successful practices, and for international comparisons on the efficacy of these practices on the accident fatality rates. Competence and professionalism are key. A competent organisation or a competent individual has all of the necessary skills, resources and authority to make the right decisions and to spread the ideas widely. Accordingly, health and safety needs to be one of the professional engineer's core competencies, not a mark of excellence.

Enjoy the book.

Ciaran McAleenan
Expert Ease International, Lurgan, Northern Ireland, UK

David A. O. Oloke
University of Wolverhampton, UK, and
Progressive Concept Consultancy (pCC) Ltd, Walsall, UK

List of contributors

GENERAL EDITORS

Ciaran McAleenan, MPhil CEng MICE
Expert Ease International, Lurgan, Northern Ireland, UK
ICE-Construction Health Safety and Welfare Registered
Chartered Civil Engineer and Chartered Health and Safety
    Practitioner

Ciaran is a civil engineer with over 30 years experience in the industry. Ciaran designed and managed a number of civil engineering projects in the water industry between 1978 and 1990 and in the period up to 2009 he worked on strategy, policy and standards development and project management within the road construction industry. He holds a Masters degree in civil engineering and is qualified as a trainer and a safety practitioner. A Chartered Civil Engineer; Ciaran holds corporate member-ship of the Institution of Civil Engineers (ICE) and the Institution of Occupational Safety and Health (IOSH). He is also a Professional Member of the American Society of Safety Engineers (ASSE). He is on the ICE's Expert Panel for health and safety, a Member of the Northern Ireland (NI) Committee of ICE, and Chair of the ICE NI Region's Expert health and safety panel.

Ciaran has received the Zurich Municipal/IOSH Supreme award for health and safety for his work as the architect of quality based safety management and in 2001 he received the National Irish Safety Organisation's Occupational Safety Award (N. Ireland Region) for consistent achievement in safety. In 2006 his own company, Expert Ease International, collected the Imhotep Good Innovative Prevention Practices prize from the International Social Security Association, for developing the operational analysis and control (OAC) model for managing safety. In 2009 he was part of the Road Workers Safety Forum (RoWSAF) team, who collected the Prince Michael of Kent's Safer Roads Award.

A Health and Safety Executive Systems Analyst and a developer of corporate systems and strategies, Ciaran has delivered lectures in North and South America, Asia, Australia and Europe. He is currently a lecturer in the School of the Built Environment at the University of Ulster and Engineering Partner in his family-run health and safety consultancy business.

David A. O. Oloke, BEng MSc PhD CEng MICE MCIOB MNSE
Construction and Infrastructure Department, SEBE, University of
    Wolverhampton, UK
Progressive Concept Consultancy (pCC) Ltd, Walsall, UK
ICE-Construction Health Safety and Welfare Registered
Chartered Engineer, Research and Development Consultant

David Oloke has vast international experience in engineering consultancy, training, research and development. He has been responsible for the planning, design and project management of a wide range of civil and structural engineering schemes. He has also managed several property damage reinstatement projects in the UK, most of which were as a result of disasters due to flooding, fire, earthquake and general impact damage. The various projects he has been involved in have totalled over £150m in value to date and have spanned across various countries including the UK, Nigeria, Canada, China, the Netherlands, France and Greece amongst others – where he has either worked or presented papers at seminars, workshops and conferences.

As a research consultant and trainer, David has been responsible for the supervision of doctorate degrees relating to studies on Construction Health and Safety management. Furthermore, he is extensively involved with the provision of Construction Health and Safety and Project Management Training to construction project managers and Construction Design and Management (CDM) Duty Holders for the University of Wolverhampton and the West Midlands Centre for Constructing Excellence. David provides consultancy support to UK Local Authorities in the West Midlands and SMEs. He is also a Principal Consultant to Progressive Concept Consultancy Ltd. David is a Registrant of the Institution of Civil Engineers (ICE) Health Safety and Welfare Register and has provided CDM coordinator and designer duties on several schemes. He has also received several international awards and certificates for his contributions as a Resource Person to the Council for the Regulation of Engineering in Nigeria (COREN), the Nigerian Society of Engineers (NSE) and the Environment, Health and Safety Committee of the London Olympic Delivery Authority (ODA). He has been a member of the ICE Health and Safety Expert Panel.

CONTRIBUTORS

Michael Battman, Gardiner & Theobald LLP, Manchester, UK

Mike has over 30-years experience in the construction industry, primarily on the contracting side. He has been involved in diverse schemes varying from the M6 Toll Road to minor utility repairs.

Mike is a Fellow of the Institution of Civil Engineers and a Chartered Member of the Institution of Safety and Health; he is currently Chair of the ICE's Health and Safety Expert Panel and was one of the founder members of the ICE's Health and Safety Register.

Mike is an Associate Director for Gardiner and Theobald, working primarily as a CDM-coordinator for a variety of clients.

David R. Bramall, Department for Regional Development Tandragee, Northern Ireland, UK

David Bramall CEng MICE MIEI MCIHT MIAT is a Chartered Engineer working for the main road authority in Northern Ireland with a particular interest in design and project management, temporary traffic management, pavement engineering and construction health and safety. He has been particularly involved in producing various standards and guidance relating to pavement engineering, health and safety and temporary traffic management. David delivers talks and training to the construction industry within Northern Ireland on these topics and works closely with the ICE Northern Ireland Region Health and Safety Expert Panel, where he is presently chairing a sub-group examining health and safety competence standards for practicing professionals.

John Carpenter, Consultant, Manchester, UK

John Carpenter CEng FICE FIStructE CFIOSH is an independent consultant and Secretary to the Standing Committee on Structural Safety (SCOSS), specialising in safety risk management. He is author of the Construction Skills CDM Guide for Designers; Health and Safety Executive research reports related to education, health, competence and building safety and coauthor of the CIRIA guide *Safe Access for Maintenance and Repair*. John is also an established training provider and author of numerous papers on the management of safety risk.

Akinwale Coker, Department of Civil Engineering, Faculty of Technology, University of Ibadan, Nigeria

Akinwale Coker PhD is a Senior Lecturer in Water Resources and Environmental Health Engineering at the Department of Civil Engineering, Faculty of Technology, University of Ibadan. For about fifteen years in the University of Ibadan, his research has focused on water and waste management engineering. His doctoral research, completed in 2002, was on the engineering applications in the management of medical waste in Nigeria. He has over 30 scientific publications and one patent to his name, and is currently the Acting Head of the Department of Civil Engineering at the University of Ibadan, Nigeria.

Peter Fewings, School of Built and Natural Environment, University of the West of England, Bristol, UK

Peter Fewings is a Chartered Builder and UWE Teaching and Learning Fellow at the University of the West of England in Bristol. He is an established author and has three previous publications which include health and safety legislation and practice. He is a specialist in construction project management

and leads a programme in that subject together with teaching and research into management subjects including health and safety. Prior to teaching Peter practiced as a construction project manager with large and medium sized contractors in the UK for 13 years. He has an ongoing interest in health and safety and supervises student research into the area.

Alistair G.F. Gibb, Loughborough University, UK

Alistair Gibb PhD BSc CEng MICE MCIOB is a Chartered Engineer, Chartered Builder and Professor of construction engineering management, joining Loughborough University in 1993, following a career in civil engineering and construction management. He is Project Director of the European Construction Institute's Safety, Health and Environment task force (ECI) and, internationally, is coordinator of the Conseil Internationale de Batiment (cib) working commission on construction health and safety. Alistair has led many health and safety research projects funded both by the UK Government and industry.

Theo C Haupt, Building Construction Science, Mississippi State University, USA

Professor Theo Haupt is the Director of Building Construction Science at Mississippi State University and has been extensively involved in construction health and safety research for several years with his PhD, gained in the United States, focusing on alternative approaches to the management of health and safety on construction sites. He has published extensively and delivered several papers on the subject in national and international journals, books and conferences. Theo has served in the capacity of convenor/organiser of, and scientific and technical committee member for, many conferences, symposia, and workshops. He regularly reviews papers on health and safety for a range of internationally acclaimed journals, serves on the editorial boards of several of these, and has been the invited guest editor for the special health and safety issues of the *Journal of Construction Research* and *Construction Industry Quarterly*, the scholarly journal of the Chartered Institute of Building. Theo is the Editor-in-Chief of the *Journal of Engineering, Design and Technology* and coeditor of the *Journal of Construction*. He is a National Research Foundation rated researcher specialising in construction-related issues, such as health and safety, and serves on the Advisory Council on Occupational Health and Safety of the Department of Labour in South Africa.

Ken Logan, Health and Safety Executive Northern Ireland (HSENI), Belfast, Northern Ireland, UK

Ken Logan BSc Dip H&S CEng MICE MIQ MIExpE is a Chartered Civil Engineer and worked for the Department of Environment for Northern Ireland Water Service for ten years. During his time with the Water Service he worked in the Central Design Unit, Operation and Maintenance and Direct Labour Organisation.

Ken joined the Health and Safety Executive for Northern Ireland in 1985 and is currently the Principal Inspector responsible for the Construction Industry. Prior to this for several years he was responsible for the mines and quarries sector and the use of explosives. Ken is appointed as an Explosives Inspector.

Philip McAleenan, Expert Ease International, Downpatrick, Northern Ireland, UK

Philip McAleenan MSc FInstLM, Professional Member ASSE, is Managing Partner of Expert Ease International and has organised and led seminars on many aspects of occupational health and safety, including safety law, principles and practices. Organiser of a symposium at the World Congress on Safety and Health at Work in Seoul, he has presented at two congresses and regularly to international professional and academic conferences. Philip has authored some 40 occupational health and safety handbooks and manuals and is codeveloper of the Organisation Cultural Maturity Index.

Ann Metherall, Burges Salmon LLP, Bristol, UK

Prior to commencing her legal career in 1998, Ann Metherall BSc CEng MICE was a British Railways civil engineer for 11 years working as a designer, resident engineer and permanent way maintenance engineer.

Ann is a Partner at Burges Salmon and part of their regulatory team, advising on contractual disputes, regulatory issues and health and safety. In particular, Ann has advised clients during accident investigations by the Health and Safety Executive and Office of Rail Regulation, at Coroner's inquests, and has defended health and safety prosecutions.

Her experience with British Railways is the inspiration for her specialism as she has been responsible for managing both worker and public safety in a relatively high risk, and certainly high profile, industry.

Brian Neale, Consultant, Cheshire, UK

Brian Neale CEng FICE FIStructE Hon IDE is an independent consultant and Secretary of the UK based Hazards Forum. He formerly worked for the Health and Safety Executive and other organisations. He chaired the drafting of BS6187:2000 *Code of Practice for Demolition* and in early 2010 he commenced chairing the newly formed committee for drafting the revision of the standard, which is due to be expanded further into refurbishment and partial demolition. As a CEN convenor, Brian chaired the drafting of one of the Structural Eurocodes which has particular relevance to the topic of demolition. He was editor of the 2009 Thomas Telford Limited book, *Forensic Engineering: from failure to understanding*, and chaired the Organising Committees of all four International Conferences on Forensic Engineering organised by the Institution of Civil Engineers.

Brian's published papers include an international dimension and his consultancy includes a training element. Other activities include being a member of the Council of Management of the Institute of Demolition Engineers.

David Porter, Rivers Agency, Belfast, Northern Ireland, UK

David Porter is currently the Director of Development with the Rivers Agency, the drainage and flood defence authority for Northern Ireland. He is responsible for the implementation of the EU Floods Directive including the mapping and hydraulic modelling requirements.

David was previously the Regional Engineer with the Rivers Agency and during this time he gained extensive practical knowledge of watercourse maintenance and the safe management of construction projects in the vicinity of rivers and the sea. He has also worked for Roads Service including three years in the bridges and structures section.

David is a Chartered Civil Engineer and an active member of both the Institution of Civil Engineers and the Chartered Institution of Highways and Transportation.

David W. Price, GexCon UK Ltd, Skelmersdale, UK

Dave Price PGDip, Fire and Explosion Engineering, is Managing Director of GexCon UK Ltd and has 18 years of experience in the field of gas and dust explosions. Until December 2007, he was an international industrial explosion protection and mitigation specialist working for a major blue chip global fire protection company. Since December 2007 Dave has headed the UK subsidiary of a leading Norwegian firm of global explosion consultants, providing accident investigations, expert witness services, probabilistic hazardous area and risk assessments, training and contract research, and dispersion/blast modeling using computational fluid dynamics for onshore and offshore facilities.

Currently a Fellow of the Institute of Fire Safety Managers (IFSM) and former President of the IFSM 2008–2010, Dave is a recognized expert and very much enjoys prescribing the explosion safety for process facilities within the biomass industry.

Delia Skan, Health and Safety Executive Northern Ireland (HSENI), Belfast, Northern Ireland, UK

Dr Delia Skan MB BCh BaO FRCP FFOM FFOM(I) MRCGP DCH is the Senior Employment Medical Adviser within the Health and Safety Executive for Northern Ireland. She has practised as an occupational physician since 1985 and has worked in the Employment Medical Advisory Service since 1987. Delia's specialist interests include the health effects of lead and asbestos, the health of doctors, health in construction, zoonoses and health inequalities. She has contributed to publications on dermatitis, carbon monoxide poisoning and GP health.

Mynepalli K.C. Sridhar, Division of Environmental Health, Department of Community Medicine, College of Health Sciences, Niger Delta University, Wilberforce Island, Bayelsa State, Nigeria

Mynepalli Sridhar PhD is currently a Professor of Environmental Health Sciences in the College of Health Sciences at Niger Delta University, Wilberforce Island, Bayelsa State, Nigeria. He obtained his PhD from the Indian Institute of Science, Bangalore, where he stayed for 13 years working on sewage treatment before moving to the University of Ibadan, Nigeria in 1977 as a professor until his retirement in 2008. While at University of Ibadan he initiated several research projects on municipal, industrial and healthcare waste management, community based waste to wealth initiatives, water quality, pollution control, lead and other toxic chemicals, environmental and health impacts, climate change and others. He has over 250 research publications in international journals and chapters in books.

Scott S. Steven, S Cubed H Ltd, Lochwinnoch, UK

Scott Steven BSc MA MEng CEng CUEW MICE MIStructE CMIOSH MaPs has worked with some of the top ten UK consultants during his 30 year career, with half of his career specialising in construction and corporate health and safety. Scott has in-depth experience in the design, contract administration, and site supervision of a wide range of civil engineering projects under most forms of procurement. His MA is in Health and Safety and Environmental Law, he has gained expert witness accreditation from Cardiff University and is a member of the ICE Health and Safety Register. Scott has sat on health and safety industry forums, professional institution boards and legal panels. Drawing on his knowledge and application of health and safety legislation, Scott has undertaken consulting engineering and safety officer roles at public outdoor and indoor international events at both temporary and permanent venues. In 2007 Scott formed S cubed H Ltd.

Section 1: Introduction

Chapter 1

Legal principles

Ann Metherall Burges Salmon LLP, Bristol, UK

doi: 10:10.1680/mohs.40564.0003

In the UK the health and safety legal framework is based on self-regulation rather than a prescriptive approach. Failure to comply with the requirements is a criminal matter.

This legal framework has a tiered approach moving from the general obligation to reduce risk to a level as low as reasonably practicable, to compliance with regulations, supported by Approved Codes of Practices and Guidance, and the development of good practice.

The reverse burden of proof placed on a defendant under the Health and Safety at Work Act 1974 emphasises the importance of being able to identify and control risks throughout the construction process but also to be aware of and implement guidance and good practice.

Although the impact of the Corporate Manslaughter and Corporate Homicide Act 2007 has increased the profile of safety, the starting point is compliance with the 1974 Act. Equally, if duty holders comply with their obligations under this Act, concerns about personal injury claims and other civil claims largely drop away.

The position in Europe and other parts of the world is different. A detailed assessment falls outside of the scope of this chapter, however some key points can be identified including the relationship between the UK regime and other parts of the world.

Introduction

What matters is that construction work is organised and done safely. Knowledge of understanding the safety legal framework is secondary but that knowledge can help identify what steps must be taken.

The purpose of this chapter is to explain in outline, but in a practical context, the principles of health and safety law. A clear grasp of those principles will make it easier to understand why (apart from the moral and social reasons associated with the prevention of accidents) compliance with the law is important. That understanding will also help you to work out what to do when dealing with innovative or unusual risk management situations.

This chapter is not a full examination of all the legal issues (that is outside the scope of the book) but the main parts of the chapter are summarised below.

In the UK the health and safety legal framework is based on self-regulation principles along with targets or goal setting, rather than a prescriptive approach which aims to give strict rules for all possible situations. However, failure to comply with the requirements is a criminal offence and can, in the worst cases, result in significant fines and even imprisonment.

> **Box 1** Legal framework
>
> The legal framework has a tiered approach moving from:
>
> - the general obligation on duty holders* to reduce risk to a level as low as reasonably practicable (ALARP);
> - to regulations (a small number of which are prescriptive (e.g. noise));
> - to Approved Codes of Practices (ACoPs) and Guidance; and finally
> - to the development of good practice which has the effect of incrementally raising the standard of health and safety performance.
>
> *Someone, or an organisation, with a legal obligation to do something (see 'Legislative framework' below)

Unusually, health and safety law places a reverse burden of proof on a defendant to any prosecution which means that if prosecuted the duty holder must prove that it reduced the risks to ALARP. This burden emphasises the importance of being able to identify and control risks throughout the construction process but in doing so duty holders must be aware of, and implement, ACoPs, Guidance and potentially good practice, to satisfy the ALARP requirement.

Although the impact of the Corporate Manslaughter and Corporate Homicide Act 2007 has increased the profile of safety obligations in the UK, the starting point is compliance with the Heath and Safety at Work Act 1974 (HSWA). Equally if duty holders comply with their obligations under the HSWA, concerns about personal injury claims and other civil claims largely drop away.

The position in Europe and other parts of the world is different. A detailed assessment falls outside of the scope of this chapter; however some key components of the legislative regime in Europe, North America and in the Middle East are identified.

The HSWA, and regulations made under it, apply in Scotland as they do in England and Wales. The Health and Safety at Work (NI) Order 1978 (HSWA NI) applies the same framework in Northern Ireland. However, the process of investigation and prosecutions do differ slightly and, where those are discussed in this chapter, the differences are highlighted.

Background

The UK's current health and safety legislative framework flows from an inquiry and subsequent report by Lord Robens as far back as 1972. At that point and in all industries, including construction, the long term downward trend in accidents, particularly fatalities, appeared to be leveling off (Hackitt, 2009). The approach to health and safety regulation was, in Lord Robens' words, 'outdated, over complex and inadequate' (Robens, 1972). It was a highly prescriptive approach that attempted to anticipate in legislation all the unsafe situations which might emerge and prescribe the control measures. As such it was incapable of tackling new issues and problems as they emerged.

This prescriptive approach derived from an earlier time where specific steps were taken to address emerging issues, often after they had led to some dreadful event. The 1966 Aberfan disaster in South Wales, where 116 children and 28 adults died following the collapse of a mine tip, is a poignant example, as Lord Robens was the Chairman of the National Coal Board at the time (Johnes and McLean, 2006). The accident led to the introduction of the Mines and Quarries (Tips) Act 1969 as the earlier Act did not provide for the management of tips at all.

The result was too much law. It was piecemeal, complex, largely incomprehensible and difficult to keep up to date. Obsolescence was rife – for example the Felt Hats Manufacture Regulations of 1902 were still in force in 1970! However, even this vast body of law did little to prevent accidents at work.

In 1966 a detailed survey of construction sites identified that of the 270 reportable accidents that happened during the survey, only 19% could be regarded as due to clear breaches of regulations. This led Lord Robens to conclude that there was a limit on the extent to which progressively better standards of safety and health at work could be brought about through negative (or prescriptive) regulation. He identified that what was needed was an effective self-regulation system. A system where high level goal setting is based on the overriding principle that those who create the risk are best placed to manage it.

This self-regulation approach of course required those who run industry to accept and exercise a high degree of responsibility as well as a legislative framework that was flexible. From his suggestion of self-regulation, the HSWA was born. Most of the key elements of the HSWA flow from Lord Robens' other recommendations including the requirements on organisations to have written statements of health and safety policy, and the creation of a national authority for health and safety at work (now the Health and Safety Executive – HSE). He envisaged that the new Act should contain a clear statement of the basic principles of safety responsibility (the general duties) supported by regulations and by non-statutory codes of practice. Lord Robens had a clear expectation that voluntary standards and codes of practice was the best way of promoting progressively better (rather than minimum) conditions of safety and health at work.

The HSWA came into force in 1974 and has remained largely unamended. Most of the regulations in place prior to 1974 have been repealed. The HSWA is generally considered to be *visionary* and the HSE, and others, consider it to be as relevant to achieving better safety performance today, as it was in 1974.

The impact of European legislation on the UK's domestic regime cannot however be ignored. Over the last 15 years, most new health and safety legislation in the UK has been introduced to implement European Union (EU) Directives. However, in principle, the EU's approach sits relatively comfortably with the UK's approach – one based on assessing and managing risk. Some of the detail which flows from the Directives has, and does, continue to cause concern and it is subject to scrutiny and negotiation usually by the HSE on the UK's behalf.

However, one impact of EU legislation is that there are more regulations than Lord Robens envisaged. He had anticipated that the detail would be contained in Guidance but although he was perhaps blessed with great foresight in recommending a self-regulation principle, he perhaps could not foresee the impact of EU legislation.

How does the HSE view the position today? Well it published a new strategy in 2009 to reset and reaffirm the direction of health and safety legislation (HSE, 2009b). It has done so in part because HSE believes that Great Britain's health and safety performance has stopped improving with this stagnation most evident in the construction industry. Although the UK's safety (and to a lesser extent health) performance continues to be among the best in Europe, it is generally recognised that, in addition to the legal reasons to improve safety performance, there are both moral and financial incentives to reassess the approach to safety management. The Robens principles survive this review and there are no proposals to make any substantive changes to the HSWA. The plan is to 'sharpen priorities' and create an environment where those in charge show leadership.

The key element of the Robens principles, and of the HSWA, is the creation of the General Duties – the first building block in the tiered legal framework.

Legislative framework

As explained above, the UK's Health and Safety Legal Framework has a tiered approach. The first level in that tier contains the General Duties placed on duty holders.

Box 2   Duty holders

The concept of 'duty holding' or 'duty holders' is key to understanding the HSWA. An explanation is therefore needed. It has no legal meaning as such but is shorthand for any person or organisation that has a legal obligation or duty to do (or not do) something. It is an extensive list and includes: employers, employees, the self-employed and some specific appointments such as the CDM Coordinator.

The General Duties are set out in sections 2 to 6 of the HSWA (sections 4 to 8 of HSWA NI). Section 2 places specific obligations on employers with respect to their employees but also, and equally important, Section 3 places an obligation on employers and the self-employed with respect to all those who are affected by their activities – commonly sub-contractors and contractors, visitors to the site and neighbours who might be affected.

Section 4 tends not to have much relevance in the construction industry because it mainly applies to the landlord and tenant relationship. There is no section 5, and section 6 focuses on the obligations of manufacturers and suppliers. What the sections all share, however, is an obligation on duty holders to identify risk and a requirement to reduce that risk to a level as low as reasonably practicable. This is an absolute duty (in other words the duty holders must comply at all times) subject only to the qualification of ALARP.

What that means in practical terms is the duty holder has to consider the scope of its activity, sometimes known as scope of its undertaking, and assess, normally, through the risk assessment process, whether it has in fact reduced the risk to ALARP. In fact, the specific requirement to assess risk flows from regulation 3 of the Management of Health and Safety at Work Regulations 1999, but the process is implied in the HSWA. If it has not reduced the risk to ALARP, the duty holder must work out what control measures it should take to do so. Elsewhere in this manual (see Chapter 6 *Establishing operational control processes*) the process of risk assessment has been explained. However, from a legal perspective there are two key issues which repeatedly catch duty holders out and lead to accidents and/or a higher risk of prosecution. Those are:

- a failure to consider the scope of activity widely enough so that some aspects of a task (often those that are on the periphery of the core activity) are not considered; and/or

- the control measures are not followed through into either procedure or policy or alternatively they are not communicated effectively to those who need to know so that they are implemented.

Whether or not something amounts to a step which is reasonably practicable, or a risk that is ALARP, is a source of much discussion and concern. However, one of the purposes of this book is to help readers assess what steps and controls might in fact be considered ALARP in a wide range of construction situations. To put the book in context it is useful to consider what must done when managing a risk.

When managing a risk there is a series of questions:

- Does the risk identified fall within the scope of the duty holder's undertaking? In other words has the duty holder either created or inherited the risk by the work it is doing?

- Are there any mandatory requirements to address this specific risk?

- If the answer is no, then what reasonably practicable steps can be taken?

Taking those three elements in more detail:

- Whether something falls within a duty holder's undertaking is, perhaps unhelpfully, defined by case law as being a question of fact (*R v Associated Octel CoLtd* [1996] 4 All ER 846). In other words, was the activity which created the risk under the duty holder's control? Usually the scope of undertaking is clear either because there is no other possible duty holder or because that responsibility has been allocated by contract or perhaps through the Construction (Design and Management) Regulations 2007 (CDM 2007).

- Having established that something does fall within the duty holder's scope of undertaking, the next question to ask is whether there are any mandatory requirements. In other words, are there any acts or regulations which require a particular step to be taken? For example, CDM 2007 requires a Principal Contractor to inform each contractor of the minimum time he will be allowed to plan and prepare before his construction work starts (regulation 22 (1) (f)).

- There are then two facets to addressing the last element of what reasonably practicable steps can be taken. The facets often operate in parallel and are interrelated. They require:

  (a) an assessment of whether there are any relevant ACoPs, Guidance or good practice; and
  (b) the risk assessment process.

The risk assessment process starts with identifying the foreseeable risks and then considering how, as far as is reasonably practicable, more risks can be eliminated or reduced. A hazard must be eliminated (or the remaining risk reduced) unless, compared with the risk, it is grossly disproportionate in terms of time, cost and effort to do so.

This explanation is the current regulatory view on the meaning of 'reasonably practicable'. Yet the potential consequence of this interpretation is that a step should be taken even if it is disproportionate (up to the point of it being grossly disproportionate) to the risk. However, new ACoPs and Guidance issued by the HSE, talk in terms of the amount of effort required being dependent on the risk. It does not talk in terms of taking disproportionate steps.

Irrespective of whether proportionate or disproportionate steps are required, most duty holders will still need guidance on how far they need go. This is where ACoPs, Guidance and relevant good practice assist as, and by complying with these, duty holders will normally have done enough to comply with the law (see 'Approved Codes of Practice, Guidance and good practice').

Routine day-to-day risk assessments are usually made on a qualitative basis using professional judgement borne out of experience and knowledge. However, in some situations a more quantitative approach, which balances the costs and safety benefits, may be required.

Quantitative Risk Assessment (QRA) normally involves a cost–benefit analysis and may be relevant where there are complex risks or controls, the costs of implementing the control measures are high, or there are a number of alternative options. This quantitative approach is largely restricted to significant safety improvements and it would be unusual for an engineer to progress a QRA in isolation from other professional safety support. However, it is important that all involved understand that sometimes risk assessments do need to be subject to a great deal more scrutiny than perhaps a risk assessment for dealing with a routine, relatively low risk, activity on site.

Regulations

The second tier of the legislative framework is Regulation. The HSWA provides the overarching legislative provisions for the control of health and safety. Beneath that sits a wide body of regulatory requirements. Despite the intentions when the HSWA came into force, and in contrast with Lord Robens' recommendations, there are now several hundred regulations made under the HSWA. It is for this reason that many organisations complain that health and safety has become incredibly bureaucratic. However, as you will see from this book, there are about 15 key construction related regulations

Although, technically, legislation which has its origins in Europe could be introduced by way of an Act, largely speaking requirements are generally brought into force by the introduction, or the amendment, of a regulation. New legislation which starts in the European Commission is debated and amended during various stages. It is during those stages that the HSE will attempt to influence the drafting by providing expert advice to those who negotiate on behalf of the government.

The HSE also leads when the implementation of European law into UK law takes place. The process is known as transposition.

The HSE normally produces a consultation document with the draft regulations and invites comments during a 3 month consultation period. Comments are analysed and discussed with stakeholders and it is normal for regulations to substantially change following stakeholder discussions. The current CDM Regulations (2007) took about 5 years to come into force and they were heavily amended from their original form. Once in final form the regulations are recommended to the UK's ministers by the HSE.

In transposing there is often criticism that the UK gilds the requirements of the European Directive. The HSE rejects that

criticism and, as part of the government's Better Regulation initiative, there is a programme of simplification. In 2009, HSE reviewed a number of requirements, including guidance to clarify the examination of equipment requirements under the Lifting Operations and Lifting Equipment Regulations 1998 with the aim of reducing the administrative burden.

Regulations are also sometimes known as statutory instruments. A statutory instrument is a form of legislation that allows the detailed requirements set out at high level in an act of Parliament (in this case the HSWA) to be subsequently brought into force, or altered, without Parliament having to pass a new act.

Construction related regulations are normally drafted by the legal department of the HSE and are generally accompanied by an explanatory memorandum. This is a good place to start if you are ever required to look at the detailed content of a regulation, as it will explain the purpose and intentions. New regulations are also subject to a regulatory impact assessment which sets out to establish whether a new regulation is expected to achieve its objectives and what effects it may have.

The method by which regulations become law varies depending on the Act which has created the provision for making them. This means that the degree of scrutiny by Parliament of a new regulation can vary enormously.

Impact of regulations

The main point to consider, however, is the nature of each regulation:

(a) Some regulations continue the theme of goal setting reflected in the HSWA and therefore steps required are usually subject to the qualification of reasonable practicability.

(b) The second type of regulation is one which effectively creates strict liability (without explaining what needs to be done). In other words the duty is absolute and there is no qualification of reasonably practicable. If the outcome occurs, the duty holder will be liable whatever steps he has taken.

(c) The third type of regulation is one which is highly prescriptive and explains in some detail how the requirements should be achieved.

Some of the regulations are specific to certain industry sectors and some have general application.

The construction-specific hazard controlling regulations are considered in more detail in other sections of this book (see Section 4 *Health hazards* and Section 5 *Safety hazards*). However, the explanation of the tiered approach would not be complete without a look at the Management of Health and Safety at Work Regulations 1999 (Management Regulations) and the key elements of the Reporting of Injuries, Diseases and Dangerous Occurrences Regulations 1995 (RIDDOR).

Management Regulations

The key elements of these regulations are the requirement to carry out risk assessments, the provision of competent health and safety advice, and to provide training. The risk assessment process is explained in outline above. There are very many books on the process as it can be complex and involved. The HSE also provides guidance on how to carry out a simple risk assessment.

The Management Regulations also require duty holders to obtain competent advice on health and safety, so that they comply with all the requirements and prohibitions which apply to the scope of that undertaking. The advice can be from an in-house expert or an external consultant but in all cases the duty holder should take steps to check that that person is competent.

RIDDOR

Following a work-related accident, duty holders must consider whether they are required to report the accident to the HSE. In fact the requirements go further than that because certain incidences of disease and dangerous occurrences (effectively near misses) must also be reported. The HSE has details on its website of where and by when the reports must be made.

The person or organisation normally responsible for making a report is the employer, but it may be someone else in certain circumstances. There is detailed guidance published by the HSE (HSE, 2008) on what to report and when. As a failure to make a report is a strict liability offence, it is usually best to err on the side of caution. Most large organisations set out the reporting requirements within their own procedures.

Approved Codes of Practice, Guidance and good practice

As explained above, the HSWA and most regulations are target setting in nature and contain little direction on how to achieve those targets. For example, the CDM Regulations require at regulation 4 that duty holders are competent. The ACoP gives detailed information on how competence might be assessed. It is the ACoPs and Guidance which form the third tier in the legislative framework and help duty holders to identify and satisfy what needs to be done to achieve ALARP.

Most of the key construction regulations are supported by either an ACoP or Guidance published by the HSE and they are referred to in some detail in this book. An ACoP has higher status than Guidance but neither should be ignored or taken lightly.

ACoPs clarify the application of the general duties and particular parts in the regulations. They also have a special status. Each ACoP contains the following statement:

> This code '... gives practical advice on how to comply with the law. If you follow the advice you will be doing enough to comply with the law in respect of those specific matters on which the Code gives advice. You may use alternative methods to those set out in the Code in order to comply with the law.
>
> However, the Code has a special legal status. If you are prosecuted for breach of health and safety law, and it is proved that you did not follow the relevant provisions of the Code, you will need to show that you have complied with the law in some other way or a Court will find you at fault.'

Each Guidance document contains this statement:

> Following the Guidance is not compulsory and you are free to take other action. But if you do follow the guidance you will normally be doing enough to comply with the law. Health and safety inspectors seek to secure compliance with the law and may refer to this guidance as illustrating good practice.

What these statements mean is if the HSE proves a duty holder has not followed a relevant provision in an ACoP, that failure will create a presumption that a breach has occurred unless the duty holder can show that they satisfied the requirement by adopting suitable alternative measures. Given the impact of an ACoP, this type of regulatory advice is generally limited to situations where there is a specific problem to deal with and where there is a strong presumption in favour of a particular method of controlling risk. ACoPs are generally therefore limited to the key issues.

In the case of Guidance there is no similar presumption. However, non-compliance is likely to mean that the HSE will consider taking enforcement action if they become aware of the omission and there is no equivalent alternative in place.

Good practice

In addition to compliance with ACoPs and Guidance, other sources of possible information on what amounts to a reasonably practicable measure is 'good practice'. It is not always clear, however, what amounts to good practice. The HSE states that (apart from all prescriptive legislation and its related ACoPs and Guidance), it would generally include standards produced by organisations such as the British Standards Institution (BSI), European Committee for Standardization (CEN), European Committee for Electrotechnical Standardisation (CENELEC), International Organisation for Standardisation (ISO), International Electrotechnical Commission (IEC) and International Commission on Radiological Protection (ICRP). However, care must be taken because these standards are not necessarily drafted with the safety of construction in mind and do become out of date.

Guidance produced by government departments may also be relevant: for example Chapter 8 of the Department for Transport's *Traffic Signs Manual* (DfT, 2009a, 2009b) which provides guidance on signage for road works, may well be a relevant standard into an investigation into a road worker accident.

Guidance produced by bodies representing a group of professionals (such as the Institution of Civil Engineers (ICE)) or trade federations may also be considered as good practice. Generally speaking such guidance has to become recognised as such for it to effectively become a requirement. However it is easy to see that what is best practice in one organisation will, in time, become tomorrow's good practice and effectively therefore a legal requirement.

This incremental development in the standards to be achieved is core to health and safety legislation in the UK.

The HSE aims to consult on the production/drafting of its ACoPs and Guidance. Most industry bodies do the same for the development and publication of good practice. Given the significant effect of ACoPs, Guidance and the development of good practice, getting involved in the consultation process may have an important benefit for you or your organisation – particularly if what is being proposed is not practical or may not reduce the risks sufficiently to justify the steps proposed.

If the situation encountered is unique and there is no relevant ACoP, Guidance or relevant good practice, then the duty holder will need to carry out a risk assessment from first principles.

Other law

Not all safety related law flows from the HSWA. For example, the Regulatory Reform (Fire Safety) Order 2005 (RRFSO) is made under the Legislative and Regulatory Reform Act 2006. Its principles are similar to the HSWA. The Road Safety Act 2006 and the very many Road Traffic Acts, are not directly related to HSWA but they do have implications for those who manage road vehicle fleets and should be taken into account when developing risk assessments and policies for managing work related driving risks and mobile phone use.

The RRFSO is enforced by local fire and rescue authorities whilst the Road Safety Act (and Road Traffic Acts) will be enforced by the police.

Criminal liability

It can come as a shock to discover that a breach of the HSWA and/or any of the regulations made under it is a criminal offence. If someone is injured then people tend to focus on compensating that person for their injuries, their ability to work in the future and any ongoing care costs. Both criminal and civil liability are important but for different reasons. A detailed review of how companies, or other statutory organisations, may have criminal or civil liability is outside the scope of this book. However, the summary below explains the principles in outline and will also help explain how the legal principles are applied when the courts get involved.

The key thing to understand is that whilst civil liability can usually be put right (or made to go away with money), that is not the case with criminal liability.

Corporate liability

HSWA

Criminal liability for a breach of health and safety legislation can come about without anyone being hurt. One of the key features of the HSWA is that it is not an outcome based Act, rather, what is relevant is whether the risk present was controlled. Another feature is that a number of duty holders (including individuals) may have liability arising out of the same incident or risk.

The HSE (in Northern Ireland (NI) the HSE for NI, HSENI) investigates accidents and incidents and makes the decision on whether to prosecute a company (in Scotland the decision on whether to prosecute is taken by the Procurator Fiscal).

Companies or organisations are generally prosecuted for a breach of health and safety legislation under sections 2 to 6 of the HSWA and/or for a breach of regulations. Simplifying slightly, a failure to reduce the risk to an employee to a level as low as reasonably practicable potentially invokes section 2, whereas the same failure to protect a sub-contractor, a visitor or a member of the public affected by its work invokes section 3.

Regulations tend to be focused towards the protection of employees and other workers but not exclusively so.

As has already been explained, both the HSWA and many regulations place an absolute duty on duty holders subject only to the qualification of reasonable practicability. The HSWA has a further little twist: where there is such a qualification, then it is for the duty holder to demonstrate that they have reduced the risk to ALARP. Technically this is known as a reverse burden of proof because in most criminal law actions it is for the prosecution to prove a defendant's guilt. Rather more emotionally (particularly for those under investigation), this provision is often described as 'guilty until proven innocent'.

One of the consequences of the reverse burden of proof is generally accepted to be a high number of guilty pleas following charge. In some cases, where the standard of safety achieved is low, this is clearly the right approach. However, even where the issues are more complex (such as whether the risk was foreseeable) then the reverse burden of proof, and the historically modest fines, meant many companies preferred to plead guilty and make a plea in mitigation to minimise the impact on their business. During mitigation, companies point to the restricted nature of the failing and their otherwise good record and systems in an attempt to keep the fine low. However, fines (particularly for large companies) are increasing and can be significant and the number of not guilty pleas appears to be increasing.

Fines

If the HSE decides to prosecute, cases are either dealt with in the Magistrates' or Crown Court. Defended cases are usually (although not always) ultimately dealt with in the Crown Court. In Scotland the District Courts deal with very minor offences whereas the Sheriff Court is the Scottish equivalent of Magistrates' Court, although it does deal with most health and safety prosecutions.

Fines in the Magistrates' Court's are a maximum of £20 000 for either a breach of the general duties under the HSWA or a breach of a regulation. The Health and Safety (Offences) Act 2008 meant that regulatory breaches after 16 January 2009 raised the fine in the Magistrates' Court from £5000 to £20 000 but kept the maximum fine for breaches under HSWA at £20 000. If the sentence is decided in the Crown Court then the fine is unlimited.

Fines can range enormously. The maximum fine levied by the Scottish Courts against Transco, following the death of four people during a gas explosion, was £15m. However, the average fine levied in 2006/07 was about £15 000.

Whichever Court decides the fine, there are some specific factors which must be taken into account. An understanding of those factors (which are drawn from the HSE's enforcement policy statement and the main health and safety sentencing case *R* v. *Howe 2* [1999] All ER 249) gives some insight into the HSE's and the Court's priorities and also emphasises the importance of responding appropriately if there is an incident. Where the offence is a significant cause of death then different sentencing guidelines apply (see below). The factors can be grouped into two areas: negative ones, which will have the effect of raising the level of fine; and positive ones, which will have the effect of reducing the level of fine:

Negative factors include:

(a) Did death occur because of the breach?

(b) What is the degree of risk and extent of danger caused by the breach?

(c) Was the extent of the breach controlled or did it continue over a long period?

(d) Was this a situation where an organisation was deliberately profiting from a failure to take necessary health and safety steps?

(e) Did it fail to heed warnings?

Positive actors include:

(a) Was there a prompt admission of responsibility and timely guilty plea?

(b) Does the organisation have a good safety record and what is its approach generally to health and safety?

(c) Were prompt steps taken to remedy the deficiencies?

(d) How far did the organisation fall short of the appropriate standard when failing to meet the reasonably practicable test?

A plea in mitigation will normally start in the Magistrates' Court. The Magistrates will consider whether they have sufficient sentencing powers and that will include whether the fine is sufficient to send a message to the shareholders. This usually means that larger companies, or companies who are subsidiaries of larger parents, will find themselves sentenced in the Crown Court.

It does not, however, automatically follow that a breach will lead to a prosecution. The HSE looks at specific criteria when making its decision in addition to the test, which applies in all criminal cases, that there must be sufficient evidence to provide a realistic prospect of conviction. The criteria largely mirror the negative and positive factors listed above but in addition the HSE will take into account:

(a) whether death was a result of a breach of the legislation;

(b) whether work has been carried out without, or in serious non-compliance, with an appropriate licence or safety case; and

(c) whether there has been a failure to comply with an improvement or prohibition notice.

In addition the HSE will also take into account the views of the victim and whether there is a requirement to prosecute to deter others. If the accident has occurred in an area the HSE is specifically targeting, such as workplace transport or falls from height, the HSE is also more likely to consider prosecution. If the HSE decides not to prosecute then it may consider other enforcement action such as an improvement or prohibition notice.

Enforcement notices

There are two types of enforcement notice: an improvement notice and a prohibition notice.

To issue an improvement notice an inspector (of the HSE) must be of the opinion that there is, or likely to be repeated, a contravention of the HSWA or a regulation. The notice will specify the time in which the contravention must be corrected. An appeal will suspend its effect.

To issue a prohibition notice an Inspector must be of the opinion that the activities involve or will involve risk of serious personal injury. The notice means that the activity must cease until the contravention is corrected or successfully appealed. An appeal does not suspend its effect.

The main grounds for appeal are:

(a) The inspector wrongly interpreted the law (i.e. there is no breach). However, to prove that, the recipient will need to demonstrate that it has done everything reasonably practicable to control the risks to employees and others affected by its activities which, as already highlighted, is a high hurdle;

(b) The recipient has been asked to control a risk which is not its to control – in other words it was not within the scope of its undertaking;

(c) The breach of law is admitted but the proposed solution is not reasonably practicable. This might arise where the inspector has been quite prescriptive about the solution and has perhaps not taken into account that the effect of reducing risk in one area increases risk in other areas; and

(d) The breach of law is admitted but the breach is so insignificant that the notice should be cancelled. In other words the risk gap (which is determined by assessing the levels of actual risk arising from a duty holder's activities) was tiny and the inspector did not act reasonably or proportionately in issuing the notice.

As it is very difficult to displace an improvement or prohibition notice once issued, it is important to work with the HSE if it is investigating an accident (or unsafe situation) to demonstrate that you are addressing its concerns. It is sometimes possible to avoid a notice if the HSE is confident that the duty holder is taking the right steps.

Corporate Manslaughter

From April 2008, companies and other organisations have also potentially been at risk of a Corporate Manslaughter or Homicide prosecution (under the Corporate Manslaughter and Corporate Homicide Act 2007). Replacing the common law offence (in other words one that was not set out in statute but in case law), the statutory offence *addresses a key defect in the law because previously [a company]could only be convicted … if a directing mind of the company was also personally liable'* (Ministry of Justice, 2007).

The largest UK company so far convicted of manslaughter had just over 100 employees. The reason is the previous requirement to demonstrate a senior 'controlling mind' who is also personally guilty of gross negligence naturally excluded most large organisations. The reality is that most fatal accidents result from a combination of different levels of fault and circumstances: the larger the organisation, the more diffuse that combination.

The controlling mind test was therefore abolished. The new law instead asks whether there has been a gross breach of duty of care and whether the organisation of activities by senior management collectively was a significant element. Effectively the offence poses the question 'just how bad was the failure and how far up did it go?'

So, in future, all those involved in the construction supply chain, including clients, will need to justify the robustness of their organisation of safety as a whole. It will no longer be enough to demonstrate that no senior individual was grossly negligent.

Summary of key aspects of the Corporate Manslaughter and Corporate Homicide Act

The Act applies to companies, partnerships, Limited Liability Partnerships (LLPs), trade unions, police, the armed forces, employer's organisations, and specified crown and public bodies within the whole of the UK (although the offence will be known as Corporate Homicide in Scotland and Corporate Manslaughter elsewhere). There are certain limitations in respect of emergency services, child protection, law enforcement and military services.

An organisation can be found guilty of the offence if the way in which any of its activities is managed or organised by its senior managers causes death due to a gross breach of a relevant duty of care. Key issues to be determined will therefore revolve around the identification of senior managers, consideration of whether breaches are 'gross' and whether relevant duties of care exist.

To be a senior manager, a person must play a significant role in making decisions or managing the organisation of the whole, or a substantial part of, the organisation's activities.

When considering whether a breach is gross, members of a jury will have to consider whether the conduct alleged amounts to a breach of that duty which falls *far* below what can reasonably be expected of the organisation in the circumstances. In doing so they must consider how serious any failure to comply with legislation was and how much of a risk of death it posed.

A relevant duty of care is defined widely and will include duties:

- to employees;

- to other persons working for or providing services to the organisation;

- as an occupier of premises;

- owed in connection with carrying on of any activity on a commercial basis;

- arising from the supply of goods and services;

- arising from construction and maintenance; and

- arising from any plant, vehicle or other things.

Most, if not all, construction activities therefore come within the scope of the Corporate Manslaughter Act.

Sentencing and consequences

If the company is convicted of corporate manslaughter, an unlimited fine could be imposed. The Sentencing Guidelines Council (SGC) published its guidelines in February 2010. It applies to sentences for a breach of HSWA and the Corporate Manslaughter and Homicide Act where the offence was a significant cause of death. The SGC suggests that the appropriate fine will seldom be less than £100 000 for breaches of HSWA, and for Corporate Manslaughter £500 000. It may be many millions.

The court can also impose a remedial order requiring the organisation to remedy any breach of duty as well as a publicity order requiring the organisation to publicise its conviction.

It is all too easy to become fixated on Corporate Manslaughter because of the name of the offence and the size of the potential fines. However a company or organisation that is complying with its obligations under the HSWA will not be guilty of Corporate Manslaughter and therefore it is probably best to focus on the HSWA requirements rather than worry about the much less likely offence of Corporate Manslaughter.

Beyond the financial consequences of the prosecution and correcting the issues, there are some other effects. Convictions, and often enforcement notices, must often be declared in tenders and are likely to affect insurance premiums.

Individual criminal liability

Individuals can also be criminally liable. Where an individual is self-employed he or she can be guilty of the section 2 to section 6 offences and can also be served with an improvement or prohibition notice. Take, for example, a sole trader who is project managing a small building or construction site where he is directing the work of sub-contractors. If that person fails to reduce the risks for those sub-contractors by a failure to coordinate the work of another sub-contractor properly, then it is possible for that person to be guilty of a section 3 offence. However, individual prosecutions under these sections are rare.

More relevant are offences under section 7 (a failure to take reasonable care), section 37 (individual liability for the corporate offence) and in the worst cases – where a death occurs – gross negligence manslaughter. Under the HSWA NI Order 1978 the equivalent sections for section 7 and section 37 are section 8 and section 34A respectively.

Section 7 – reasonable care

Section 7 of the HSWA places a duty on an employee at work 'to take reasonable care for the health and safety of himself *and other persons who may be affected by his acts and omissions at work*' (emphasis added). Section 7 also places a duty to cooperate with his employer in the discharge of the employer's obligations.

There is no case law on the meaning of reasonable care. However, HSE notes (HSE, 2009a) that the extent to which risk could or should have been foreseen will be an important factor in defining the existence of the duty. The question of whether someone did in fact take reasonable care will be one for members of the jury to decide and they will be invited to consider whether the employee in practice followed the procedures and method statements in accordance with his or her training.

Liability under section 37

Directors and other similar officers will be guilty of an offence where the organisation has committed an offence and it is due to their 'consent or connivance' or is 'attributable to any neglect'. What this means is an individual is potentially liable for a corporate offence – usually breaches of sections 2 or 3 of the HSWA.

The meaning of consent and connivance has not been specifically considered by the courts. However, consent is generally understood to mean the director was aware that the possibility of danger existed, that there was a reasonably practicable step, which could have been taken, but nonetheless he agreed to carry on with the activity without taking such a step.

Connivance probably falls just short of consent but is usually taken to mean that a director knows what is going on and gives tacit agreement by neither encouraging the approach nor taking any steps to stop it.

The meaning of neglect has been considered by the Court of Appeal (R v. P Ltd and another [2007] EWCA Crim 1937). It arises where a director should have taken action, not based on what he knew, but what he should have known. Although there are no statutory health and safety duties placed on directors, there is Guidance set out in the document published by the HSE and the Institute of Directors (IoD) – Leading Health and Safety at Work (IoD and HSE, 2007) – which gives information on the collective and individual responsibilities of directors and board members. This guidance is therefore likely to be relevant to establishing the standard of safety management directors should achieve and therefore what they should have known.

Whether or not someone is a director or similar officer does not depend on their job title. What matters is the degree of control they exercise and whether they are 'the decision makers within the company who have the power and responsibility to decide corporate policy and strategy' (R v. Boal [1992] 2 WLR.890). The scope is 'to catch those responsible for putting proper procedures in place; it is not meant to strike at underlings'.

These offences are prosecuted rarely and almost never involve directors of medium or large companies. Between 2002 and 2007, 33 company directors or senior officers were prosecuted. During the same period about 1100 workers have died at work (Rapp. R Black or White SHP August 2009). However, as one of the key elements of a Corporate Manslaughter offence is whether there has been (collective) senior management failing, it is easy to see that, with directors and senior managers

generally coming under more scrutiny, there may be more section 37 prosecutions in the future.

Gross negligence manslaughter

Equally, prosecutions of individuals following work related deaths for gross negligence manslaughter are relatively rare. This is a common law offence and the principles of the offence are set out in a House of Lords case R v. Adomako. The case involved an anaesthetist. He was convicted of gross negligence manslaughter following the death of a patient which flowed from Mr Adomako's failure to notice that a tube had become disconnected from a ventilator and the patient had stopped breathing.

The principles are similar to the test for corporate manslaughter:

(a) there must be a breach of duty of care to the victim who has died;
(b) the breach must cause the death; and
(c) the breach of duty amounted to gross negligence and therefore a crime.

Where the two offences differ is in establishing what the standard of care is and then by how much below the standard the offender must fall.

The standard of care is based on the principles of the law of negligence (see below). The test of whether the breach was gross negligence is more difficult. The test is a little circular: was, having regard to the risk of death (not just injury), the conduct of the defendant so bad in all the circumstances as to amount to a criminal act or omission? The jury is asked to consider whether breach showed such a disregard for the life and safety of the fatally injured person as to amount to a crime. In assessing the level of disregard it is for the jury to set the standards. Mere oversight or civil negligence is not enough. The disregard for safety must carry obvious risks of death.

In Scotland the nearest equivalent is culpable homicide. It is, like its English equivalent, an offence under common law. It is committed where the accused has caused loss of life through wrongful conduct, but where there was no intention to kill or 'wicked recklessness' (The Scottish Government, 2009). While the offence charged remains the same there can be a great variation between individual cases including whether or not the act causing the death was voluntary or involuntary.

Consequences

It has always been technically possible to be sent to jail for a breach of the HSWA. However, the circumstances were very limited – carrying out a licensed activity such as the removal of asbestos or failure to comply with a prohibition or improvement notice. Since the introduction of the Health and Safety (Offences) Act 2008, a breach of section 2, section 3 and section 37 raises the possibility of a custodial sentence.

At the date of writing, no Guidelines on when the courts should consider a custodial sentence have been published. However, in a rare step, a statement about the intention of the new penalty was issued by a government minister (Lord McKenzie): he said custodial sentences powers are intended to be used 'only where very serious circumstances applied'.

Convictions for gross negligence manslaughter (culpable homicide, in Scotland) can lead to a custodial sentence. The maximum sentence is life imprisonment although the terms tend not to exceed 3 to 4 years except in exceptional circumstances.

Civil liability

Anything more than the briefest summary of civil liability principles is outside the scope of this chapter. However to understand how the HSWA and regulations interact with an employer's or individual's civil liability, a quick overview is required.

Civil liability arises in two main ways – as a breach of contract or in negligence. Contracts can contain a clause which will make a breach of a party's health and safety obligations a breach of contract, and therefore give the other party a right to recover its losses which flow from the breach. However, the main area where compliance to health and safety obligations is relevant is in negligence and in particular a claim arising out of personal or fatal injury.

To bring a claim for personal or fatal injury the claimant must show that he was owed a duty of care, that there was a breach of that duty, that the breach caused the injury and that the outcome was foreseeable.

It is also possible to bring a claim for breach of statutory duty. Claims cannot be made for a breach of the HSWA but can for a breach of regulations unless the regulation specifically excludes it. Regulations which allow breach of duty claims tend to be ones which are focused on the protection of workers (such as the Control of Noise at Work Regulations 2005 and the Personal Protective Equipment at Work Regulations 1992).

That is not to say that HSWA, regulations, ACoPs and Guidance have no role in personal injury claims as they are often highly relevant in deciding the standard of care expected by the defendant.

If an injury comes about because of the behaviour of another employee at work then the employer will normally be (vicariously) liable for the employee's actions and therefore will be required to pay compensation.

It can be difficult to successfully defend a claim for personal injury because there is a natural bias towards the person injured. However, to do so the defendant will need to show that it either did not owe a duty of care or it did not breach the standard of care. If it did breach the standard of care, the defendant would have to show that the breach did not cause the injury. There are often arguments about whether the injured person's own actions had some impact on the outcome and a defendant will often suggest that there was some contributory negligence. To make that suggestion then the employer will need to show that the individual was well trained, competent and supervised in the task he was undertaking.

Where there has been an act of negligence, the purpose of damages is to place the claimant in the position they would have been had the act not taken place. In the case of contract damages the aim is to generally put the claimant in the position that they would have been in if the contract had been performed.

In personal injury claims, however, the basis for assessment of damages is a complex area. Personal injury damages are divided into special and general damages. Special damages are all past quantifiable loss (loss of earnings, past medical expenses, cost of damaged property and travel expenses) and general damages are all other losses (future loss of earnings, future medical expenses, amount for pain, suffering and loss of amenity, and loss of earning capacity). General damages therefore are extremely difficult to quantify.

In the event of a claim it is important to remember to notify the insurer in accordance with the terms of the relevant policy.

Before issuing a formal claim, the claimant will normally make a detailed request for relevant documents such as the accident report, risk assessments and training records. It is therefore sensible to collate these documents at the time of the accident because it can be many months or up to 3 years (5 years in Scotland) after the incident before the request is made. On construction sites that can mean that the records are difficult to locate and therefore hamper the ability to defend the claim, if that is appropriate.

Inquests

If there has been a work related accident there will normally be a coroner's inquest (in Scotland a Fatal Accident Inquiry by the Procurator Fiscal). This will be in addition to any HSE or police investigation although evidence collected by these bodies will form part of the coroner's inquiry.

Unlike criminal proceedings, an inquest is an inquisitorial process rather than an adversarial one. What this means is that it is the coroner who directs the overall approach to the inquest, and legal representatives for the interested parties are there to assist the coroner's inquisition.

The purpose of the coroner's inquest is not to blame anyone for the death. Rather it is to establish the cause of death which is confined to four relatively narrow questions, which are:

(a) who has died;
(b) where;
(c) when; and
(d) by what means and in what circumstances?

The first three questions are usually (but not always) straightforward. The fourth question can be more problematic. It is sometimes seen as a way to widen the scope of the inquest and attribute blame. The approach adopted by each coroner can vary enormously which means in practice some inquests are narrowly confined and in others there can be a detailed review of not just the relevant risk assessments, method statements and training but the entire approach to safety management on a site or within a business.

At the end of an inquest there will be a verdict. Traditionally these were in short form: including accident, open and unlawful killing. It is however more common today for a narrative verdict to be given which describes the circumstances in more detail. Additionally the coroner may make a report to prevent deaths (based on the evidence heard at the inquest). The organisation must respond to the report, although it is not under any obligation to act on it.

In Scotland Fatal Accident Inquiries (FAIs) are held under the Fatal Accidents and Sudden Deaths Inquiry (Scotland) Act 1976. An inquiry must be held in cases of death in custody or as a result of an accident at work but may be held in other cases of sudden, suspicious or unexplained death, or death in circumstances that give rise to serious public concern. The purpose of an FAI is not to apportion blame for the death in either the civil or criminal sense but is essentially a fact-finding exercise carried out in the public interest. The purpose of an FAI is to determine:

■ where and when the death took place;

■ the cause of the death;

■ reasonable precautions whereby the death might have been avoided;

■ the defects, if any, in any system of working which contributed to the death or any accident resulting in the death;

■ any other relevant facts relevant to the circumstances of the death.

Insurance

In contrast with civil liability, where compensation is made to put right the thing which has gone wrong, it is not possible to insure against fines, although it is possible to insure against the cost of defending a prosecution. This is for public policy reasons.

The most likely way the costs of defending a health and safety or corporate manslaughter prosecution will be covered is through legal expenses insurance (either as a stand-alone policy or as part of a more general policy). Legal expenses insurance can be bought by a company or an individual (it is also included in some household insurance). Although it will be dependant on the precise wording, the costs of legal advice during the investigation stage, and also through the inquest process, may be covered. The Insurance Companies (Legal Expenses Insurance) Regulations 1990 are relevant to this type of insurance and covers, amongst other things, the insured's right to choose the solicitor to advise him.

The primary purpose of public liability or project insurance is to cover the sums the insured becomes legally liable to pay to third parties.

Employer's liability (EL) insurance is also another possible route whereby a company's defence costs, and even the costs of representation at an inquest, will be covered. However, the main reason for EL insurance is to cover claims by employees during the course of their employment.

EL insurance is a statutory requirement (Employer's Liability (Compulsory Insurance) Act 1969). This Act requires all employers to take out and maintain insurance for injury and disease sustained by an employee arising out of and in the course of that employment in Great Britain. An employee will normally be covered unless the employee has wilfully ignored safety instructions.

The question of whether someone is an employee for the purposes of EL insurance often arises – normally someone working as a labour only sub-contractor will fall under the scope of the EL policy (and the insurer should be notified how many people are in this position). However it is not always clear-cut.

The HSE is responsible for enforcing compliance to the Act. If an employer does not have cover in place, it commits a criminal offence and may be fined up to £2500 for each day it is not insured. Unlike cover provided by the Motor Insurers' Bureau, for uninsured drivers involved in accidents, there is no safeguard for employees whose employer has not taken out insurance.

Directors' and officers' (D&O) insurance is designed to cover the liability of company directors rather than the company. In some cases the D&O policy may be extended to cover key employees acting in the capacity of directors where they are involved in a senior management function of a company.

If either a company or an individual is aiming to invoke an insurance policy to cover its defence costs there are some key points to bear in mind:

■ Does the wording of the policy in fact cover the investigation and defence costs? For example a D&O policy can exclude illegal acts and so the director may need to repay defence costs if he is convicted. In legal expenses insurance it is not uncommon for coverage to only kick in at the point at which the company or individual is charged.

■ With the exception of an EL policy, you are usually required to notify insurers of accidents and claims immediately or within a short time limit. A failure to do so may mean cover will not be provided.

■ Written consent to incur the defence costs will almost always be required. It is important therefore to involve the insurer and confirm approach at the earliest opportunity.

Other jurisdictions

The earlier part of this chapter has explained the principles behind the health and safety legislative framework in the UK. The position in the rest of world is on the whole very different even if the intended outcome (the reduction of accidents) is the same. The HSWA framework is adopted in Eire and Hong Kong. The principle of ALARP is being looked at for adoption in some other countries but in the rest of Europe the position is different.

Europe

Like the UK, all EU members must ensure that their domestic legislation complies with the requirements of the EU legal framework. In the case of health and safety the main piece of legislation is the Council Directive 89/391/EEC, which in addition to some provisions on employers and employees to encourage improvements in the safety and health of workers at work, also provides the basis for the rather large number of specific safety directives. It lays down general principles concerning the prevention and protection of workers against occupational accidents and diseases. It contains principles concerning the prevention of risks, the protection of safety and health, the assessment of risks, the elimination of risks and accident factors, the informing, consultation and balanced participation and training of workers and their representatives. This framework hints at an HSWA style regime elsewhere in Europe.

How the directives are in fact implemented in each country varies reflecting the differences in domestic legislation. For example in Spain, the Framework Directive is incorporated into its legal framework through a specific statute, the Prevention of Occupational Hazards Act 1995 and a number of general statutes. In addition there are further regulations issued by the regions or Autonomous Communities.

Essentially, the Spanish framework adopts a preventative approach although (like many countries) there are regulations which take a prescriptive approach. Enforcement of safety law is done by the Labour Inspection Service and Social Insurance Institution. Like the HSE they have a dual role providing information and advice on how to comply with obligations. The National Commission of Safety and Health at Work draws together the work of all those involved: labour inspection service, employers' federations, trade unions and the obligatory insurance organisations.

Unlike the UK, Spanish companies cannot (as a matter of general law) be held criminally liable. Only those responsible for the running of a company can be prosecuted where they were acting with guilty intent or negligence. Companies though can be liable (in a civil sense) if the crimes are committed by employees in the execution of their duties. What that means is that routine breaches of health and safety law are not criminalised as they are in the UK, but when something does go wrong there tends to be a greater focus on individuals than in the UK.

United States of America

The position is more complicated in United States of America (USA) where there are federal and (potentially) state and city requirements. The approach is generally more prescriptive than in the UK. The Occupational Safety and Health Act 1970 (OSH Act) created the Occupational Safety and Health Administration (OSHA). OSHA has a similar role to the HSE, establishing and enforcing standards and providing assistance to employees.

The OSH Act imposes on employers a duty to provide a safe and healthy place of employment – the General Duty Clause. This means that an employer must keep its place of employment free from recognised (foreseeable) hazards that are causing or likely to cause death or serious physical harm to its employees.

This means employers are required to control safety risks that are not governed by specific regulations. There does not need to have been an accident, only that the circumstances are likely to cause death or serious injury.

In addition, the OSH Act requires employers to comply with OSHA's safety and health rules. There are four major industry groupings of OSHA standards which are published in the Code of Federal Regulations including construction. Some of the standards are very prescriptive. There are also horizontal standards which apply to a number of industries. For example, training and education is a specific duty applicable to all employers and requires that safety training programmes be tailored to the specific needs of the workplace.

Enforcement will be by OSHA unless the state has an OSHA approved state plan at which point enforcement transfers to the state. The state plan is required to be at least as effective as the federal OSHA requirements. Compliance officers can issue citations for any specific violations of the Act or regulations.

OSHA can impose civil penalties against employers but not employees. The level will depend on the type of violation. So a 'de minimus' notice (one which is considered to be a minor violation) will attract no penalty but a 'failure to abate' notice could result in a penalty as much as $7000 per day.

Federal criminal charges against the employer can be brought under the OSH Act only for willful violations of OSHA standards or regulations that cause death and not under the General Duty Clause obligations. In contrast to the HSWA regime therefore, serious criminal charges are levied if there has been a specific breach of a rule and not because the organisation has failed in its general duties to control foreseeable risks.

Technically there can be a federal prosecution of an individual for homicide which flows from a work related death and (if there were aggravating features and involvement by senior management) homicide charges against the company. In some states there can also be criminal liability where a corporate offence occurs because of the conduct of an employee. In some cities, it is possible to be prosecuted for homicide or assault in addition to penalties for breaches of local building codes.

Middle East

Generally there is no specific health and safety law in the Middle East countries. The requirements are included in general labour law at both federal and state levels and traditionally if accidents are investigated then it tends to be by the police.

For example, in Abu Dhabi, there is federal legislation (which applies to all seven emirates) and emirate specific law. Some of the law is quite prescriptive but it is not supported with a general duties clause we would recognise in the UK or any supporting guidance. However there is, it appears, a move towards a more goal setting approach for health and safety with the publication of a code of practice for construction projects which includes detailed obligations for on-site health and safety.

This whistle stop tour around the rest of the world demonstrates that whilst the goal of a reduction in accidents is likely to be the same, countries will go about it in different ways. If you are working overseas then you will need to take steps to find out about the legal regime in place and what both the corporate, and your personal, obligations are.

If you are working with an international company then it is likely that the system for managing safety adopted will be based on the system used in its home country but customised to reflect specific local regulations. Increasingly, when countries with developing safety legal regimes are letting tenders, they expect designers and contractors to adopt international best practice and given that the UK's record in health and safety performance is good, the UK regime including the principle of ALARP may be adopted.

Summary of main points

The HSWA creates a self-regulation framework where its target is to see risks managed or controlled to a level that is as low as reasonably practicable. It is underpinned by a process of risk identification, assessment and management. That process is not one that is carried out by duty holders in isolation. Duty holders can obtain extensive assistance from ACoPs, Guidance and industry good practice when preparing their risk assessments.

The reverse burden of proof in most prosecutions means it is for the duty holder to show that it has in fact discharged its risk management obligations and has reduced risks to ALARP. Awareness and compliance (where relevant) to the ACoP, Guidance and industry good practice is therefore a vital component of discharging safety obligations. This book provides a useful route map through that body of advice.

References

Department for Transport (DfT). *Traffic Signs Manual Chapter 8 – Part 1: Design. Traffic Safety Measures and Signs for Road Works and Temporary Situations,* 2nd edition, 2009a, London: The Stationery Office.

Department for Transport (DfT). *Traffic Signs Manual Chapter 8 – Part 2: Operations. Traffic Safety Measures and Signs for Road Works and Temporary Situations,* 2nd edition, 2009b, London: The Stationery Office.

Hackitt, J. *Leadership in Health and Safety – The Essential Role of the Board.* Annual Rivers Lecture on the 18 March 2009 by Judith Hackitt CBE, HSE Chair. Transcript available online at http://www.hse.gov.uk/aboutus/speeches/transcripts/hackitt180309.htm

Health and Safety Executive. *A Guide to the Reporting of Injuries, Diseases and Dangerous Occurrences Regulations 1995,* 3rd edition, 2008 (reprinted with amendments 2009), London: HSE Books. Available online at: http://www.hse.gov.uk/pubns/priced/l73.pdf

Health and Safety Executive. *Prosecuting Individuals OC 130/8 Version 2,* 2009a, available online at: http://www.hse.gov.uk/foi/internalops/fod/oc/100-199/130-8.htm

Health and Safety Executive. *The Health and Safety of Great Britain: Be Part of the Solution.* (06/09 C100), 2009b, London: Health and safety Executive. Available online at http://www.hse.ov.uk/

strategy/strategy09

Institute of Directors and Health and Safety Executive. *Leading Health and Safety at Work. Leadership Actions for Directors and Board Members,* 2007, London: Health and Safety Executive. Available online at: http://www.hse.gov.uk/pubns/indg417.pdf

Johnes, M. and McLean, I. *The Aberfan Disaster,* 2006. Available online at: http://www.nuffield.ox.ac.uk/politics/aberfan/home.htm

Ministry of Justice. *A Guide to the Corporate Manslaughter and Corporate Homicide Act 2007,* 2007. Available online at: http://www.nio.gov.uk/guide_to_the_cmch_act_2007_web.pdf_oct_07-3.pdf

Robens of Woldingham Alfred Robens Baron chairman, Great Britain Committee on Health and Safety at Work, Great Britain Department of Employment. *Safety and Health at Work: Report of the Committee 1970–72. (Cmnd. 5034),* 1972, London: HMSO.

The Scottish Government. 2009. Information for Bereaved Families and Friends Following Murder or Culpable Homicide, 2009, Edinburgh: The Scottish Government. Available online at: http://www.scotland.gov.uk/Publications/2009/02/24125340/0

Referenced legislation

Construction (Design and Management) Regulations 2007. Statutory instruments 320 2007, London: The Stationery Office.

Control of Noise at Work Regulations 2005. Statutory instruments 1643 2005, London: The Stationery Office.

Corporate Manslaughter and Corporate Homicide Act 2007. Elizabeth II – Chapter 19, London: The Stationery Office.

Employer's Liability (Compulsory Insurance) Act 1969.

European Council Directive 89/391/EEC of 12 June 1989 on the introduction of measures to encourage improvements in the safety and health of workers at work. *Official Journal of the European Union L183,* 29/06/1989 P. 0001–0008.

Fatal Accidents and Sudden Deaths Inquiry (Scotland) Act 1976. Elizabeth II, 1976. Chapter 14, London: HMSO.

Health and Safety at Work Act 1974. Elizabeth II. Chapter 37, London: HMSO.

Health and Safety at Work (Northern Ireland) Order 1978. Statutory Instruments 1978 1039, London: HMSO.

Health and Safety (Offences) Act 2008: Chapter 20, London: The Stationery Office.

Insurance Companies (Legal Expenses Insurance) Regulations 1990. Statutory Instruments 1990 1159, London: HMSO.

Legislative and Regulatory Reform Act 2006. Elizabeth II – Chapter 51, London: The Stationery Office.

Lifting Operations and Lifting Equipment Regulations 1998. Statutory Instruments 2307 1998, London: The Stationery Office.

Management of Health and Safety at Work Regulations 1999. Statutory Instruments 1999 324, London: The Stationery Office.

Mines and Quarries (Tips) Act 1969, London: HMSO.

Occupational Safety and Health Act 1970 (USA).

Personal Protective Equipment at Work Regulations 1992. Statutory Instruments 1992 2966, London: HMSO.

Prevention of Occupational Hazards Act 1995 (Spain).

Regulatory Reform (Fire Safety) Order 2005. Statutory Instruments 1541 2005, London: HMSO.

Reporting of Injuries, Diseases and Dangerous Occurrences Regulations 1995. Statutory Instruments 1995 3163, London: HMSO.

Road Safety Act 2006. Elizabeth II – Chapter 49, London: The Stationery Office.

Cases referenced

Rapp. R Black or White SHP August 2009
R v. *Adomako* [1995] 1 AC 171
R v. *Associated Octel Co. Ltd* [1996] 4 All ER 846
R v. *Boal* [1992] 2 WLR.890
R v. *P Ltd and another* [2007] EWCA Crim 1937
R v. *F Howe & Son Engineers Ltd.* [1999] 2 ALL ER 249

Further reading

Websites

British Standards Institution (BSI) http://www.bsigroup.com
European Committee for Standardisation (CEN) http://www.cen.eu

European Committee for Electrotechnical Standardisation (CENELEC) http://www.cenelec.eu
HSE risk management pages http://www.hse.gov.uk/risk/index.htm
Institution of Civil Engineers (ICE), Health and safety website http://www.ice.org.uk/knowledge/specialist_health.asp
International Organisation for Standardisation (ISO) http://www.iso.org
International Electrotechnical Commission (IEC) http://www.iec.ch
International Commission on Radiological Protection (ICRP) http://www.icrp.org
Sentencing Guidelines Council (SGC) http://www.sentencing-guidelines.gov.uk

Chapter 2

doi: 10:10.1680/mohs.40564.0017

Recognising health hazards in construction

Delia Skan and **Ken Logan** Health and Safety Executive Northern Ireland (HSENI), Belfast, Northern Ireland, UK

CONTENTS

To date, safety has been the dominant driver for health and safety performance. Forward looking construction companies agree that this drive for excellence in safety performance must continue but it must be matched with an equal drive for health. Rita Donaghy in her 2009 report into the underlying causes of construction-related accidents focuses on safety but also acknowledges that ill health in construction remains a serious issue. Initiatives such as BuildHealth and Constructing Better Health seek to provide a framework to enable construction companies to actively manage health risks and to adopt an holistic approach to health.

This chapter provides background information on the importance of sustaining the health of our construction workers through prevention of injury and ill health, and ensuring access to support and rehabilitation. It identifies relevant work and individual related health risks, and provides a model to make more meaningful to construction workers how the workplace can influence their wellbeing. The paucity of data on the health of construction workers is acknowledged.

Introduction

Construction work is varied and offers opportunities for creativity and innovation. It also poses risks to both health and safety. In the case of risks to health many of these cause disease after moderate to long latency periods. Examples include noise induced hearing loss and asbestos related diseases. Similarly heavy physical work and manual handling may cause cumulative harm to the musculoskeletal system. Studies show that construction workers are more likely to suffer from work-limiting conditions and to retire early. Need in this sector for health intervention is high. The construction industry presents unique opportunities for health improvement in that it provides access to a population from broad socio-economic backgrounds which is largely male, often itinerant, and which may be reluctant to seek medical advice. Whilst some aspects of the work organisation may be perceived as providing health promotion opportunities, there is no doubt that the working environment and the patterns of work present particular challenges for meaningful health interventions.

Background

In the UK, deaths caused by asbestos-related diseases outnumber considerably those arising as a consequence of workplace injuries (HSE, 2009). The proportion of these within construction and related trades is high. Asbestos with silica and lead are 'invisible hazards'. Perceived as integral to the fabric of buildings their presence is not easily identified, they are not labelled and consequently their health risks may remain unrecognised. Similarly, manual handling and noise may be seen as risks that 'go with the job' and their control and management may remain unaddressed.

Construction workers themselves are also known to have relatively unhealthy lifestyles with evidence that the prevalence of smoking and drinking to excess is high when compared with other sectors. In Europe a number of studies show they have a high incidence of permanent work incapacity mainly as a consequence of musculoskeletal and cardiovascular diseases. That construction workers stop work at an earlier age has been shown in a number of different countries (Brenner and Ahern, 2000). The negative health consequences of the unemployment which results are issues for all of society. Therefore interventions to rehabilitate construction workers back to work and prevent their unnecessary early retirement and associated worklessness are likely to be in the best interests of the health of the individuals, their families and the community as a whole.

Set against the above, however, is an industry which traditionally has struggled to support workers and where efforts to take a planned management approach to ill health prevention and workplace risk management is particularly challenging. The barriers relate to systems of work as well as to misperceptions. As regards the former short contracts, temporary workers and sub-contractors and the lack of accessible occupational support are only some of the reasons. Likewise many of the work-related conditions have medium to long latencies, i.e. exposure today will result in health consequences several years later. This delay may have the effect of allowing health-related risks to have a lower priority and to be relatively neglected. Paucity of data on ill health from this sector contributes to misperceptions of risk. Similarly, misunderstandings about the true meanings of health and occupational health make it difficult for employers to conceptualise how workplaces can influence wellbeing. It is against this background that this chapter has been developed.

In 2006 in Northern Ireland the authors endeavoured to explore this issue in greater detail and conducted planned visits to a number of large construction employers with the aim of exploring their approach to health at work. There was a consistency in our findings in that large construction companies had very well developed policies, they had a sound understanding of the legal, ethical and moral arguments with regard to health and they had excellent systems for rehabilitation, support and health surveillance for their own employees. However, they acknowledged the challenge of health management at the subcontractor interface with issues such as fitness standards for safety critical work, arrangements for health surveillance and access to competent occupational health advice also emerging. Nonetheless they were committed to the idea of investing effort and resource towards improving the health, not only of their own workforce but also that of sub-contractors. 'Build-Health' (see below) was founded based on a 'health champions' approach and supported by a web based resource of the same name.

Concept of wellness

Early discussions focused on what was meant by occupational health and the need to distinguish it from occupational health services. Consideration was given to the factors that influence health at work including regulations, the behaviours of individuals themselves, their engagement and the culture of an organisation. Analogies were made with public health models which describe the determinants of health and which recognise the importance of education, engineering and regulation in ill health prevention. Through such dialogues it was possible to enhance 'literacy' on occupational health, acknowledge the role of individuals themselves in health improvement and ensure clarity of purpose.

Wellbeing definition

The idea of wellbeing at work was developed. This was seen as a positive attribute or resource which could be influenced by both the employer and the worker. It was informed by the World Health Organisation (WHO) definition of health as a 'state of complete physical, mental and social wellbeing and not merely the absence of disease...', and its definition of mental health as a 'state of wellbeing in which the individual realises his or her own abilities, copes with normal stresses of life, works productively and fruitfully and makes a contribution to his or her community'.

It also drew from the concept of healthy working lives as put forward by the Scottish Executive (2004) which defines it as follows:

> a healthy working life is one that continuously provides working-age people with the opportunity, ability, support and encouragement to work in ways and in an environment which allows them to sustain and improve their health and wellbeing. It means that individuals are empowered and enabled to do as much as possible, for as long as possible, or as long as they want, in both their working and non-working lives.

Mapping workplace interventions

Workplace interventions which could have a positive impact on wellbeing are made up from the following strands:

- workplace risk management to include organisational and individual risk;

- occupational health support and rehabilitation (interface between the construction industry and an occupational health provider); and

- health promotion.

Integral to the shamrock model shown in **Figure 1**, which intends comprehensively to address workplace wellbeing, has to be leadership and a health advocacy culture. Senior management commitment and employee involvement and engagement at all levels are essential.

Each of the modalities will now be considered in turn and the influence of culture discussed.

Managing risk

Organisational health risks

Organisational health risks, i.e. those created by the construction work itself, are myriad. Some are classified and presented in **Table 1**.

Individual health risks

Health and safety legislation (see Chapter 1 *Legal principles*) requires that employers put in place a safe system of work.

Figure 1 Shamrock depicting modalities involved in workplace wellbeing

ICE manual of health and safety in construction © 2010 Institution of Civil Engineers

Type	Examples	Diseases
Physical	Noise	Hearing loss and tinnitus
	Vibration: hand, arm	Hand–arm vibration
	Whole body	Back pain
	Ultraviolet radiation	Premature skin ageing and skin cancer
	Ionising radiation	Range of effects
	Dusts:	Chronic obstructive pulmonary disease
	Wood dust	Asthma
	Asbestos	Asbestos related diseases
	Silica	Silicosis and lung cancer
Chemical	Cement/grout/resins	Dermatitis
	lead	Lead poisoning
	Carbon monoxide	Carbon monoxide poisoning
Biological	Leptospira (rat urine)	Leptospirosis
	Tetanus (spores in the ground)	Tetanus
Ergonomic	Manual handling risks	Musculoskeletal problems, e.g. back pain, knee pain
Psychosocial	Work-related stress	Negative health consequences
	Job insecurity	
	Long working hours	
	Physically demanding environment	
	Absence of support	
	Drugs and alcohol	

Table 1 Classification of construction related health risks

Fitness of an employee for a task is integral to such systems. Construction employers must therefore consider whether illness or its consequences could foreseeably affect such systems. In doing so an employer will find it useful to have validated standards for specific tasks and to understand the concept of safety critical work.

In 2008, fitness standards for construction related trades were published under the Constructing Better Health banner (see latest publications on their website) along with an easy to use matrix (Constructing Better Health, 2009).

The document quoted from the Faculty of Occupational Medicine's (2006) *Guidance on Alcohol and Drugs Misuse in the Workplace*, which defined safety critical workers or roles as:

> those involving activities where because of the risks to the individual concerned or to others the employees need to have full unimpaired control of their mental or physical capabilities.

Constructing Better Health identified the following construction related jobs as safety critical:

- HGV / LGV drivers
- all mobile plant operators (including crane drivers)
- rail and road n track side workers
- asbestos licensed workers
- tunnellers or those working in confined spaces
- tasks carried out at height where collective preventive measures to control risk are not practicable
- others may be identified during the risk assessment process.

Input from occupational health professionals is helpful in defining standards and will be necessary in reaching decisions on fitness in individual cases taking account of the standards and the nature of work. Employees themselves will find it useful to understand the concept of safety critical work and the rationale which has informed decisions on fitness standards. It is important that all understand that standards are not intended to pose barriers to employment and to realise the implications for employment of the Disability Discrimination Act.

Support and rehabilitation of the ill employee and role of occupational health services

Work is generally good for our health and wellbeing (Black, 2008). During working age, the main conditions which require that time is taken off work are 'common health problems', e.g. mental health (e.g. 'stress'; depression), musculoskeletal (e.g. back pain; arthritis), or cardio-respiratory (e.g. hypertension; asthma). Whilst these conditions cause distress and can affect many of us at one time or another, for many there is no serious underlying disease or lasting harm. Many of these conditions are not well managed at work and therefore can cause long term disability (Waddell and Burton, 2006).

For such conditions there is a strong evidence base which supports the benefits of early intervention and early return to work (Waddell et al., 2008). Likewise, adjustments to tasks have been shown to reduce disability, delay retirement and reduce sickness absence. Many of the adjustments to task are relatively simple such as allowing more time to complete a task or changing the work schedule. Interventions may also include input from a case manager, physiotherapy or cognitive behavioural therapy. Given the evidence that many construction workers suffer from work limiting conditions, early intervention is particularly relevant in this sector yet few studies have been conducted. In one study an educational and counselling intervention for musculoskeletal disorders was implemented in one group of workers and the findings compared over a 26 month period with those in a control group. Work ability of those in the intervention group improved slightly over time, but the number of work disability pensions were unaffected (Stenlund, 2005). In the social security system in Sweden, a model for rehabilitation and prevention has been developed for workers in this sector.

Ill employees and sickness absence

Many construction employers do not have regular access to occupational health support. Thus the approach to the management of sickness absence is particularly difficult. There is a range of sources of information including from the Health and Safety Executive (HSE) (HSE, 2004) and the Chartered Institute of Personnel and Development (CIPD). In larger firms a policy is an essential first step and is also recommended for smaller organisations. This can explain the range of actions an organisation will take and should refer to the requirements of the Disability Discrimination Act (DDA) and to the principle of 'reasonable adjustments' contained within it. External assistance from occupational therapists, occupational health professionals or general practitioners may be required. In the UK advice and guidance are also available from disability employment advisers based in job markets.

Many construction employers will not have regular access to occupational health support and when a person becomes ill there may be circumstances in which he or she wishes to seek advice from the employee's general practitioner. Requests from employers for reports are covered by the Access to Medical Reports Act 1988 and require appropriate consent. Employees can refuse to give consent. In requesting such reports employers should request advice on the functional consequences of illness and provide a list of the duties and demands of the job. Employers should understand the requirements of the DDA.

Role of occupational health services

Occupational health providers aim to prevent ill health and promote health at work. Their skills and competencies can encompass risk identification and management as well as traditional occupational heath services at the worker interface. At a strategic level their contribution to policy development is valuable as is their role of health advocates.

Fitness assessments and statutory health surveillance

Functions carried out by occupational health services include pre-employment assessment, determination of fitness for work following sickness absence with, where necessary, determination of fitness for safety critical work as well as specific statutory assessments, e.g. audiometry, hand–arm vibration screening and night workers' assessments. **Table 2** provides examples of legislation which refers to health surveillance and the circumstances in which it is considered appropriate. Statutory *medical* surveillance will be required if construction work poses a risk of significant lead exposure, or as required under the Control of Asbestos Regulations 2006 or the Work in Compressed Air Regulations 1996. Where there is a risk of dermatitis, a trained responsible person can carry out the necessary skin inspections when required under the Control of Substances Hazardous to Health (COSHH) Regulations. Occupational health staff can in these circumstances provide the necessary training. Employ-

Regulations	Circumstances (examples)	Usual methods
The Working Time Regulations 1998 (amended 2009)	Night workers as defined	Confidential medical questionnaire with, where necessary, an assessment
The Control of Noise at Work Regulations 2005	Regular exposure exceeds 'upper exposure action value' and in circumstances where an individual is particularly sensitive to noise	Questionnaire and audiometry to meet Health and Safety Executive (HSE) standards
The Control of Vibration at Work Regulations 2005	Risk assessment indicates a risk to the health of the individual Or Employees are likely to be exposed to vibration at or above an exposure action value	Questionnaire and assessment as described by HSE
Control of Substances Hazardous to Health Regulations 2002	Risk of dermatitis or asthma in circumstances of use of a substance	Questionnaire and visual assessment for dermatitis Questionnaire and lung function tests for asthma
The Control of Lead at Work Regulations 2002	Significant lead exposure	Medical examination (appointed doctor) and relevant biological monitoring
The Work in Compressed Air Regulations 1996	Workers in compressed air	Medical examination (appointed doctor)
The Ionising Radiations Regulations 1999	Classified people and others as defined in the Regulations	Medical examination (appointed doctor)

Table 2   Statutory health surveillance: examples and methods

ers can check the registration of doctors and nurses using the registration system within the General Medical Council and the Nursing and Midwifery Council respectively. It is important that such statutory surveillance in which employee attendance is required is distinguished from health promotional activities such as measurement of blood pressure or blood cholesterol (which are voluntary). The latter aim to contribute to the determination of individual cardiac risk and do not target the health consequences of organisational risks.

Health promotion

This is described by the World Health Organisation as the 'process of enabling people to increase control over their health and its determinants, and thereby improve their health'. Workplace health promotion is seen as: 'the combined efforts of employers, employees and society to improve the health and well-being of people at work'.

Workplaces are seen as priority settings for health promotional activities such as dietary and exercise advice, advice on the harmful effects of drugs, alcohol (Northern Ireland Government, undated) and cigarettes and guidance on stress, etc. The construction workplace provides access to large groups of workers (as described above), many of whom do not regularly access health services. Health promotional activities aim to empower workers to take actions to improve their health. In turn this empowerment may prompt a greater uptake of measures to safeguard health and safety in the workplace.

Culture

Safety culture and climate are discussed extensively in the *National Construction Agenda* issued by the National Occupational Research Agency (NORA) in 2008. It refers to Pidgeon and O'Leary's work (2000) which argues that a 'good' safety culture may reflect and be promoted by four factors:

1 senior management commitment to safety;
2 realistic and flexible customs and practices for handling both well defined and ill defined hazards;
3 continuous organisational learning through practices such as feedback systems, monitoring, and analysis;
4 a care and concern for hazards that is shared across the workforce.

In the UK the HSE (2005) suggests five indicators which are known to influence safety culture as follows:

1 leadership
2 two-way communication
3 employee involvement
4 learning culture
5 attitude towards blame.

Senior managers and workers

Recently an approach in Northern Ireland (NI), under the Build-Health initiative has seen the emergence of construction health

champions, committed to health, with visible strong leadership and commitment from the start of the project. Empowerment and employee involvement are actively encouraged and supported. The Health and Safety Executive for Northern Ireland (HSENI), through its mentoring role, encourages dialogue on health at all levels.

Senior managers have a clear leadership role. They set the agenda, decide on resource allocation and are major influencers of culture in workplaces. In areas such as communication and employee involvement and the development of trust they have key roles to play. It is perhaps too easy to neglect workers themselves as key influencers in the health and safety agenda. In effect, the sharing with them of healthy working practices and their experience of cultures which have a health focus are key to enabling their advocacy and involvement. These are particular challenges, however, for the construction industry where approaches to worker empowerment are still relatively underdeveloped and worker transience is particularly high.

Organisational health risk management

This section describes the more common organisational health risks and their management. It emphasises the roles of designers in risk prevention and the priority of the collective over the individual approach.

Prevention: roles of designers

Designers have a major influence on health risks in the sector and upstream design decisions with a focus on injury and disease prevention have the potential to impact positively and at a collective level on downstream health. Their specifications on materials, size of blocks and the design of plant rooms and construction vehicles are but some examples. 'Designing out' of scabbling prevents risks from noise, vibration, dust and silica. Likewise these risks are removed when hydraulic shears are used for pile cropping as an alternative to hand operated jack hammers.

Specific hazards

An overview of health hazards within the construction industry is given below. For further detail on these hazards and the design issues associated with them see Chapter 8 *Assessing health issues in construction*.

Dermatitis

Dermatitis is an inflammation of the skin caused by contact with external agents. Skin irritants such as paint thinners, cement, detergents and wet work are the most common cause. However, sensitisers specifically in the case of construction workers when working with chromium (contained in cement) and epoxy resins, are also recognised causes. Exposed areas of the skin are most at risk, with the hands being most commonly involved. Symptoms include soreness of the skin with an itch. Typically in the early stages, these symptoms get worse as the working week progresses and improve over weekends. However, as the disease progress this improvement may disappear.

The skin appears dry or scaly and it may have tiny blisters. Prevention relies on choice of safest materials, collective measures to prevent exposure such as engineering solutions and limitation in use of substances, appropriate personal protective measures – gloves, etc. – and proper washing facilities with the use of conditioning creams where appropriate. Where a risk exists, health surveillance will be required.

Noise

Construction work is generally noisy as a consequence of the use of heavy equipment, power tools and vehicular movement. Research has shown that, after a lifetime of construction work, hearing loss averaged 60% among all trades and in some trades was up to 80% (Dement et al., 2005). Noise induced hearing loss is irreversible but can easily be prevented. The consequence of excessive noise is not just hearing loss but also tinnitus (ringing in the ears), which has the potential to be a much more intrusive disabling condition.

The following list provides information on typical noise levels in construction machinery:

- jackhammer: 102–111 dB
- excavator: 109 dB
- earthmover: 87–94 dB
- front end loader: 86–94 dB
- ready mix lorry: 112 dB.

Noise is frequently associated with the use of hand held vibration tools and thus construction workers are exposed to several risks from the use of one tool, for example in scabbling or the use of jackhammers.

Hearing loss caused by noise is manifest in the higher frequencies, typically reduced hearing at 4 kHz. Hearing loss is gradual and early symptoms may be an inability to distinguish high frequency consonant sounds.

Reducing exposure

Elimination of risk at source

'Designing out' and elimination of risk at source are preferred options. Issues such as technical measures to reduce noise, segregation and reduction of period of exposure, rest periods also require consideration. Legislation describes the circumstances in which hearing protection has to be offered and when it has to be worn. This is an inferior control measure given the problems with constant wearing and that bystanders are not protected

Employee engagement

It is important to tell employees about the risk and to explain the rationale for hearing protection. If the second action level of

85 dBA is reached an audiometric assessment will be required. Noise induced hearing loss has a medium latency and therefore it may be difficult to convince workers that control measures should be implemented. They should be reminded that the loss is irreversible and to share with them demonstration audios of hearing loss amongst sufferers.

Hand–arm vibration syndrome and carpal tunnel syndrome

Hand–arm vibration syndrome is a condition involving the nerves, blood vessels and muscles of the upper limb. Symptoms include whitening or blanching, numbness and tingling.

Assessment

The first step in managing this risk is to recognise the particular activities likely to be associated with the risk and to determine vibration exposure levels using available data in preference to direct measurement of the vibration. In Great Britain a daily exposure action value of 2.5 m/s^2 A(8) applies whilst the daily exposure limit value is set at 5 m/sec^2 A(8).

Prevention

Efforts should be made to reduce exposure below the daily exposure action value. Measures include design of the building (to avoid work methods likely to pose vibration risks), purchasing policies to ensure that low vibration equipment is preferentially bought, limitation of period of use and systems for tool maintenance. Where risks to health are likely health surveillance will be required (see **Table 3**). Carpal tunnel syndrome (a condition in which the median nerve at the wrist is compressed) can also occur as a consequence of vibration exposure. Both conditions are statutorily reportable in the UK.

Asbestos exposure

Asbestos minerals are compound metallic silicates which have crystallised as long thin particles. There are three types: blue asbestos (crocidolite), brown (amosite) and white (chrysotile). Risks to health of blue and brown asbestos are greater than those from the white form.

Asbestos in buildings

Very many buildings constructed in the 1950s to 2000 contain materials in which asbestos was combined with other building materials to form asbestos cement products or asbestos-containing insulating products. Such materials are commonplace in our buildings and are not easily identifiable as such. Asbestos which is in good condition does not generate fibres and there is no risk posed to health. However, damaged asbestos or work on existing asbestos-containing materials, particularly with the use of power tools, will generate significant amounts of airborne asbestos fibres.

Typical locations of asbestos in buildings include:

- lagging or insulation of boilers or pipes
- sprayed asbestos used for insulation fire protection and ducts

Management plan	Indicator
1 Identify health risks where possible at design and planning stage and design these out so far as is reasonably practicable.	Evidence that methods statements have been reviewed and that where possible risks to health have been designed out.
2 Assess risks to health from work activities and communicate these and their means of control to all at risk. (Many sites will pose a risk of dermatitis, manual handling, dust exposure and noise. The BuildHealth website will direct you to the more common risks and their means of control.)	Evidence that: health risks have been assessed and that controls have been implemented. those at risk are ware of the risks and the proper controls. (See also 7 below.)
3 For the ill employee develop a basic rehabilitation plan and communicate this to the workforce. Describe the role of the GP, DEA and occupational physician. See advice.	Employees know the range of supports and adjustments available and understand the role of the Disablement Employment Adviser, and how their GP can work best with their employer.
4 Identify safety critical work. See website: standards.	All sites have a list of safety critical jobs and standards for these tasks are identified.
5 Using the resources on the website agree the fitness standards for safety critical work with occupational health providers.	All safety critical workers are aware of the key fitness standards for their work and understand the need to communicate health changes/problems to management. The employer can demonstrate that safety critical workers reach the relevant standard of fitness.
6 Health surveillance as described in COSHH, the Control of Noise at Work, and the Control of Vibration at Work Regulations is in place. Statutory Medical Surveillance as described in the Control of Lead at Work Regulations and the Control of Asbestos Regulations is carried out according to risk.	Health surveillance records available and are properly kept.
7 There is a planned programme of training/awareness raising events on workplace health risks.	Employees understand the range of health risks on site and the controls which are available.
8 Arrangements are in place to allow workers themselves to draw attention to concerns regarding health at work as well as safety.	A procedure exists at the worksite for workers to comment on gaps in health and safety practices and such comments are actively pursued.
9 Sites undertake a review of the general health needs of workers and make plans to address them	Workers are encouraged to attend a series of promotional events to help them gain a greater understanding of the steps they can take to better their own health.
10 Company directors include worker wellbeing as a critical measure of success alongside other corporate targets.	Annual reports, press releases, speeches show reference to wellbeing of workers.

Table 3 BuildHealth Management Plan
Reproduced with permission © BuildHeath (http://www.buildhealthni.com), with acknowledgement to Mr Jim Leitch.

- asbestos insulation board
- ceiling tiles
- soffit boards below roofs.

Removal or stripping of these materials or work that could lead to high levels of fibres being released is likely to require a licensed contractor. It is important at the outset that employers consider if risks to health could arise, through a comprehensive risk assessment.

Asbestos: health effects

The health consequences of asbestos exposure include:

- asbestosis: a condition in which the lungs become more rigid
- lung cancer
- mesothelioma: a tumour of the lining of the lung or of the abdomen
- pleural thickening
- pleural plaques.

Employee awareness

A 2008 survey conducted by the British Lung Foundation (2008a, 2008b) has shown that awareness among construction tradespeople of the hazard posed by asbestos is low. The survey of 400 tradespeople found that:

- less than a third were aware that asbestos exposure can cause cancer;
- 12% knew that asbestos exposure could kill them;
- 30% wrongly believed most asbestos has been removed from UK buildings;
- 74% have had no training in how to deal with asbestos.

Prevention and employee engagement

Within the construction industry it is extremely important that all workers are aware of the extensive use of asbestos and that they know how to take steps to ensure that any building in which they work is free of asbestos. The statutory requirement which includes, in the case of non-domestic premises, the requirement to have an asbestos register, should ensure that workers are alerted to its presence.

Dust and crystalline silica exposure

Many construction related activities are associated with dust generation. Concrete and mortar along with certain rock types contain crystalline silica and work activities such as concrete scabbling, cutting, drilling, demolition, façade work, stone masonry and blast cleaning are among those in which exposure can occur.

Prevention

The health risks of silica exposure include silicosis and lung cancer. Assessment of the risk will be required. Design and choice of materials at the planning stage should help eliminate the risk. Dust suppression and personal protective wear will also require consideration. As is the case with asbestos, employee engagement to ensure adequate knowledge of the risk and of the appropriate control measures is essential.

Many construction sites are dusty. Such heavy dust exposure has been linked to the development of chronic obstructive pulmonary disease. This serves to emphasise the need to implement measures to suppress all dusts.

Back pain and musculoskeletal disorders

Musculoskeletal disorders are associated with various factors both physical and psychological. Construction work is physically demanding involving manual materials handling, lifting, bending, thrusting and risks of whole body vibration. Such activities are linked to increased reports of back pain, aggravation of symptoms and 'injuries'. The construction industry is associated with high rates of musculoskeletal disorders as evidenced by a Swedish study showing that 72% of all sick leaves longer than 4 weeks was due to musculoskeletal problems (Holmstrom et al., 1992) and a further study which showed that early retirements due to these conditions were more common in construction workers than other men. Designers through their specification as well as work methods which reduce the amount of bending can help prevent these conditions. Brick, block, kerb and paving slabs laying, moving and installation of plasterboard and mechanical and electrical equipment at height are some examples of construction-related activities which pose risks to the musculoskeletal system.

Risk assessment of such tasks will be necessary and options to avoid or reduce the collective risk such as the use of mechanical lifting equipment should be given priority. Likewise tasks can be altered and work load or the work environment changed to help reduce the risk. Training on safe systems is important and consideration should be given to measures of effective implementation of interventions (van der Molen et al., 2005). Employees' understanding of the means to self-manage back pain should it occur can be; promoted through the use of information booklets such as the *Back Book* (Bigos et al., 2002). Evidence supports the use of early intervention in the management of simple back pain. Some temporary adjustment to work activities may be necessary but it is important to emphasise that recovery is expected in the majority of people who suffer from simple back pain.

Unnecessary worklessness is to be avoided given the association between unemployment and poor health, particularly psychological ill health.

Other risks: carcinogens and biological agents

Work activities within construction pose risk of exposure to established human carcinogens including ultraviolet light, asbestos and silica as well as possible carcinogens such as diesel fume. Information on cancer, its prevention and workplace management, has been developed (HSENI, 2009). Certain construction work activities such as those in sewers may pose risks from leptospirosis, a bacterial disease transmitted through contact with rat urine. Tetanus is also a potential risk where organisms in soil contaminate a wound.

Summary of main points

Construction sites present many opportunities to protect and improve health. Leadership and employee engagement are essential elements of a healthy workplace culture. Systems to address organisational risks and to support and rehabilitate workers when illness supervenes need further development in this sector as does health promotion. Benefits to workers' health and wellbeing, productivity and overall population wellbeing are likely to accrue.

Disclaimer

The views expressed in this chapter are those of the author alone and not necessarily those of Health & Safety Executive for Northern Ireland.

Box 1 Case Studies

Health champions sites: Northern Ireland approach

Under the BuildHealth initiative, Health Champion sites were identified that are intended to serve as centres of excellence for occupational health practice and to promote dissemination of health messages to a range of employees including smaller sub-contractors and their employees. A Management Plan which contains all the modalities is used as the vehicle to address health on such larger sites; a copy is provided in **Table 3**.

BuildHealth website http://www. buildhealthni.com/

A central feature of the BuildHealth initiative was the creation of a website. The purpose of the BuildHealth website is to provide information on the range of risks within all aspects of occupational health as described above. The core of the website is a series of fact sheets relating to organisational risks, fitness for work and safety critical work and health surveillance, health promotion, and drugs and alcohol. These are presented on the web as easy steps to implement with added information being provided through web linked tools.

References

Bigos, S., Roland, M., Waddell, G., Klaber Moffett, J., Burton, A. K. and Main, C. J.. *The Back Book,* 2nd edition, London: The Stationery Office.

Black, Dame Carol. *Working for a Healthier Tomorrow. Dame Carol Black's Review of the Health of Britain's Working Age Population,* 2008, London: The Stationery Office. Available online at: http://www. workingforhealth.gov.uk/documents/working-for-a-healthier-tomorrow-tagged.pdf

Brenner, H. and Ahern, W. Sickness Absence and Health Retirement on Health Grounds in the Construction Industry in Ireland. *Occupational Environmental Medicine* 2000, **57**, 615–620.

British Lung Foundation. *Tradespeople Asbestos Awareness Survey,* 2008a. Available online at: http://www.lunguk.org/Resources/British% 20Lung%20Foundation/Website/Media%20and%20Campaigning/ Documents/Survey%20report%20final_formatted.pdf

British Lung Foundation BLF Survey Reveals Alarming Ignorance of Asbestos Risk Amongst Tradespeople, Press Release, February 2008b. Available online at http://www.lunguk.org/ media-and-campaigning/media-centre/archive-press-releases-and-statements/feb2008/BLFsurveyrevealsalarmingignoranceo-fasbestosriskamongsttradespeople.htm

Constructing Better Health. *Heath Assessment Matrix,* 2009. Available online at: http://www.constructingbetterhealth.co.uk/wcore/showdoc.asp?id=599

Dement, J., Ringen, K., Welch, L., Bingham, E. and Quinn, P. Surveillance of Hearing Loss Among Construction and Trade Workers at Department of Energy Nuclear Sites. *American Journal of Industrial Medicine* 2005, **48**, 348–358.

Donaghy, R. *One Death is Too Many: Inquiry Into The Underlying Cause of Construction Related Accidents. Rita Donaghy's Report to the Secretary of State Department of Work and Pensions,* 2009, London: The Stationery Office. Available online at: http://www. dwp.gov.uk/docs/one-death-is-too-many.pdf

Faculty of Occupational Medicine. *Guidance on Alcohol and Drug Misuse in the Workplace,* 2006, London: Faculty of Occupational Medicine of the Royal College of Physicians.

Health and Safety Executive (HSE). *Asbestos Health and Safety,* 2004. Available online at: http://www.hse.gov.uk/asbestos/index.htm

Health and Safety Executive (HSE). *RR367 – A Review of the Safety Culture and Safety Climate Literature for the Development of the Safety Culture Inspection Toolkit,* 2005, London: HSE Books. Available online at: http://www.hse.gov.uk/research/rrhtm/rr367.htm

Health and Safety Executive (HSE). *Managing Sickness Absence and Return to Work,* undated. Available online at: http://www.hse.gov. uk/sicknessabsence

Health and Safety Executive Northern Ireland (HSENI). *Advice on Cancer for Construction Workers,* Belfast: HSENI. Available online at: http://www.hseni.gov.uk/advice_on_cancer-2.pdf

Holmstrom, E., Lindel, l. J. and Moritz, U. Low Back and Neck/ Shoulder Pain in Construction Workers; Occupational Workload and Psychosocial Risk Factors. Part 2 Relationship to Neck/ Shoulder Pain. *Spine* 1992, **17**, 672–677.

National Occupation Research Agenda (NORA) Construction Sector Council. 2008. *National Construction Agenda: For Occupational Safety and Health Research and Practice in the US Construction Sector.* Available online at: http://www.cdc.gov/niosh/nora/ comment/agendas/construction/

Northern Ireland Government, Department of Health, Social Service and Public Health. *Workplace Drugs and Alcohol Policies,* undated. Available online at: http://www.dhsspsni.gov.uk/workplace_ policies1.pdf

Pidgeon, N. and O'Leary, M. Man-made Disasters: Why Technology and Organisations (Sometimes) Fail. *Safety Science* 2000, **34**, 15–30.

Scottish Executive. *Healthy Working Live: a Plan for Action,* 2004, Edinburgh: Scottish Executive, Available online at: http://www. scotland.gov.uk/Publications/2004/08/hwls/0

Stenlund, B. The Galaxen Model – A Concept for Rehabilitation and Prevention in the Construction Industry. *Scandinavian Journal of Work Environmental Health* 2005, **31** (suppl. 2), 110–115.

Van der Molen, H. F., Sluiter, J. K., Hulshof, C. T. J., Vink, P. and Frings-Dresen, M. H. W. Effectiveness of measures and implementing strategies in reducing physical work demands due to manual handling at work. *Scandinavian Journal of Work Environmental Health* 2005, **31** (suppl. 2), 75–88.

Waddell, G. and Burton, A. K. *Work and Health: Changing How We Think About Common Health Problems,* 2006, London: The Stationery Office. Available online at: http://www.mindfulemployer.net/ Work%20and%20Health.pdf

Waddell, G., Burton, A. K. and Kendall, N. A. S. (Vocational Rehabilitation Task Group – Industrial Injuries Advisory Council). *Vocational Rehabilitation: What Works for Whom and When?,* 2008, London: The Stationery Office.

Referenced legislation

Access to Medical Reports Act 1998. Elizabeth II. Chapter 28, London: HMSO.

Control of Asbestos Regulations 2006. Reprinted November 2006, January 2007 and March 2007. Statutory Instruments 2739 2006, London: The Stationery Office.

Control of Lead at Work Regulations 2002. Statutory Instruments 2676 2002, London: The Stationery Office.

Control of Noise at Work Regulations 2005. Statutory Instruments 1643 2005, London: The Stationery Office.

Control of Substances Hazardous to Health Regulations 2002. Reprinted April 2004 and March 2007. Statutory Instruments 2677 2002, London: The Stationery Office.

Control of Vibration at Work Regulations 2005. Statutory Instruments 1093 2005, London: The Stationery Office.

Disability Discrimination Act 1995. Elizabeth II – Chapter 13, London: The Stationery Office.

Ionising Radiations Regulations 1999. Statutory Instruments 1999 3232, London: The Stationery Office.

Work in Compressed Air Regulations 1996. Statutory Instruments 1996 1656, London: HMSO.

Working Time Regulations 1998. Statutory Instruments 1998 1833, London: The Stationery Office.

Working Time (Amendment) Regulations 2009. Statutory Instruments 1567 2009, London: The Stationery Office.

Further reading

Carpenter, J. *Series of Health Guidance Sheets for Designers,* 2010. Available online at: http://www.ice.org.uk/knowledge/specialist_ health.asp

Introduction and explanation; Site set-up; Groundworks; Concrete works; Steelwork; Buildings; Housing; Bridges; Demolition; Refurbishment; Drainage and services.

Websites

BuildHealth http://www.buildhealthni.com

Chartered Institute of Personnel and Development (CIPD) http://www.cipd.co.uk

Constructing Better Health http://www.constructingbetterhealth.co.uk

Faculty of Occupational Medicine of the Royal College of Physicians (FOM) http://www.facoccmed.ac.uk

Health and Safety Executive http://www.hse.gov.uk

Health and Safety Executive Northern Ireland (HSENI) http://www.hseni.gov.uk

HSE, Control of Substances Hazardous to Health (COSHH) http://www.hse.gov.uk/COSHH/index.htm

Institution of Civil Engineers, Health and Safety http://www.ice.org.uk/knowledge/specialist_health.asp

The National Institute for Occupational Safety and Health (NIOSH), USA government site http://www.cdc.gov/niosh/

World Health Organisation (WHO) http://www.who.int/en/

Section 2: Workforce issues

ice | manuals

Chapter 3

Responsibilities of key duty holders in construction design and management

doi: 10:10.1680/mohs.40564.0029

David A. O. Oloke Construction and Infrastructure Department, SEBE, University of Wolverhampton, UK and Progressive Concept Consultancy (pCC) Ltd, Walsall, UK

CONTENTS

Construction often involves a complex set of operations that culminate in the delivery of a product. Several varying human and environmental factors occur throughout the process and these often generate several independent and/or inter-dependent hazards. The process of health and safety risk management will thus need to be innovative and on a par with the life cycle of the project. It is therefore important that all participants are aware of the responsibility thrust upon them individually and acting jointly throughout the entire process.

The Construction (Design and Management) Regulations 2007 place specific duties on key participants on any construction project. This chapter describes the key Duty Holders whilst also giving an overview of their main responsibilities. Some interrelationships of these roles are also examined, with a view to enhancing the effective understanding and implementation of one of the basic requirements of the Regulations – Communication. Construction Design and Management is always evolving and it is essential that Duty Holders update their knowledge, not only on emerging technology and concepts, but also on processes and systems that will enhance effective health and safety management. A case study which presents some emerging innovative techniques for managing health and safety on a typical construction procurement cycle is discussed.

Box 1	Key learning points

- To define key Duty Holders under the CDM Regulations 2007.
- To describe the main responsibilities of each key Duty Holder under the CDM Regulations 2007 in line with the Approved Code of Practice (ACoP) and a wide range of available Industry Guidance.
- To examine emerging trends and discuss innovative techniques for accomplishing these roles.

Introduction

Construction can be an intricate process that often involves a complex network of operations that culminate in the delivery of a product. More often than not, this product (a building, a highway or other infrastructure) is unique even when the original intention was to replicate an existing prototype. This attribute of uniqueness of the product happens as a result of the varying human and environmental factors that are at interplay. These factors introduce a set of often interrelated hazards that imply that risk management will need to be innovative and in step with the life cycle of the project. It is therefore important that all participants are aware of the responsibility thrust upon them individually and acting jointly throughout the entire process.

Since the advent of the Construction Design and Management (CDM) Regulations 2007 on 6 April 2007, the United Kingdom (UK) construction industry has been witnessing a variety of efforts aimed at giving publicity to the Regulations – which replaced the CDM 1994 which had been in force up to that moment. Particularly, it was considered most important that those who would be entrusted with responsibilities under the Regulations be made to understand clearly what their responsibilities were. Also, where these had either been non-existent or were a wide variant from the CDM 1994 roles, it is considered necessary that those differences are clearly understood by all involved.

The CDM 2007 recognises Duty Holders as those with specific health and safety roles to play in the procurement and delivery of construction projects. These include: Client, CDM Coordinator, Designer, Principal Contractor and Contractor.

The Approved Code of Practice (ACoP)

The ACoP (HSE, 2007) gives useful information to Duty Holders, which enables them to understand their roles and duties under the CDM 2007 Regulations. Hence, the ACoP should help Duty Holders fulfil their various duties. It highlights the need for them to understand the legal duties placed on them as Clients, CDM Coordinators, Designers, Principal Contractors, Contractors, self-employed and staff workers. It also enables Duty Holders better to understand the circumstances in which domestic Clients do not have duties under CDM 2007 – but how the Regulations still apply to those doing work for them. The ACoP elaborates on the variants of the 'new' role of CDM Coordinator, who under the CDM 2007 is a key project adviser for Clients and also responsible for coordinating the arrangements for health and safety during the planning phase of larger and more complex projects.

Furthermore, it is important that Duty Holders know which construction projects need to be notified to the Health and Safety Executive (HSE) before work starts and the ACoP gives information on how this should be done. Notifiable projects are those that would last for 30 working days or more or involve more than five person days (HSE, 2007). The CDM Coordinator is responsible for this process and is to inform the HSE through the completion of the HSE Form 10. Guidance is also given on how to assess the competence of organisations and individuals involved in construction work and on how to improve cooperation and coordination between all those involved in the construction project and with the workforce. Finally, a major thrust of CDM 2007 is the emphasis on reducing unnecessary paperwork; the ACoP emphasises what essential information needs to be recorded in construction health and safety plans and files, as well as what shouldn't be included.

The duties of the individual Duty Holders are now discussed in summary.

The Client

A Client is defined in CDM Regulations as the person who, in the course or furtherance of a business, seeks or accepts the services of another which may be used in the carrying out of a project for him/her or a person who carries out a project him/herself.

Due to the fact that Clients have a substantial influence on the way the project is run, it is very important that the decisions they make are carefully considered as these directly affect resource control, project team composition and coordination.

According to the CDM 2007 Guidance (Construction Skills and HSE, 2007a), Clients include local authorities, school governors, insurance companies and project originators on Private Finance Initiative (PFI) projects. The range of projects commissioned/undertaken by these Clients varies enormously in size and complexity. It should be noted, however, that Client duties are fundamentally the same.

Before the project commences, it is important to identify who within the Client organisation will be the single point of contact, i.e. 'the Client's representative', for the project. This is not required by the Regulations but is considered best practice and can avoid confusion about 'who was going to do what and when' that may arise as the project develops. The Office of Government Commerce (OGC) has produced good practice examples for Client organisations in the public sector. These

may be useful for private sector Clients who undertake a large project too.

Some large and experienced Clients would normally be in a position to nominate a 'representative' from the membership of their in-house staff. However, if relevant competencies are not available, a third party can be appointed to act in this role for you. It is a worthy caution to note, however, that CDM 2007 does not allow the Client to transfer their duties to a third party, as was the case in CDM 1994 for the 'Client's agent' role. Inexperienced Clients can, however, find consolation in the fact that the appointment of an external agent to carry out the Client's duties is still acceptable. Albeit, in such cases, the Client retains the responsibility to ensure that their duties under these regulations are met and can only delegate the tasks. It is thus essential that the arrangements for the reporting and monitoring of this role are carefully considered by the Client.

Generally, Clients are now required to check competency of Designers, Contractors and individuals doing the work and to give as much information as possible or as required about the site and/or premises in advance of the works. It is also pertinent that Clients take reasonable steps to ensure that their arrangements for managing their own duties on a project, as well as those of others with duties under CDM 2007, are suitable. (As previously stated, however, Clients may seek professional advice in ensuring that this and other duties are satisfactorily performed.)

In terms of operational considerations, Clients need to factor in the time for setting up the proper procedures prior to commissioning work and generally cooperate with the Contractor to allow them to discharge their duties. Appropriate lead-in time should be allowed to account for the unique requirements of the project. Their own work has to be coordinated so it does not affect the safety of other Duty Holders, especially those doing the job on site, and they need also to ensure that the Contractor has appropriate welfare in place before work starts. They should make sure there are arrangements in place so that what is built complies with, amongst others, the various Workplace Health, Safety and Welfare Regulations. It is also the Client's responsibility to ensure that they are provided with information about what has been built (the Health and Safety File) to enable future management of health and safety in the building over its life and ultimate demolition (including partial demolition). Hence, they should make sure that a Health and Safety File is prepared, ready for handover at the end of the project, and kept readily available for future work/new owners.

In addition, and where projects are Notifiable, Clients are also expected to appoint a competent CDM Coordinator, who will assist them with their duties and other legal functions, and also appoint a competent Principal Contractor to plan and manage the work. The Client should not allow the work to start until the Contractor has prepared a suitably developed plan to manage health and safety and installed welfare facilities.

However, Clients do not have to plan or manage construction projects or specify how work should be done – that is provide method statements – although they need to be clear about the brief. They do not need to provide welfare facilities

ICE manual of health and safety in construction © 2010 Institution of Civil Engineers

or be expert in determining what constitutes good or adequate welfare. Clients are also not expected to: check designs to see that they comply with Designer duties and that the designs satisfy all the Workplace Regulations; visit the site to supervise or check construction work standards; employ third-party assurance advisers to monitor health and safety standards on-site or subscribe to a third-party competence assessment scheme.

Where work is not Notifiable, the appointment of a CDM Coordinator and Principal Contractor is not a required duty of the Client. By implication, therefore, they are not expected to have a health and safety plan in place before construction work starts and receive a Health and Safety File. However, good practice would dictate that similar arrangements – that is providing health, safety and welfare facilities – on the smaller projects, although not a statutory requirement, would make good sense.

The CDM Coordinator

A CDM Coordinator (CDM-C) is the Client's adviser in matters relating to construction health and safety. The role involves advising and assisting the Client in undertaking the measures needed to comply with CDM 2007, including, in particular, the Client's duties both at the start of the construction phase and during it. It is important that any person or organisation acting in this capacity has a firm understanding of the general requirements of the project.

According to the CDM 2007 Guidance (Construction Skills and HSE, 2007b), the CDM-C's role is most vital to the success of a project. A CDM-C is responsible for providing proactive advice and practical help to the Client in response to Client and project demands. He or she should also provide specific advice, systems or support to the Client on compliance with regulation 4 and Appendix 4 of the ACoP (HSE, 2007) relating to health and safety resources and competence. It is also the responsibility of the CDM-C to support the Client in identifying and ensuring suitable arrangements for the project, how they will be delivered by the team to achieve project safety and other related Client–project benefits, and how they will be reviewed and maintained throughout the life of the project. It is understood, however, that some Clients may have arrangements in place already, which may mean that less advice and assistance from the CDM-C is required. The Client may also have specific requirements that will need to be implemented by other Duty Holders.

Information flow is a key requirement of the successful implementation of CDM 2007. The CDM-C is the key Duty Holder who is thus responsible for developing a strategy with the team for maintaining the flow of relevant health and safety related information throughout the lifetime of the project to make sure that what is needed reaches the right people at the right times. This includes information required by Designers, pre-construction information, whenever it is required, and information for the Health and Safety File. This further implies that the CDM-C should promote the suitability and compatibility of designs and actively seek the cooperation of Designers at all project phases when dealing with the risk consequences of

construction and workplace design decisions.

Traditionally, the Client (under the older CDM Regulations) had a statutory duty to engage an Agent for the purposes of providing technical support through the CDM process. However, CDM 2007 no longer recognises the Agent's role. Under the current Regulations, the CDM-C is responsible for providing support to the Client and advising on the suitability of the Principal Contractor's construction phase plan. The CDM-C is also further responsible for encouraging and developing links between permanent and temporary works Designers and actively liaising with the Principal Contractor to ensure safe design. This role is highly essential on a variety of projects – increasing with complexity. On large scale redevelopment projects, for example, the CDM-C provides the much needed interface between Designers where substantial elements of temporary works design are involved in the development. On such projects, design requirements can be dramatically influenced after opening-up works. In addition, demolition can reveal the need to reconsider assumed load paths. Invariably, therefore, the design of both temporary and permanent works and the role of the CDM-C in ensuring that all requirements for health and safety must be adequately considered throughout the entire process.

The Designer

A Designer is defined in the CDM Regulations as any person who in the course or furtherance of a business prepares or modifies a design and/or arranges for or instructs any other person under their control to do so.

The above definition makes the role of a Designer much broader than the historical/traditional role of a Designer prior to the introduction of the CDM Regulations. It implies that the Client, Principal Contractor or others could be undertaking the Designer duties if any of their work or their contribution to the project are encompassed by the definitions above. However, for a Designer it is imperative to consider how their designs can be built, used, maintained and demolished without causing harm to the health and safety of construction or maintenance workers or to those who use the buildings or structures.

In respect of the discharge of their duties for all projects, therefore, Designers must make sure that they are competent and adequately resourced to address the health and safety issues likely to be involved in the design. Considering that, in a large number of cases, they are also the first to be approached by the Client, they will also have to check that Clients are aware of their duties.

Furthermore, when carrying out design work, Designers are required to avoid foreseeable risks to those involved in the construction and future use of the structure. In doing so, they should eliminate hazards (so far as is reasonably practicable, taking account of other design considerations) and reduce risk associated with those hazards that remain. One way they can achieve this is to ensure that they provide adequate information about any significant risks associated with the design. It is important to stress that CDM 2007 places emphasis on significant risks as opposed to the very obvious risks which would normally be accounted for in the Contractor's Method Statement, and Designers are particularly required to ensure this is complied with. Finally, Designers are required to coordinate their work with that of others in order to improve the way in which risks are managed and controlled.

These duties, when carried out with due diligence, help to mitigate the health and safety risks associated with the construction processes resulting from the design. **Tables 1** and **2** highlight these hazards in the light of safety and health issues respectively.

It behoves Designers, therefore, to ensure that the implications of their designs are carefully considered, utilising the most appropriate risk assessment techniques. One of the major thrusts of CDM 2007 is the need to reduce the amount of unwarranted paperwork which appeared to have characterised the implementation of CDM 1994. It is therefore essential that Designers ensure that they undertake and report Design Risk Assessments that are very specific to the requirements of the project as informed by their designs.

Also often neglected by Designers is the need to consider the project's whole life. The design choices that are made must be such that they do not only look at buildability/constructability matters in isolation, but that they extensively consider maintenance and demolition of the facilties at the end of their useful life as well.

The Principal Contractor

A Principal Contractor (PC) is responsible for Contractors and workers on site and for planning, managing and monitoring the construction phase in a way which ensures that, so far as is reasonably practicable, it is carried out without risks to health or safety.

The CDM 2007 Guidance (Construction Skills and HSE, 2007d) states that Principal Contractors must satisfy themselves that Clients are aware of their duties to ensure that for notifiable projects a Coordinator has been appointed and HSE has been notified before they start work. They need to ensure

Working at height	Falling from height is the biggest (single) cause of fatalities. There are steps that can be taken at the design stage to eliminate or mitigate this, e.g. prefabrication, maintenance strategies
Vehicles and other transport	The likelihood of being struck or crushed by construction (or 'in use') vehicles is reduced by strategic consideration of circulation, separation and space. During construction the Contractor is responsible for the detailed implementation, but the Designer can sometimes facilitate this by appropriate consideration during the design phase. 'In use' issues will need to be discussed with the Client
Power cables and electrical installations Structural instability	The risk of electrocution emphasises the need for good information (from surveys if necessary) and avoidance of unnecessary activity in the vicinity of electrical supplies. This is particularly important on refurbishment/extension projects
Slips, trips and falls	Risk of collapses typically applies to buildings and trenches. Be extra vigilant when refurbishing buildings. Consider carefully the need for deep trenches, and their excavation, if adjacent to other works. These account for large number of injuries and are very disruptive and costly overall
Others	Project specific hazards, e.g. significant fire risks arising from the design

Table 1 Safety hazards that Designers influence
Reproduced, with permission, from Construction Skills and Health and Safety Executive, 2007c. © Construction Industry Training Board 2007.

Musculoskeletal	This is one of the most common causes of ill health. Designers should consider lifting (e.g. choice of unit size), operating space, and the ergonomics of relevant activities
Noise-induced hearing loss	Current Regulations require significant reductions in the level of exposure to workers from those previously tolerated. If noisy or vibration-prone activities unnecessarily result from the design, this may result in additional project costs
Hand–arm and whole body vibration	Designers can obtain useful advice from Contractors and suppliers of equipment
Dermatitis and other skin related problems	Designers should consider whether there are alternatives to materials or processes which cause particular problems
Asbestos related diseases	This is a major issue on refurbishment projects. Influence can be exerted through adequate information provision and careful consideration of survey information and the management plan
Other	Project specific, e.g. presence of vermin and bird excreta, specific materials, dusts, sprays, contaminated land, lead

Table 2 Health hazards that designers influence
Reproduced, with permission, from Construction Skills and Health and Safety Executive, 2007c. © Construction Industry Training Board 2007.

ICE manual of health and safety in construction © 2010 Institution of Civil Engineers

that they are competent to address the health and safety issues likely to be involved in the management of the construction phase and that the construction phase is properly planned, managed and monitored, with adequately resourced, competent site management appropriate to the risk and activity.

The PC on a project will also need to ensure that every Contractor who will work on the project is informed of the minimum time they will be allowed for planning and preparation before they begin work on site in addition to ensuring that all Contractors are provided with the information about the project that they need to enable them to carry out their work safely and without risks to health. Requests from Contractors for information should be met promptly. Where the assistance of the CDM-C/Client is required to facilitate this, it is essential that appropriate steps are taken promptly.

It is a key duty of the PC to ensure safe working and coordination and cooperation between Contractors and that a suitably developed Construction Phase Health and Safety Plan ('the Plan') is prepared before construction work begins and that this is developed in discussion with, and communicated to, Contractors affected by it. The Plan should be implemented and kept up to date as the project progresses and should be reviewed prior to any significant changes being made. The PC has the benefit of examining arrangements put in place by the Client to ensure that they satisfy themselves that the Designers and Contractors they engage are competent and adequately resourced.

The PC is further required to ensure that suitable welfare facilities are provided from the start of the construction phase. These should include sanitary conveniences, washing facilities, drinking water, changing rooms and secure storage, facilities for rest such as suitable arrangements to ensure meals can be prepared and eaten and a means of boiling water. They should also take reasonable steps to prevent unauthorised access to the site by preparing and enforcing any necessary site rules in addition to displaying the completed project notification.

It is essential for the PC to provide access to relevant parts of the Plan and other information to Contractors, including self-employed persons, in sufficient time for them to plan their work. In addition, effective liaison with the CDM-C on design carried out during the construction phase, including design by specialist Contractors and any implications this may have on the Plan, is essential. Communication with parties under the control of the PC must be directed via the PC, who should promptly provide the CDM-C with any of such information relevant to the Health and Safety File.

On workforce matters, it is the PC's responsibility to ensure that all the workers have been provided with suitable health and safety induction, information and training and that they are effectively consulted about health and safety matters.

The Contractor

A Contractor is defined in the CDM Regulations as any person (including a Client, PC or other Duty Holder) who, in the course or furtherance of a business, carries out or manages construc-

tion work. Part 2 of CDM 2007 stipulates that for all projects – including domestic and non-Notifiable projects, Contractors must carry out certain specific duties.

Specifically, Contractors must check that Clients are aware of their duties and must satisfy themselves that they and anyone they employ or engage are competent and adequately resourced. The must also cooperate with others and coordinate their work with others working on the project. If a Contractor is overseeing the work for a domestic Client, then they should ensure that the Contractors' work is properly coordinated, and that there is good cooperation and communication.

All Contractors need to plan, manage, and monitor their own work and that of their workers to make sure that they are safe when starting their work on site, that it is carried out safely and that health risks are also addressed. The effort invested in this should reflect the risk involved and the experience and track record of the Contractor's workers. Where unsafe practices are identified, appropriate remedial action must be taken to redress the situation.

It is essential for Contractors also to ensure that all of their Sub-contractors are informed of the amount of time that they will have for planning and preparation before they are expected to begin construction work. Their workers (whether employed or self-employed) should also be provided with any necessary information, including site induction, training, information from risk assessments, and relevant aspects of other Contractors' work (where not provided by a Principal Contractor), to enable them to work safely.

Furthermore, under Part 4 (regulations 25–44, setting out the general duties on construction sites) of CDM 2007, for all projects, including domestic and non-Notifiable projects, Contractors are required to ensure that the site is a safe place to work – there should be suitable and sufficient access, egress, and working space. The site should be properly maintained. Risk assessments should take account of machinery and the risks they pose to individuals. The site should be clean, tidy and secure from trespass. There is a specific requirement not to leave timber and other materials with projecting nails where someone could step, trip or fall on them. Contractors are also responsible for recommending regular site inspections.

They should always ensure that as part of a safe system of work, structures do not collapse and are not overloaded or misused. Work involving demolition and dismantling should be planned and hazards managed. Arrangements for this should be recorded in writing before work begins. Explosives should be stored, transported and used safely and securely without endangering anyone through blast or debris.

The use of adequate supports or other methods to prevent collapse, in addition to the use of edge protection and other appropriate measures to prevent anything falling into excavated areas, should be adopted. Such areas should be inspected at the beginning of each shift, and after any event which may potentially cause damage, particularly after rain. In particular, it is essential that heavy plant is not allowed to operate near the sides, as this may weaken the excavation. The safety of cofferdams and caissons must be ensured and they must be suitable, well maintained, and appropriate, with good escape routes. They must also be inspected at the beginning of each shift and after any event which may potentially cause damage. If the Contractor is required to inspect an excavation, cofferdam or caisson then the relevant Regulations from Part 4 will apply. Where the person carrying out the inspection is not content that work can be carried out safely, they must provide a report to the person for whom the inspection was carried out. These reports are similar to those used in other construction-related legislation (see Chapter 1 *Legal principles* for further information on legislation), such as the working at height regulations (see Chapter 14 *Working at height and roofwork* for further coverage of working at heights). The person who receives the report must keep a copy available for inspectors on site for three months after that work is completed.

On-site responsibilities of the Contractor also demand that the safety of energy distribution installations, including underground, overhead, or concealed temporary or permanent supplies, should be ensured. Electric generators and cables particularly need to be safely managed and hazards highlighted with warning signs where they cannot be eliminated.

Appropriate steps should be taken to ensure the prevention of drowning where likelihood is high, in addition to ensuring that adequate provision is made for rescue equipment. Transport of any persons over water should be safe, that is, any vessel used should not be overloaded or over-crowded. See Chapter 21 *Working on, in, over or near water.*

It is the responsibility of the Contractor also to ensure that the risk of traffic and vehicular movement accidents is minimised as far as is reasonably practicable. They are to ensure the safety of site traffic routes and the safe use of vehicles on site. The vehicles must be ensured to be loaded, operated, unloaded or towed in a manner which does not put the safety of driver, passenger, pedestrians or other individuals at risk. This includes prevention of unintended movement of a vehicle, and taking steps to prevent the vehicle's fall into an excavation, pit or water, or it overrunning the edge of any embankment or earthworks. See Chapter 19 *Transportation and vehicle movement* for further detail.

Other responsibilities of the Contractor include ensuring that there is adequate provision made for the prevention of risk from fire, explosion flooding or asphyxiation and that the emergency procedures are suitable and sufficient for any foreseeable hazard. In addition, they should also ensure that the emergency routes and exits are suitable and sufficient for any foreseeable hazard, as well as ensuring the provision of adequate temperature and weather protection and the provision of adequate lighting (see Section 5 *Safety hazards* for further coverage of the hazards mentioned here).

The Guidance (Construction Skills and HSE, 2007e) also states that a report of inspection must include the following salient items of information: name and address of the person on whose behalf the inspection was carried out; location of the place of work inspected; description of the place of work, or part of that place, inspected (including any work equipment and materials); date and time of the inspection; details of any

matter identified that could give rise to a risk to the health or safety of any person; details of any action taken as a result of any matter identified; details of any further action considered necessary; and the name and position of the person making the report.

In addition to these specific duties, Contractors are also obliged to ensure that any design work they do complies with regulation 11 of CDM 2007. The regulations applicable to Contractors which apply to Notifiable projects only are contained in regulation 19.

However, for Notifiable projects, in addition to the above, Contractors are also required to: check that a CDM-C has been appointed and HSE notified before starting work (having sight of a copy of the notification of the project to HSE (Form 10) with the appointments detailed in it is normally sufficient). Where a firm is adequately resourced enough to provide the CDM-C role (in addition to their Designer duties), it is advisable to ensure that the CDM-C role is fulfilled by persons independent of the actual design team. However, as Contractors they will need to cooperate with the Principal Contractor, CDM-C and others working on the project. They are under obligation to tell the Principal Contractor about risks to others created by their work. This includes anything, for example from Risk Assessments and written systems of work, which might justify a review or update of the construction phase plan. Also, a relationships protocol could be considered to be a useful document in the plan, where more than one contractor is on site, setting out who does what and when.

In furtherance of the above, therefore, they are to also comply with any reasonable directions from the PC and with any relevant rules in the construction phase plan. Any problems with the plan or risks identified during their work that have significant implications for the management of the project should be highlighted. This includes anything, for example from Risk Assessments and written systems of work, which might justify a review or update of the construction phase plan. Contractors have to tell the PC about accidents and dangerous occurrences. Incidents occurring under the Reporting of Injuries, Diseases and Dangerous Occurrences Regulations 1995 (RIDDOR) must be notified to the PC, so that they can monitor compliance with health and safety law and, if necessary, review the arrangements for the management of health and safety. Generally, information for the Health and Safety File must be provided in good time. In addition, where Contractors are involved in design work, including temporary works, they also have duties as a Designer and this also applies to non-Notifiable projects.

As part of measures to ensure a robust and efficient health and safety management system, Contractors must generally ensure that work does not commence until they are sure that the Client is aware of their duties under CDM 2007 and that the names of the CDM-C and PC have been made known – with the HSE being notified of these arrangements where the project is Notifiable. Contractors need to provide access to the parts of the construction phase plan which are relevant to their work. It is their duty also to check that suitable and sufficient welfare facilities have been put in place by the PC and that suitable and sufficient precautions have been taken by the PC to prevent access to the site by unauthorised persons.

In addition, on all Notifiable projects, a Contractor will be working under the direction and within the management arrangements of a PC. However, if some of the work is sub-contracted to another firm, it will be the Contractor's responsibility to ensure that any Sub-contractors engaged abide by the same rules that they themselves are adhering to. The Sub-contractors must do likewise for any Sub-contractors that they in turn may take on.

It is necessary to stipulate in the contract with the Sub-contractors that they should insist on the same criteria from any of their Sub-contractors that is insisted on from them, including: assessing the Sub-contractors' competence; passing on relevant information; controlling Sub-contractors and ensuring that they work as planned and that they comply with the site rules. They should also ensure that there is coordination between their Sub-contractors, their own employees, other Contractors and the Principal Contractor. Hence, they should coordinate communications between their company and other Duty Holders, including Sub-contractors, where not already provided for by a PC. This should include information on who speaks to whom, when and what about; whilst they must also ensure the setting-out process for submission of Method Statements and Risk Assessments to the principal or main Contractor is made in good time. Contractors should ensure that their Sub-contractors suitably instruct, supervise and train their workforce and that they comply with the site rules and any induction processes – allowing Sub-contractors sufficient time to prepare and carry out the work. It is also required that they inform the Principal Contractor about whom they have sub-contracted.

The Sub-contractor

Sub-contractors generally carry out specific duties under the Contractor. However, all Sub-contractors must: adhere to their own safety policy and observe any additional duties imposed by their health and safety policy or that of any principal or main Contractor in addition to any instructions given by persons enforcing the safety standards imposed by such duties. They should be properly insured against all relevant risks and receive their health and safety monitoring and/or inspection reports and, where necessary, take appropriate action to correct any matters brought to their attention.

Sub-contractors should ensure that their employees have the necessary experience, knowledge and training to carry out their duties and clearly understand their responsibilities. They should understand and adhere to the arrangements, rules and so on and should not start work until they are aware of what is required of them.

In a similar vein to Contractors carrying out work directly, Sub-contractors must allocate sufficient resources to ensure they manage any risks arising out of their own work activity effectively – considering all significant hazards, undertaking risk assessments of their work activities and establishing the

control measures necessary to provide safe systems of work. They should provide information to their employees, including details of risk arising out of a work activity and the safe system of work to be implemented as a result, and cooperate with the PC, other Contractors and their Sub-contractors, whilst following any reasonable directions from the PC.

Sub-contractors are to inform the PC of any death, injury, ill health or dangerous occurrence (via the Contractor if possible, directly if necessary) but provide the PC with information, as requested, to be included in the Health and Safety File (via the Contractor, if possible, directly if necessary). All employees of the Sub-contractor must be provided with the appropriate personal protective equipment (PPE) and discipline. They should be willing and able to remove from site any of their employees that breach site health and safety rules and inform the Contractor if they sub-contract work to others.

As part of competence assessment measures, the prospective Sub-contractors may be asked the following questions:

- Can you demonstrate that you have sufficient insurance?
- Can you demonstrate your company's commitment to health and safety?
- Can you demonstrate that your company's duty to provide your employees with health and safety training has been carried out?
- Can you demonstrate how your company gets to know when any of its employees have accidents, and how you follow them up?

The key word to note is 'demonstrate'. Any assessment of competence must include an examination of demonstrable evidence.

Duty Holder responsibilities and information flow structure

The key Duty Holders described above will need to perform their duties within a project in a cooperative and communicative manner. It is essential that each party gains a full understanding of their duties whilst at the same time they also understand how their roles impact on other Duty Holders and the requirements for deliverables relating to health and safety management.

Figure 1 illustrates a typical Notifiable project and the relationship between key Duty Holders and information flow path as required under CDM 2007.

Innovation in CDM coordination – a case study of London 2012

The significance of the role of the CDM-C cannot be overemphasised. Working to ensure that the vital links between all other duty holders are established as effectively as possible during the project, the CDM-C is also ultimately responsible for collating the health and safety information that will outlive the construction programme – that is the Health and Safety File. The role thus has to be properly managed to ensure efficiency and productivity.

Fulfilling this duty effectively can be a challenge on large and complicated projects. For example, the long term regeneration

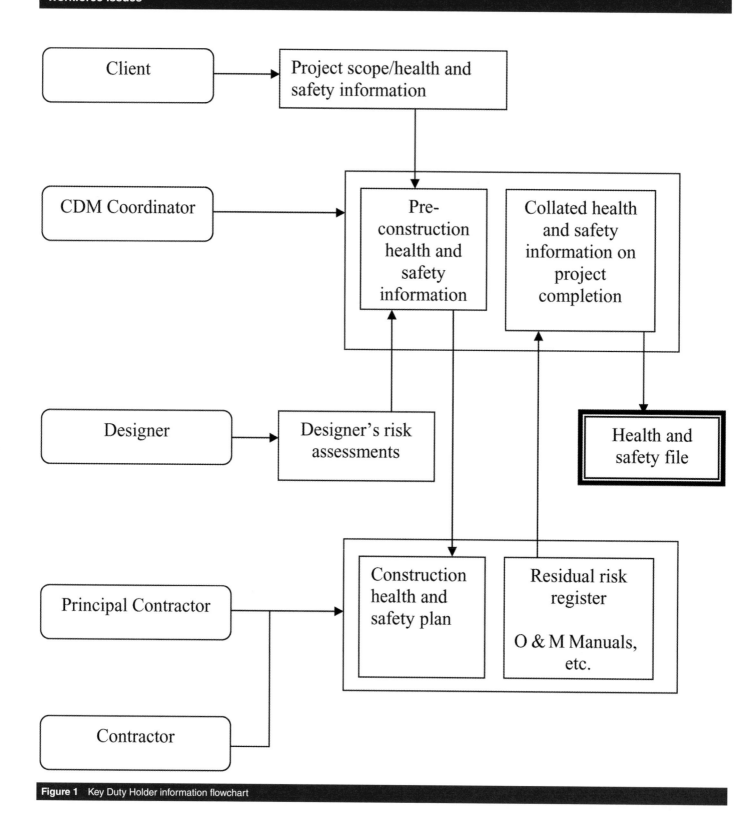

Figure 1 Key Duty Holder information flowchart

of the 2.5 km² of the proposed 2012 Olympic Park in London was considered too big for a single CDM-C appointment (Scopes, 2009). This is because the project has about 9000 people working at the same time across 38 projects. The Olympic Delivery Authority (ODA) thus decided to procure CDM-C Services for all its multiple design and build packages. A unique CDM Integrator role was then appointed to manage the large and diverse group of CDM Coordinators.

The arrangement did not only lead to a significant reduction in fees but it also meant that a uniformly high standard of service and a consistent approach resulted. Reportable accidents in the first year of construction, for example, were only around 7% of industry average.

Summary and conclusion

This chapter has highlighted the specific duties of key participants on construction projects. Key Duty Holder responsibilities were given and some interrelationships of these roles were also examined as a basis for enhancing the effective understanding and implementation of one of the basic requirements of the CDM 2007 – Communication. Construction Design and Management always evolves and this trend creates attendant challenges which mean that Duty Holders will need to consistently update their knowledge on emerging technology, concepts, processes and systems that will enhance effective health and safety management. A case study which presents some emerging innovative techniques for managing health and safety on a typical construction project was also discussed.

References

Construction (Design and Management) Regulations 2007 (CDM 2007) (S.I. 2007 No. 320). London: HMSO.

Construction Skills and Health and Safety Executive. *Construction (Design and Management) Regulations 2007 Industry Guidance for Small One-off and Infrequent Clients,* 2007a, King's Lynn: Construction Skills. Available online at: http://www.cskills.org/supportbusiness/healthsafety/cdmregs/guidance/Copy_3_of_index.aspx

Construction Skills and Health and Safety Executive. *Construction (Design and Management) Regulations 2007 Industry Guidance for CDM Co-ordinators,* 2007b, King's Lynn: Construction Skills. Available online at: http://www.cskills.org/supportbusiness/healthsafety/cdmregs/guidance/Copy_2_of_index.aspx

Construction Skills and Health and Safety Executive. *The Construction (Design and Management) Regulations 2007 Industry Guidance for Designers,* 2007c, King's Lynn: Construction Skills. Available online at: http://www.cskills.org/supportbusiness/healthsafety/cdmregs/guidance/Copy_5_of_index.aspx

Construction Skills and Health and Safety Executive. *The Construction (Design and Management) Regulations 2007 Industry*

Guidance for Principal Contractors, 2007d, King's Lynn: Construction Skills. Available online at: http://www.cskills.org/supportbusiness/healthsafety/cdmregs/guidance/principal.aspx

Construction Skills and Health and Safety Executive. *The Construction (Design and Management) Regulations 2007 Industry Guidance for Contractors.* 2007e, King's Lynn: Construction Skills. Available online at: http://www.cskills.org/supportbusiness/healthsafety/cdmregs/guidance/Copy_4_of_index.aspx

Health and Safety Executive (HSE). *Managing Health and Safety in Construction: Construction (Design and Management) Regulations 2007 Approved Code of Practice* (HSE L144), 2007, London: HSE Books. Available online at: http://www.hse/gov.uk/pubns/books

Scopes, P. J. London 2012: A New Approach to CDM Co-ordination, 2009. *ICE Proceedings, Civil Engineering* 2009, **162**(2), 76–86.

Referenced legislation

Construction (Design and Management) Regulations 2007. Statutory instruments 320 2007, London: The Stationery Office.

Work at Height Regulations, Statutory instrument 2005, No. 735, London: The Stationery Office.

Reporting of Injuries, Diseases and Dangerous Occurrences Regulations 1995 (S.I. 1995 No. 3163), London: HMSO.

Workplace (Health, Safety and Welfare) Regulations 1992. (S.I. 1992 No. 3004), London: HMSO.

Further reading

Construction Skills. *Construction Site Safety – Health, Safety and Environmental Information* (GE 700/09), 2009, King's Lynn: Construction Skills.

Websites

Construction Skills http://www.cskills.org

Health and Safety Executive (HSE) http://www.hse.gov.uk

HSE Forms: Notification www.hse.gov.uk/forms/notification/

HSE Publications http://www.hse.gov.uk/pubns/index.htm

HSE, RIDDOR http://www.hse.gov.uk/riddor/

Institution of Civil Engineers (ICE), Health and safety http://www.ice.org.uk/knowledge/specialist_health.asp

Office of Government Commerce http://www.ogc.gov.uk

Chapter 4

Managing workers' conditions

Philip McAleenan Expert Ease International, Downpatrick, Northern Ireland, UK

doi: 10:10.1680/mohs.40564.0039

CONTENTS

The Seoul Declaration of 2008 recognises that occupational health and safety is a common responsibility taken on by all societal players rather than solely the responsibility of government, employers and occupational health and safety institutions. With the aim of moving away from reactive accident prevention activities towards a positive culture of enhancing the well-being and welfare of workers the signatories committed the participating bodies to enhancing worker safety on the international level.

At national and local levels there are many examples of good practice adopted by tripartite government, employer and employee groups that provide models for the continual improvement of welfare and health: the BuildHealth and BuildSafe initiatives in Northern Ireland established minimum standards of safety and health conditions on site for companies tendering for government contracts; the various skills programmes establish minimum levels of occupational health and safety competences for all who enter or work on sites, and in support of this, statutory regulations direct Clients to design, build, maintain and eventually demolish structures in a manner that is non-injurious to workers, users and public alike.

Box 1   Key learning points

Readers will be introduced to the principal responsibilities of the key Duty Holders in relation to assessing hazards to health, safety and welfare that face workers on construction projects. Guidance will be provided that will enable designers and contractors to put in place appropriate controls to ensure a safe and healthy site.

Introduction

Britain's construction industry employs 2.2 million people. In the past 25 years, more than 2800 of these workers have died from injuries that they have received at work. In 2007/2008 there were 72 deaths out of a total of 229 for all workplace fatalities, giving a fatality rate of 3.4 per 100 000 workers compared to 0.8 for all. This trend extends across to a higher rate of injury in construction compared to other industries, with a rate of 599.2 per 100 000 for injuries (517.9 for all industries), or 136 771 injuries reported (Office for National Statistics, 2009). According to the Labour Force Survey there were 299 000 reportable injuries (rate 1000 per 100 000) in construction in 2007/2008, indicating almost 49% under-reporting (HSE, 2010).

With regard to comparing illness rates, for statistical purposes these are not significantly different when compared with all industries (Health and Safety Executive (HSE)). However the British Government's White Paper, *Choosing Health: Making Healthier Choices Easier* (Department of Health, 2004) showed that 2 million people suffer ill health that they believe was caused by or made worse by their work (this figure has not been broken down by industry sector). Such ill health covers, amongst other things, compromised lung and heart functions, stress and psychological illnesses, musculoskeletal problems, and skin conditions.

In Northern Ireland (NI), with a much smaller population and approximately 75 000 employed in the industry, the numbers of fatalities have been in the order of 4–5 per year, with a drop to 2 in 2008/2009 and major accidents rising year on year from 68 in 2000/2001 to 103 in 2008/2009. The over three-day lost time accidents have fluctuated around the 150 to 170 mark during the same period (Health and Safety Executive Northern Ireland (HSENI)).

Overall some 34 million days were lost in 2007/2008 (28 million in respect of illness and 6m in respect of injuries) (HSE). In social terms this means that somewhere in the region of 1 million people (family and friends) were directly affected by or knew someone injured at or who had become ill through work, and in terms of wage income, this represents an approximate loss of £443 million, based on the average wages for males in the UK in 2008 of £521 per week (Annual Survey of Hours and Earnings (ASHE) (ONS, 2008)). (This figure does not take account of any social welfare benefits injured persons and their families may have received.) However this represents only a small fraction of the costs of workplace accidents and a more complete analysis of social and economic costs can be obtained from the *Costs to Britain of Workplace Accidents and Work-related Ill Health* report (HSE, 1999).

The Bilbao Declaration

The problems and difficulties with health and safety in construction are not an exclusively British or Northern Ireland problem, and in November 2004 the European Agency for

Safety and Health at Work (EU-OSHA) organised a construction safety summit as part of its annual European Week for Health and Safety at Work. The participants included European federations representing the construction and building industries and the engineering and architectural sectors, with the importance of the event being signified by the presence of the Minister for Social Affairs and Employment from the Dutch Presidency.

The outcome from the summit was the Bilbao Declaration, Building in Safety (EU-OSHA, 2004). The summit called upon 'all relevant parties in the construction sector to commit to resolute actions to achieve the permanent improvements that are required by the EU health and safety strategy, notably through a full and effective application of the national legislation transposing Directive 92/57/EEC' (Council Directive on Implementation of Minimum Safety and Health Requirements at Temporary or Mobile Construction Sites, EEC, 1992).

The declaration went on to elucidate a number of important concepts which have subsequently been incorporated into legislation and strategies for health and safety.

Regard for health and safety is not just for the construction phase but is applicable to the whole life of the project from the concept and design stage, through construction to use and maintenance and finally to the demolition at the end of life of the structure. This acknowledgement is part of the Construction (Design and Management) (CDM) Regulations 2007) that require all parties to the construction project to work together to integrate health and safety into the planning and management of projects from the design concept onwards. Additionally, whilst the law has been moving towards it, often through case law, the declaration calls on the design community in Europe to maximise to their full potential the safety and health aspects of design which are integral to the construction process. The CDM Regulations recognise this and place a duty on Designers to design out risk and to inform others where there is a residual risk so that they may effectively manage it.

The declaration states that 'reputable and sustainable occupational health and safety standards can only be secured within an overall context of high quality standards being achieved'. The growing use of integrated management systems is not just a cost-effective way of managing, but also integrating the various requirements and standards leads to improved quality and safety in the process and on the outcome (see Chapter 7 *Occupational health and safety management systems*).

Quality, according to the declaration, is achieved through cooperation between all the competent partners in the process, and this is recognised in the CDM Regulations. The duty to appoint coordinators is to ensure that the different partners do work together at all the relevant times (see Chapter 3 *Responsibilities of the key duty holders in construction design and management*).

The procurement process has a central role to play in the safety of workers in construction in that Clients and Contractors, etc. should be examining the workers' conditions in the supply chain as part of the assessment of bids. This relates directly to EC Directive 2004/18/EC, Coordination of Procedures for the Award of Public Works Contracts, Public Supply Contracts and Public Service Contracts (EEC, 2004).

The Bilbao Declaration is linked to the Seoul Declaration, signed at the 2008 World Congress of Occupational Safety and Health (ILO, 2008), which reiterates the concept of prevention contained within the International Labour Organisation's (ILO) framework documents on prioritising prevention in the workplace, and to the Quebec Protocol (ISSA, 2003) which advocates the integration of safety and health issues into vocational and professional training from the outset. The laws in Britain and Northern Ireland (NI) cover much of the above concepts and requirements whilst specific regulations and codes of practice, many dealt with in detail in this book, add substance to the principles.

Workers' conditions

The conditions under which workers are required to carry out their daily work activities can be taken as an indication of how much respect they have been afforded and how much they really are seen to be the main asset of a company. The HSE reported (June 2009) that almost 90% of employers regard their workforce as such, with 65% of employees stating that good health and safety practices make them feel valued. Furthermore, being the creatures that we are, work is integral to our wellbeing and the Choosing Health Report (Department of Health, 2004) confirms that in general our health is improved by being in work. The fact that this is far from the case when the conditions in which people work are unhealthy or unsafe goes without saying. But what is worth saying is that it is not simply a matter of the healthfulness of and the control of hazards in the workplace that impacts upon the safety and wellbeing of workers, but also the nature of the work itself, the role the worker has in the decision making process, especially in relation to his or her own activities, excessive hours or work demands, and the rewards (remuneration and promotion) attained as a result of effort.

Some of these factors, when negative, will lead to ill health: the risk of heart disease for example increases the greater the imbalance between effort and reward or the lack of job control, whereas monotony may give rise to reduced concentration and thereby to an increased risk of accidents occurring. The relief of boredom may be achieved by some by 'spanner in the works' behaviours: small, deliberate acts designed to break up the tedium of the day which may in some cases have larger, unforeseen consequences.

This range of factors that impact upon workers' conditions suggests that more is going to be required than designing out the major risks and controlling the remedial risks through on-site hazard assessments and operational controls. There is a need to add into the equation the psychological effects of the chosen work processes, the managerial structures and the hierarchies of decision making, and the employment contract with its impact in industrial relations.

Added to this there is the fact, raised by a number of the keynote speakers at Seoul 2008, that the right to good working conditions is a fundamental human right as per Article 23 of the Universal Declaration on Human Rights (United Nations, 1948), in which case the protections must be guaranteed by the state and this in turn should take into consideration the efficacy of law making and enforcement.

Once we have an understanding of those factors that affect workers' conditions we can begin to design projects that take into account the whole range of requirements necessary to ensure compliance with our duty to ensure a safe and healthy workplace.

Consideration, as is highlighted by various contributors to this book, commences at the earliest stages when the concept is first tabled. There are questions about how the idea, when translated into a practical reality, will affect the conditions of those who are tasked with building it or those charged with maintenance after it goes into use. Though the answer to such questions may not be within the ambit of the Client's abilities, it is for the Designer to take on board the concept and to advise the Client of its limitations and effects on health, safety and welfare.

Client

Informed by the Designer, the Client must tailor his demands and allocate his resources to maximise the quality, safety and health and environmental aspects within the context of getting a quality structure capable of meeting the requirements of the end users.

There are limitations in what can be done, some of which are absolutes, e.g. a fixed budget, in which case it is imperative that all the critical elements of the project are properly costed and presented in respect of each design, including (of course) the costs for the design element. Everyone hopes to come in under cost but in reality cost overruns are common, whether in small or large scale projects: Arsenal's Emirates Stadium is hailed as an exemplar of coming in on budget and on time (Kitching, 2006), whereas the Wembley Stadium, completed 10 months after its due date, is still beset with legal problems and claims for over £200m which could see the eventual costs exceed £1 billion. Legal costs at the time of writing stand at about £65 million with an estimate of almost a year before the case will get into court (Rowson, 2009). (An interesting fact that was noted in the *New Civil Engineer* article was that photocopying for the Wembley Stadium came in at £1 million.) At a more extreme end of the scale is Boston's Big Dig, which was initially costed at $2.8 billion but which eventually came in at $28 billion. But equally, a £60 000 overrun on a domestic project is as likely to drive the Client into bankruptcy, leaving the project unfinished, or the contractors with substantial fees owing to them.

Budget considerations in the context of workers' health and safety conditions should take into account the provision of adequate welfare facilities from the outset, including sufficient toilets and washrooms, canteen and rest area facilities, first aid and changing rooms, clothes drying areas, training and induction areas (this author has trained workers in dirty workshops with boxes and crates for seats as there were no other facilities, and such circumstances are not yet a thing of the past). It is

axiomatic that if you can't afford the health, safety and welfare requirements for a project you can't afford the project.

The budget should include adequate remuneration for the whole workforce, whether employed directly or sub-contracted in. To this end it may be useful to have a policy on wage and salary rates which would apply throughout the supply chain. One point to note is the illegality of paying below the minimum wage – and it is not good practice to pay at or just above it. This latter point is particularly relevant when you consider that in order to obtain a competent workforce which will transform the designs into a quality structure in a safe manner you will need to ensure that they receive remuneration commensurate with their competence.

It is also important to ensure that wage rates are consistent for all workers and that workers who are non-nationals are not discriminated against; again Seoul 2008 highlighted this as a scourge of globalisation and of recession.

On the matter of remuneration, there is a case for a larger workforce to meet time targets, than a smaller one that is doing excessive overtime. Notwithstanding the European Working Time Directive (EEC, 1993), excessive overtime leads to fatigue, which leads to reduced concentration and thus to accidents and poor workmanship. Overtime will be required from time to time, but it should not be a factor of most workers' daily lives on site. There is the added social benefit in that more people in properly paid work is economically desirable, less of a drain on the public purse and, in human terms, meets that fundamental need to work.

Also include in the budget sufficiency for technological solutions to safety issues, whether for construction or for maintenance. Some of these will be engineering solutions to problems such as working at height (see Chapter 14 *Working at height and roofwork*), or mechanical solutions to handling operations. It takes but one serious accident to wipe out any cost savings that would have been obtained from using manual solutions, and statistically up to 95% of the cost of an accident will be uninsured (HSE, 2005).

It is not unreasonable for the Client to want to maximise the potential for any site that he has, but unless he is able to lease sufficient land immediately adjacent to the construction site, the space available must be allocated according to the requirements of the construction process as well as those of the finished project. This means that the site must have space sufficient for the welfare facilities detailed above, as well as for the ease of movement of plant and people around the site, storage, fabrication and so on. Cramped working conditions, poor manoeuvrability, inadequate storage all increase the hazards and contribute significantly to the reduction in quality of working conditions.

Connected with this element is the requirement to ensure that the infrastructure necessary for the project is in place before the main construction phase commences: routeways, water, sewage and power suitable for the peak levels of workers on site are to be in place or constructed to take additional capacity as and when needed. The Designer and Principal Contractor may be responsible for determining what these requirements are likely to be, but the Client must ensure that the financial and time resources are allocated in time for them to be carried out.

Time resources is another critical factor in determining whether the conditions of work are high quality or inadequate. Realistic time scales must be established between the Designer and the Client at the earliest stages of the project. There is a degree of flexibility and projects can be completed in a shorter time scale if and only if adequate human and material resources are provided to do so without compromising the safety and health of the workers and the quality of the structure. However, there are absolute time limits, sometimes as a result of technical constraints, which the Client should be aware of and which he should respect and allow for.

Some things can be calculated at an early stage, e.g. quantities of steel and therefore fabrication and delivery times, likewise for quarry products, cement, concrete, hardcore, etc. Other things cannot be timed until surveys have been carried out, tests on new designs conducted; even the process of tendering and assessing the whole gamut of bids may take an undefined amount of time that in turn may be affected by political considerations. It has been shown that the insufficient allocation of time at the design and pre-construction phase can lead to poor working conditions during the construction phase as time and cost overruns loom (Dann and Levitt, 2009).

Resourcing the project is the responsibility of the Client: if he has a fixed budget then the project must be tailored to come within that budget; if the project is determined then the finances to realise it must be provided. If there is room to take from Peter to pay Paul it must be done within the fundamental tenet that the project must be designed, built, maintained and demolished in such a way as to afford no harm to workers or to those who are affected by the structure at every stage of its life.

Before moving on it is worth noting the comments of David Orr in his inaugural address as President of the Institution of Civil Engineers (ICE):

> Clients can become blind to the true cost of the work through haste, uncritical enthusiasm and lack of professional input. They can come up with an unrealistically low initial estimate through a combination of insufficient costs for the known scope and omission of key items.
>
> And if the initial estimate is unrealistically low, projects become burdened with the perception that 'costs have spiralled out of control'. But of course, the problem is often not real increases in cost; simply that the initial published estimate was so low it was never realistic in the first place. (Orr, 2007)

The Designer

The role of the Designer in advising the Client on the project he wishes to have is important in determining whether the workers are able to construct the project in a safe and healthful manner. Some of those responsibilities follow on naturally from what has been stated in respect of the Client's responsibilities, others emerge from the competence and professionalism of the Designer to highlight what is possible and what cannot be done with respect to legality, current technological advances and functionality.

Architecture, when unfettered by physical and social constraints, is as creative as any art form and there are more designs on paper than ever were translated into practical projects (though not all because they were not physically or humanly possible). But when it comes to translating an idea to a design to a practical reality, the Designer (architect, engineers, etc.) must determine what the absolute limitations are with respect to geography, environment, technology, etc. and must ensure that the Client is aware of them. He or she must insist upon the time necessary to conduct surveys and tests in order to determine what is possible, what the actual requirements will be to make it possible and how these will impact upon the conditions that the workers will experience. Planning and scheduling the construction phase is more successful if it commences with, and is integral to, the design process. It is a requirement on Designers to be aware of hazards in their designs and to eliminate these at this stage, if possible, and to inform the Contractors of any residual risks that they will need to consider and control. And these hazards should not be limited to the physical hazards that might lead to accidents, but also to such hazards that may lead to ill health (including mental health) or which create adverse conditions for the workers. By way of example, the construction of a high rise building will create different problems depending on where it is to be built, e.g. on the coast or inland? How many seasons will it take to construct? What effect will it have on the workers' environment as it progressively rises through those seasons. Will they, for example, be constructing the 30th or 40th floor in summer or winter, each of which brings different problems for the workers? Will the project be surrounded by low or high rise building, and what effect will this have on wind patterns and updrafts?

These same types of questions should be applied to the maintenance of the finished structure. How do you maintain the exterior of a high rise building, windows, lights, signs, inspections, etc.? How will different developments around the building in the future affect what is to be done now on the building?

With each iteration of the design these questions are asked and solutions arrived at and put to the Client. That it takes time to do this is undoubted, but time and cost at this stage will be offset by a well managed project at the construction phase, which, all things being equal, will come in on time, on budget and having met the safety, health and welfare requirements of the workforce. This of course entails the Principal Contractor meeting his duties and requirements.

Contractor

In practical terms it is up to the Principal Contractor to ensure that the facilities necessary for the welfare of workers and for their protection from harm are in place for when they are required. Before commencement, safe systems for working are to be agreed and notified to all who will be working to them, all existing hazards are to be identified and appropriate control measures put in place, competent workers are to be employed and competent Sub-contractors capable of identifying and controlling the hazards associated with their elements of the project are to be engaged.

Whenever possible the Client should engage the Principal Contractor at an early enough stage so that they can work with the design team in assessing the buildability of the project and the requirements for safeguarding the workforce. There is valuable practical experience that the Principal Contractor can bring to the table and this opportunity must not be lost by discounting the contribution they can make. (By the same token, bringing young engineers and graduates onto the project at an early stage will provide them with good experience and an excellent learning opportunity.)

Information relevant to workers' welfare includes details of the actual requirements for the welfare facilities based on the number of workers and duration of the construction phase, the hazards associated with construction in general and the project in particular, and indeed information of construction methods or alternative materials that will ameliorate safe and healthy processes.

In turn the Contractor will work with the Sub-contractors at an early enough stage to obtain their specialist knowledge as regards safety and to ensure their effective integration into the construction process whenever they are due on site.

The specialist knowledge of Sub-contractors should not be downplayed and subordinated to generic rules for safety, especially when they appear to conflict. Where there are conflicting requirements for the protection of workers, both the Principal Contractor and the Sub-contractor should agree a satisfactory solution and communicate this to all affected, especially where there is a deviation from a 'normal' rule, with the reasons fully explained and understood by all the relevant workers.

The Contractor is responsible for ensuring that appropriate site inductions are available to all who come onto the site. Inductions are an important aspect of on site communications that should be designed to reflect the nature of the specific project and flexible enough to adapt to ever changing conditions. They are not some generic, broad based health and safety talk to be given to everyone and anyone so that a box on a checklist can be ticked. Inductions should therefore be tailored to the needs and experience of the inductees; a directly employed site worker there for the duration of the project will have a different induction requirement than the person who is an infrequent or a one-off visitor. People whose trade or profession is in construction will not generally (though this is not always the case) require an induction to generic construction hazards: what they require is information on the site-specific conditions and controls, whereas someone outside the industry or with substantially less experience may well require a general overview of construction hazards in order to comprehend the site-specific hazards and controls. The principle here is that the workers employed by Designers, Contractors and Sub-contractors are competent in their field and its application to construction and will therefore bring with them knowledge of the health and safety issues. At the recruitment and/or assessment stage this

will have been determined and from this a range of induction programmes can be developed.

In respecting the competence of those who are authorised to be on site, the Principal Contractor is establishing the groundwork for cooperation between all on site and reducing the opportunity for disgruntlement and disaffection which, over prolonged periods of time, may lead to increased negative and unsafe behaviours and reduced performance. Respect is a two-way channel and if you expect respect then it must be given in return.

When it comes to the development of site rules, the Principal Contractor must be aware of the effect that rules, especially generic rules, have on those who are required to follow them. The CDM Regulations empower the Principal Contractor to include in the Health and Safety Plan any rules necessary for the management of the construction phase, the operative words being 'necessary for the management'. This means that the need for the rule must be determined and its extent agreed upon before bringing it into effect; for example, the generic personal protective equipment (PPE) rules that adorn the entrances to so many sites cause concern for many workers, especially Sub-contractors or professionals with a legitimate reason to be on site. What the signs indicate is that the notion of PPE being a control measure for specific hazards that are not capable of being controlled reasonably practicably by other measures has not been considered properly. Whilst it may be true that on most of the site there are hazards that may cause head injuries, this is not always the case at all times, e.g. when floor tilers are working in an all but completed area with no overhead hazards. Indeed the requirement for these workers to wear hard hats may impinge upon their ability to lay the tiles to the quality they would otherwise achieve.

Likewise a generic rule to wear gloves or eye protection does not advise the person who has to wear them of what hazard they are to be protected from and therefore what type and standard of glove/eye protection they must select.

Rules must not be for convenience: they must require of the worker a behaviour that is necessary for their safety and for the effective management of the site. Rules for mere convenience encourage rule breaking and when that happens all rules, including those that do happen to be necessary, are liable to be broken, thereby decreasing the level of safety on site.

At this point the discussion raises two further areas for consideration, the first regarding hazard identification and controls and the other relating to the question of whether generic controls or site-specific assessment and controls are preferable. These are not mutually exclusive issues and for no particular reason the issue of assessment and controls is treated first.

Hazards assessment and controls

The preceding paragraphs have highlighted some of the issues that the various parties to the construction project must consider in order to identify and control those matters that will impact upon the conditions under which employees will work. The assessment process, the main purpose of which is to

Figure 1 Generic safety signs

improve work conditions on site, will be much more detailed and specific, gathering information relevant to the project and devising and implementing effective solutions for controlling the identified hazards and hazardous conditions.

The identification and consideration of controls is an intellectual process which is continuous throughout the lifetime of the design and construction phase as well as at the use, maintenance and demolition points in the finished structure. It is carried out by all players in the project from the Client, through the Designers, to the Principal and Sub-contractors and to the workers and others who are legitimately involved in any aspect of the project. It is a process which informs action (without which it is a waste of time) and action in

its turn prompts further assessment to ascertain its efficacy in achieving the desired objective (output in terms of product and safety) and whether it has given rise to new seen and previously unforeseen hazards.

The legislation talks of risk and risk assessment and this has become the mantra of many, creating an onerous and massive paper monster for industry to the extent that it has been investigated by one Parliamentary Working Party (UK Parliament Works and Pensions Committee, 2008) and has led the HSE and HSENI to advocate 'sensible risk assessment'.

Risk assessment without a doubt is a central feature of the process of assessing and controlling hazards, but in and of itself it has limitations. Risk is the probability of unwanted outcomes

ICE manual of health and safety in construction © 2010 Institution of Civil Engineers

occurring; it is not capable of predicting when such an event will occur. No matter how remote the probability, it could and may occur the next time the activity is undertaken.

The existence of a risk of whatever probability means that there is a corresponding reduction in the control of the activity/process or of the final structure. Some things can be controlled with absolute certainty – in that all is known about the hazard, what is required to conduct the activity safely in face of the hazard, and the workers having the necessary competence and being compliant in carrying out the activity in the safe manner. Other things cannot be controlled with the same confidence of absolute certainty because it is not possible to know all that is necessary to make the decision – for example, the quality of steel may and should be of a high standard, but the minutiae of detail about the internal structure cannot be known to the extent that its collapse point can be predicted with complete certainty, rather than merely estimated. What is known is the probability of a failure based on past experience and that such a probability will increase with time. Thus the controls in the construction are backed up by appropriate inspection and test regimes and emergency procedures which in combination should ensure the safety of workers and users.

The second aspect of current risk assessment practice is the determination of severity of outcome. This is valuable information because it will allow the assessors to prioritise the actions that will be required to safeguard workers.

A word of caution though – a risk assessment is not an end product but a thought process that leads to action, whether this is a modification of the design, a change to the technical methods or technologies used in the construction process or an awareness by and alteration in workers' behaviours.

Nor is it necessary to produce a written risk assessment. This is a misinterpretation of the legislation which requires that significant findings of the assessment be recorded. Any such finding that compromises the safety and health of workers or the integrity of the process must be dealt with appropriately so that the works proceeds safely and working conditions are not compromised.

What may be required as the output of a risk assessment is a written set of controls, but this should be an infrequent practice rather than the norm. Competent workers, by dint of that competence, will know and apply with the appropriate degree of flexibility the control measures for most of their routinely carried out activities. Unusual hazards or particularly complex procedures may require a written control document which, respecting workers' competence, should provide sufficient information to allow them to proceed safely but not be so much as to be too onerous to read or comprehend nor should such a document contain an undue amount of irrelevant information that would tend to obscure the relevant, and finally it should not negate the authority of the worker to make appropriate decisions. A good assessment will be carried out by the workers and team who are to undertake the task.

Thus engineering will come up with designs that offer safer solutions to meet the project requirements, or may specify engineering controls in respect of aspects of the construction process or alternative methods to avoid or reduce such issues as working at heights, manual handling, etc. Contractors will develop safe methods for those elements of the project taking into consideration the activities of other workers and trades on site at the same time, and workers will detail specific controls on permits in respect of high hazards activities.

Coordination of the activities of the different parties involved in the project and cooperation across all disciplines and levels of the hierarchy are essential to ensure that all the hazards are clearly identified and understood and the most effective controls agreed upon and implemented to ensure the safety of all concerned. The approach is to view the whole gamut of controls that are possible and available, from design, via safe methods to mechanical and PPE, as a matrix which when used in an appropriate combination will provide effective and guaranteed safety. This is more suited to the 'reasonably practicable' principle that underpins United Kingdom and Northern Ireland health and safety law without negating the principle that some controls afford higher security than others and should therefore be considered before these others. Regardless of what controls are to be put in place, projects must be safe to start and should be safe to execute.

This leads neatly to the question as to whether safe and healthy conditions are best achieved through generic or project and task specific controls. Of course the answer is neither one nor the other, but in recognising that competent workers and teams require flexibility to react to the ever-changing circumstances on a construction site, rules that are in the form of broad guidance to workers are likely to achieve greater safety. Sometimes an absolute rule is essential, for example the setting of traffic routes around a site (see Chapter 19 *Transportation and vehicle movement*), the creation of exclusion zones around craning operations, or the separation of volatile substances from each other (see Chapter 20, *Fire and explosion hazards*). At other times a broad rule that encompasses the flexibility necessary for individual decision making affords better protection – for example a rule that requires all confined space operations to be conducted under the permit to work system will, in being so limited in guidance, allow the competent confined spaces worker, fully cognisant of best practices, to make and establish specific controls for each confined space he is working on (see Chapter 16 *Confined spaces*).

Case studies

On the route to achieving improved working conditions for workers in the industry, various groups and conglomerates have established policies to reduce accidents and lost time injuries (LTIs); others have come into being specifically to deal with particular matters or aspects of industry health and safety.

In Northern Ireland in the early 2000s the Government Construction Client Group in conjunction with the HSENI and the member bodies of the Construction Industry Forum (NI) established the BuildSafe NI Initiative which over a period of five years set a target for the reduction of major injury accidents by 50% based on the 2002 level and established requirements for Contractors that were designed to assist them in achieving this target. These included a requirement for minimum levels of health and safety competency supported by training that would continually increase the competence of workers, and a requirement that companies have third party accreditation of the health and safety management systems.

At the end of this period, when the BuildSafe NI steering group ceased to exist, a new group, BuildSafely, was established within the construction sector, with regional groups which each run bi-annual seminars for directors and senior managers of companies on many aspects of safety including showcasing best practice by other companies in the industry. These seminars generally take place during targeted safety weeks during which a mobile training unit visits sites within the region giving practical training to workers on working at heights, slips, trips and falls, and manual handling. These prove to be very welcome by both large and small companies and the seminars are well attended.

The UK Contractors Group (UKCG), formally the Major Contractors Group, have established targets to reduce year on year fatalities and serious injuries in construction, and though their specific target of 10% annual reduction has not been met as of publication, improvements are noticeable with an average of 5% year on year reduction achieved. They have a target of a fully trained workforce and in support of this have established an Accepted Records Scheme, a list of the various programmes of training and certification available to construction workers.

They have also developed and published an occupational health strategy that highlights the keys areas where workers' health suffers and details plans for tackling the issues, such as skin conditions and hand/arm vibration and for the promotion of health surveillance strategies.

Likewise the Northern Ireland BuildHealth programme, an initiative between government clients, the HSENI and contractors to tackle specific disorders by establishing targets, for example to introduce mechanical handling for kerbing, and which includes raising health issues at site inductions and the use of health champions on large projects.

A similar initiative exists in Britain: the Constructing Better Health project, which aims to improve the management of occupational health by setting standards for health issues and health provision, collecting and transferring data to and from employers and the provision of specialist support for return to work schemes.

Summary of main points

These initiatives, in their various ways, are in line with the various declarations and protocols on occupational health and safety that have been addressed elsewhere in this volume, including the International Labour Organisation framework document and the Seoul Declaration of 2008 on preventions, the Bilbao Declaration of 2004 on improved occupational health and safety in construction through quality, and the

Québec Protocol on occupational health and safety training. But as we have seen in the figures at the commencement of this chapter – 72 deaths, 136 771 reported injuries and 28 million lost days through ill-health in a workforce of 2.2 million – there is still a long way to go in the industry in order to ensure that the conditions under which our workers are engaged are safe and free from hazards to physical and mental health.

To achieve more, all that is required is a commitment at all levels to improving workers' conditions and a willingness to develop and implement improved assessments and controls.

References

Dann, C. and Levitt, R. *Cost and Schedule Risk Management of Large Capital Products*, Stanford University, June 2009 (seminar presentation available online at: http://www.adepp.com/files/csra_pdf.pdf)

Department of Health. *Choosing Health: Making Healthier Choices Easier*, 2004, London: The Stationery Office.

European Agency for Safety and Health at Work. *Bilbao Declaration – Building in Safety*, European Construction Safety Summit, Bilbao, November 2004.

Health and Safety Executive (HSE). *HSE HSG101 The Costs to Britain of Workplace Accidents and Work-related Ill Health in 1995/96*, 1999, London: Health and Safety Executive.

Health and Safety Executive (HSE). *Ready Reckoner*, 2005. Available online at: http://www.hse.gov.uk/costs/costs_overview/costs_overview.asp

Health and Safety Executive (HSE). *Health and Safety Statistics*, 2010. Available online at: http://www.hse.gov.uk/statistics/

International Labour Organisation. *Seoul Declaration of Safety and Health at Work,* 2008, Seoul, Korea.

International Social Security Agency. *Quebec Protocol for the Integration of Occupational Health and Safety Competence into Vocational and Technical Education*, 2003, Quebec, Canada: ISSA, International section for education and training for prevention.

Kitching, R. Keeping it Simple. *New Civil Engineer* 2006, 3 August. Available online at: http://www.nce.co.uk/keeping-it-simple/482674.article

Office for National Statistics. *Annual Survey of Hours and Earnings*, 2008, London: ONS.

Office for National Statistics. *Self-reported Work-related Illness and Workplace Injuries in 2007/08: Results from the Labour Force Survey*, 2009, London: Health and Safety Executive. Available online at: http://www.hse.gov.uk/statistics/lfs/lfs0708.pdf

Orr, D. *At the Heart of Society*, Presidential address, ICE, 2007.

Rowson, J. Ready to Rumble. *New Civil Engineer* 2009, 22 January. Available online at: http://www.nce.co.uk/ready-to-rumble/1972517.article

UK Parliament Works and Pensions Committee. *Evidence on Health and Safety*, February 2008. Available online at: http://www.publications.parliament.uk/pa/cm200708/cmselect/cmworpen/uc246-iii/uc24602.htm

United Nations. Universal Declaration on Human Rights, UN, 1948.

Referenced legislation

Construction (Design and Management) Regulations 2007. Statutory instruments 320 2007, London: HMSO.

ICE manual of health and safety in construction © 2010 Institution of Civil Engineers

EEC. Council Directive 2004/18/EC on the Coordination of Procedures for the Award of Public Works Contracts, Public Supply Contracts and Public Service Contracts, 31 March 2004.

EEC. Council Directive 92/57/EEC on the Implementation of
Minimum Safety and Health Requirements at Temporary or
Mobile Construction Sites (eighth individual Directive within
the meaning of Article 16 (1) of Directive 89/391/EEC), 24 June
1992.

EEC. Council Directive 93/104/EC – Working Time Directive of
the European Union. 23 November 1993. Amended by Directive
2000/34/EC of the European Parliament and of the Council of
22 June 2000.

Further reading

Health and Safety Executive (HSE). ALARP 'at a Glance', undated.
Available online at: http://www.hse.gov.uk/risk/theory/alarpglance.htm

Health and Safety Executive (HSE). Sensible Risk Management,
August 2006. Available online at: http://www.hse.gov.uk/risk/
principles.htm

Joyce, R. *CDM Regulations 2007 Explained*, 2007, London: Thomas
Telford.

Websites

BuildHealth http://www.buildhealthni.com

Constructing Better Health http://www.constructingbetterhealth.co.uk/

European Agency for Health and Safety at Work (EU-OSHA) http://
osha.europa.eu

Health and Safety Executive (HSE) http://www.hse.gov.uk

Health and Safety Executive Northern Ireland (HSENI) http://www.
hseni.gov.uk

Institution of Civil Engineers (ICE), Health and safety http://www.
ice.org.uk/knowledge/specialist_health.asp

International Labour Organisation (ILO) www.ilo.org

Office for National Statistics, Labour Force Survey http://www.
statistics.gov.uk/statbase/Source.asp?vlnk=358

UK Contractors Group http://www.ukcg.org.uk/

UK Contractors Group (UKCG), Accepted Records Scheme http://
www.ukcg.org.uk/UK-Contractors-Group-UKCG-Accepted-
Records-Scheme-c651e72

Section 3: Managing occupational health and safety in construction

Chapter 5

The different phases in construction – design in health and safety to the project life cycle

Scott Steven S Cubed H Ltd, Lochwinnoch, UK

doi: 10:10.1680/mohs.40564.0051

CONTENTS

This chapter will discuss the integration of health and safety into the design life cycle of a construction project and explore the difficulties in establishing a baseline for acceptable risks or tolerable risks – what are suitable and sufficient safeguard measures.

The discussions will underpin the principle that the effort expended in planning and managing the health and safety aspects of the project should be proportionate to the risk and complexity of the project. The chapter will consider the issues to be tackled to eliminate hazards and to control the risks inherent in the design of the works to be constructed. It will consider the importance of identifying hazards early on in the project process to enable them to be properly managed. The discussions will promote the added value to be gained by embedding health and safety management into the procurement strategy and investment decisions from the project inception. Practical examples will be used to illustrate the matters being discussed and legal cases, when appropriate, to provide an understanding of what is an acceptable standard. When discussing these matters consideration will be given to the factors that will influence the construction risk management process to allow informed judgements to be made with a degree of confidence to miminise threats and maximise opportunities in a cost effective way.

Box 1   Key learning points

From reading the chapter you will gain an in-depth appreciation of the hazard elimination and risk reduction process in the project design life cycle and tools to:

- Confirm that there is sufficient time and resources to comply with designers duties under health and safety legislation.
- Facilitate good communication through proactive management of information flow.
- Integrate health and safety into the design, from project inception, to enable the construction works emanating from the design to be constructed without harm to those involved or may be affected by the works – injury free workplace.
- Coordinate health and safety aspects of the design work through design reviews relating to the hazardous operations in the design process and the construction, use, maintenance and demolition of the works.
- Liaise with the contractor regarding ongoing design of the permanent and temporary works.
- Embrace the notion that all injuries are preventable – no injury is acceptable.
- Why we should not set goals indicating that some numbers of injuries are expected and thus acceptable.
- The pledge to an injury free workplace solves the schedule, cost or production versus health and safety dilemma.

The important thing to remember is people in the industry are judged on the strength of their daily decisions so they need to be robust. Projects are successfully delivered by people working together making a genuine contribution to the success of the project. The processes will increase the opportunity to deliver consistently and enable the hazard elimination and mitigation measures to control the impact of the hazards inherent in the design to be carried out in a coherent systematic manner. The outcome will be projects delivered efficiently, effectively and safely with all those involved remaining fit for life.

Introduction

This chapter will discuss the integration of health and safety into the design life cycle of a construction project and promote early consideration of the fundamental health and safety issues relating to the use, buildability, cleaning and maintaining of the asset. The approach promotes that health and safety is considered as an integral part of the project design and decision making process and is carried out in a systematic, structured and timely manner tailored to the context of the construction environment. To improve the effectiveness of the approach the overall framework needs to be dynamic, interactive and responsive to change.

The emphasis of the approach promoted is early intervention of designing out hazards in design – 'avoid the hazard' – the first step that is most important and beneficial in the health and safety in design hierarchy. Once the opportunity to 'plan out the hazards' has been fully explored and exhausted the emphases will focus on the second step of the health and safety in design hierarchy –'reducing and controlling' the impact of the hazards remaining in the design and inform those using the design of the residual risks inherent in the design. The third step and last resort will be to rely on good working practices and personnel protective equipment (PPE) which the designer will have little control over regarding its use. The early intervention and systematic approach will avoid wasted effort at later stages in the design process – once the design has been frozen – to make fundamental changes in an attempt to eliminate hazards and control residual health and safety risks in the design. The approach enables the Designers to set the health and safety tone for the entire project from day 1 and to connect the designing and planning with the construction, maintenance, use and eventual demolition. Unfortunately all too often the different phases are disconnected.

One of the aims of the chapter is to embark the reader on a journey or continue them on their journey which fosters and nurtures a culture of caring for the health and safety of others and aligns their beliefs and attitudes to zero harm.

The chapter will explore the difficulties in establishing a baseline for acceptable risks or tolerable risk in construction, use, maintenance, cleaning, dismantling and eventual demolition of the works and the integration of health and safety into the design to promote an injury free workplace. The focus will be on the actions a Designer can take – suitable, sufficient and proportionate – within the context of the construction works being carried out to eliminate hazards and mitigate the impact of the hazards remaining in the design throughout the project life cycle in a systematic, structured and timely manner. In other words, the actions a Designer can take that will make a difference. The Designer's actions should promote that the construction works emanating from the design can be carried out in safe conditions – a safe workplace – and promote that the working activities can be carried out using safe acts – without compromising the safety behaviours of the workforce. These discussions will consider the most common causes of accidents and ill health in construction, clarify the use of designer risk assessments and reinforce the need for the provision of relevant information on the unusual risks and the risk most likely to be difficult to manage in the design. This will include making available the right information to the right people at the right time in a form suitable for the end user and the use of internationally recognised signage to visibly highlight the dangers where they exist on the project – a pictorial story!

Why? Because the Designers have a unique opportunity to assist the construction Team and play their part to rid the industry of the appalling accident and ill health record that currently exists. The ill health and injuries in many instances are devastating the lives of those who work in the industry and the ones that they support. The industry needs to embrace the philosophy that people can go home from work healthier than they arrive – the designers need to focus on zero harm to the workforce and those that may be affected by the works.

The discussions will dispel the myth that injuries and ill health – a 'slow accident'– are an inherent part of the construction industry. Setting health and safety performance targets acknowledges that a certain number of injuries and harm will occur and sends the message that injuries and ill health up to these targets is okay. This is unacceptable and should be discouraged. The discussions will embrace the need to nurture a caring culture to eliminate unsafe conditions – physical working conditions – and unsafe acts – unsafe behaviour – to realise injury free workplaces. Focus on zero harm.

Opportunity management of hazards in design available to Designers – key role to play

So why should we be so interested in understanding how to integrate health and safety into the planning and design process? Simply because research has demonstrated that the root cause of fatal construction accidents can attribute 35% to design, 28% to planning and 37% to construction. Therefore half of the fatal construction accidents should have been prevented before the works start on site, leaving less than 40% to be avoided during construction. In other words we need to change the mind set from accepting accidents will happen on the construction project to predicting what might occur from the certain and uncertain occurrences on the project, and making an informed judgement on what action can be taken to prevent the next accident or incident – deliver incident and injury free projects.

It is important that ill health – the 'slow accident' – is not overlooked when considering the construction industries' performance. These accidents are not sudden events obvious on site and thoroughly investigated at the time they occur. They generally happen due to exposure to unsafe acts and conditions over a long period of time, with a time delay. They have far reaching effects with a significant proportion of the workforce over 50 years of age having to leave the construction industry due to ill health, including hearing loss, asbestosis, mesothelioma, not able to use their hands properly, etc. The quality of their life has been significantly deteriorated and in some cases shortened.

The Designers are in a unique position to identify and eliminate hazards and reduce risks that arise during construction work and provide information in the design. The Construction Design and Management (CDM) coordinator provides the 'safety net' that the designer has carried out the design such that the works can be carried out safely. Together with the client they should be setting the tone for the work activities to be carried out in an injury free workplace. The legislative responsibilities of the duty holders are fully discussed in Chapter 3

Responsibilities of the key duty holders in construction design and management.

The mind set within the construction industry over the decades from the 1900s has changed from accepting that a fatality will occur on a major construction project with an estimate of probability of one fatality per mile of tunnel construction, one death per floor of skyscraper construction, 13 deaths for the construction of the major viaduct, to a mind set intolerable of any level of injury. It is evident from inspection of the industry's health and safety performance that, despite a step change in mind set of what is morally and socially acceptable, there has not been a commensurate change in the industry's accident and ill health performance.

This reinforces that the new goal setting legislation we currently operate under and the threats of prosecution in the event when something goes wrong alone (see Chapter 1 *Legal principles*) have not created the desired outcome of injury free projects or embeded and driven health and safety excellence.

We need to nurture this mind set of incident and injury free projects into reality by the construction team working together to make it happen and take up the unique opportunity available to designers within the construction industry. The Designers must play their part in enabling everyone in construction to go home every day healthy and safe.

The start is to change the concept that health and safety in design is tolerable of injury up to a number is acceptable. To avoid work related injury the designer needs to eliminate from the design occurrences that will lead to unsafe acts or unsafe conditions which contribute to a person suffering from ill health and/or resulting in a person being injured.

Construction risk management process – decision making framework

To promote confidence and assurance that the designer will consistently manage the health and safety aspects of the project proportionate to the risk and complexity associated with the project, a framework is required to focus on eliminating hazards associated with the design and developing design solutions that will reduce the impact of the hazards remaining in the design – a decision making process to undertake the risk mitigation management.

In the other chapters of this book the focus has been on either occupational risks or other safety risks associated with construction work activities. The intention in this chapter is to consider how the decision making process can be applied to address the hazards associated with occupational health risks and safety risks in the design of a construction project. The emphasis here is to provide a framework to enable the construction risk management process to be structured so that the most effort is targeted on the overall health and safety design of the project where it can do most good in order to avoid, mimimise or absorb the effects. The emphasis will be on identifying the context of the working environment, and securing collaborative working with others involved in the delivery of the project and ensuring the control measures remain relevant throughout the design development and construction process. One approach in achieving this is by considering the risk management process contained in the new international risk management standard discussed below.

Risk management process

The new standard for international risk management, ISO 31000: 2009 Risk management – principles and guidelines, primarily details the principles, framework and process to improve the identification of opportunities and threats, and increase the likelihood of achieving the health and safety objectives for the project. The International Organisation for Standardisation (ISO) has stated that ISO 31000: 2009 is intended to be used to harmonise risk management processes in existing and future standards by providing a common approach in support of standards dealing with specific risks and/or sectors, not replacing the standards. ISO 31000: 2009 is not intended for the purpose of certification.

The standard contains the risk management five process steps and two continuing steps to ensure effective management of the risk and is shown on **Figure 1**. The figure is not a start to finish process as demonstrated by the two-way arrows and this is actively encouraged. The power of the tool is you can start at any point and go forwards and backwards. Some start with the communication and consultation stage and others start with context.

The Health and Safety Executive (HSE) publication *Five Steps to Risk Assessment* (HSE, 1996) provides a simple approach to assessment but makes no specific mention of context, nor the need to communicate and consult. The HSE five steps to assess the risk in the workplace are:

- Step 1 – Identify the hazard.
- Step 2 – Decide who may be harmed and how.
- Step 3 – Evaluate the risk and decide on the precautions.
- Step 4 – Record your findings and implement them.
- Step 5 – Review your assessment and update as necessary.

In the other chapters of this book, risk assessment approaches are discussed and reference should be made to them for further knowledge (see Chapter 6 *Establishing operational control processes* and Chapter 7 *Occupational health and safety management systems* for more details).

Under ISO 31000: 2009 the definition of risk management in the context of health and safety in the construction industry is to coordinate activities to direct and control the construction works with regard to risk. If the output from the risk assessment – identifying, analysing and evaluating the 'risk' – will have negative consequences – it should be referred to as a threat. 'Risk' with positive consequence should be called an opportunity. The opportunity will have a positive impact on cost or provide a time saving or health and safety performance enhancement.

The risk identification is about asking what could happen when, how, why and who will be involved. A variety of

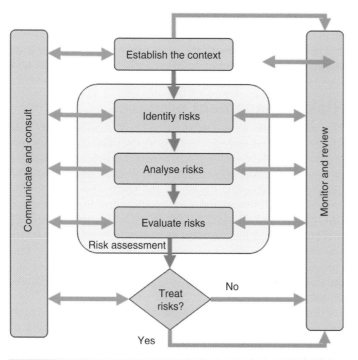

Figure 1 Risk management process
Reproduced, with permission, from ISO 31000: 2009. © International Organisation for Standardisation

mechanisms exist to identify risk, at all levels and this will be discussed later.

In risk analysis, the identified risk is to be assessed in terms of the likelihood of it occurring, the associated magnitude of the impact and the current controls evaluated for effectiveness. The general principle of the model that we will study is its application to increase the understanding of the decision making process. It is also useful to try to assess the consequence and likelihood of the controls failing in part or completely. The difference between the uncontrolled and controlled risk severity will be a very good indication of how critical the controls are to the safeguard measures required to carry out the works without harm being realised. When undertaking the assessment, consideration should be given to the population affected giving collective protective measures priority over individual protective measures. The output will translate into an indication of the relative importance of the risk to enable consistent comparison and prioritisation.

Risk evaluation involves making decisions about risks and how they have to be effectively managed. This requires identifying a risk owner with the authority to determine the degree of mitigation to be applied to make the risk tolerable and acceptable to include in the works. During construction this will be the contractors and once the works are passed over this will be the owner/occupier of the asset – i.e. the structure.

In some instances, additional mitigation measures will be required by the asset owner/occupier–controller of the premises – and the degree of the residual risk evaluated. For example, in a process plant designed to the appropriate safety standards the acceptable operational condition of the piece of equipment, once constructed, may result in it running at a temperature that could not be touched, requiring the operator to put in place additional safeguard measures – an exclusion zone – to protect those using the place of work, including the workforce carrying out planned maintenance and repairs.

The treatment of the threat will result in the risks identified, details of owner, and control measures to be implemented to be compiled to form a risk register. The significant and unusual risks could be supported with relevant risk mitigation action plans.

The mitigation measures must remain effective. This may require a maintenance regime or control systems to be put in place to keep the safeguard measure in good condition, which in itself may introduce further risks and a balance will need to be struck between performance, whole life cost, timescale, reliability and easy to maintain throughout the asset's in-service life. The risk represents value for money, e.g. confined space with mechanical ventilation. The maintenance introduces further confined space working which is a high risk activity but operationally may significantly reduce the residual risk.

If the number of safeguard measures employed to degrade the likelihood of the risk occurring is significantly increased, the risk owners will have a responsibility to assure themselves that the mitigation measures remain proportionate and effective through an inspection or monitoring programme to reflect the nature and complexity of the risk, e.g. anchor point for a fall arrest lanyard.

It will be paramount at each stage of the process to communicate and consult with the stakeholders, and monitor and review the whole process denoted by the two-way arrow flow in **Figure 1** – dynamic, consultative and responsive to change.

The concept has been introduced as it is a very powerful tool and assists in developing a good understanding of informed decisions based on sound information within the context that the works will be undertaken. This will be of great assistance when making professional judgements on acceptable risk under the current legislative framework which we will consider shortly.

However, before this we need to explore some of the terminology further.

Issues, uncertainty and risks

Within the Standard the definition of risk is 'uncertainty on objectives'. The objectives can have different aspects, such as financial, health and safety and environment. In the context of this chapter we would be selecting safety objectives, e.g. continuous improvement, incident and injury free projects, and setting a target the designer can achieve.

The term 'uncertainty' relates to lack of information, or understanding or knowledge of either the initiating event, or change in circumstances, or several of these. They may represent either a threat or an opportunity to the project. In terms of an example relating to excavation works the uncertainty may include level of contamination, extent of ground contamination or gaps in legacy historic information on use of the site.

Uncertain occurrences are events that may or may not occur and the construction risk control measures will depend on their significance. There will be significant health and safety uncertainties, and these will need to be addressed within the risk management process and included on the project hazard elimination and risk reduction register.

There will be insignificant uncertainties which will not require specific construction risk management action; however they may be required to be addressed within the project's planning if they have a high probability of occurring so they do not impact the overall execution and safety of the project.

The key is to keep asking about uncertainty and the effect it could have on the objectives, in this instance health and safety. To successfully deliver projects the decision process must have resilience – the capacity to adapt to change in a complex and changing environment.

If the probability of an uncertain occurrence happening reaches 100% then it becomes an issue and will become part of the main programme of works – a certain occurrence.

Certain occurrences, as you would expect, are occurrences that will happen and the term 'issue' is used to describe these types of events. The issues irrespective of their occurrence being significant or insignificant must be accounted for in project planning and scheduling activities with control measures commensurate to the level of risk.

The process requires that we establish the context relating to the risk assessment. If the construction works are being carried out on a greenfield site or a brownfield site, then the risk assessment related to excavations on each site is significantly different. Good working practices to undertake excavation on the greenfield site most likely will not be sufficient to control the risk on the brownfield site where there is a high probability that the ground may be contaminated. This will necessitate that the designer communicates and consults with the client to obtain available information on the ground conditions together with information on the historic use of the site, to establish the types and level of ground contamination that may be present. If there is no information then a ground investigation would be required to obtain the necessary information to enable the risks to be identified. It is not acceptable to leave it to the contractor doing the work. In both instances assumptions on the ground conditions will be required over the footprint of the works. However, if the excavation works related to repairs or preventative maintenance to an asset on a site operated under a permission and licensing regime then the context of the risk assessment will be significantly changed again due to the activities on the site involving significant operational hazards with high levels of risk which gives rise to wider public concern.

Take for example, the foundation level of an old masonry harbour wall founded on rugged rock head. Most likely the level of the rock head will be highly variable along the length of the structures. If the project involved dredging to accommodate vessels with larger drafts the uncertainty of the foundation levels may have a significant impact on the design of the harbour works. To accommodate the uncertainty of significant underwater underpinning works the costing and planning and scheduling would need to take cognisance that there is a probability of the occurrence of lower foundation levels

and make provision within the project for the risk to be managed. This could be in a variety of ways from the client holding financial sums to pay for the underpinning works if they occur, to provisions being included within the Works Contract for the contractor to manage the risk by making an allowance within the contract price. Irrespective of how the costs are managed the foundation levels being higher than the dredge depth would result in the foundations being undermined by the dredging works and significantly more construction risk management to develop the control measure to safeguard the structural integrity of the masonry harbour walls. This risk is foreseeable and must be included in the design.

Risk can be identified through formal or informal methods, including experience, previous project history and brainstorming.

Target resources

In terms of construction health and safety risk management, the effort to manage the risk must be targeted where it can do most good in order to avoid, mimimise or absorb the effects. To meet this guiding principle, the hierarchy of effort that should be targeted by the designer can be broadly broken down as follows:

- construction activities with *significant* health and safety risk:
  - significant construction activities – certain occurrences
  - significant uncertain occurrences with high probability
  - insignificant certain occurrences
  - insignificant uncertain occurrences with high probability.

- construction activities with *insignificant* health and safety risk which will be effectively managed by relevant good industry practices:
  - significant certain occurrences
  - significant uncertain occurrences with high probability
  - insignificant certain occurrences
  - insignificant uncertain occurrences with high probability.

Foreseeable risk

Based on the context and assumptions of the project and the construction work activities we will have significant certainties, insignificant certainties and insignificant uncertainties. These will be able to be included in the planning and scheduling for the project and the hazard elimination and risk reduction process due to the designer's knowledge and experience of managing the information available and health and safety control measures available in good practices. With a degree of confidence the construction risks emanating from these instances will be foreseeable. This is very important under the legislation, as designers are required to manage foreseeable risk.

Omissions

From the Designer's perspective of discharging his legal duties, the significant uncertainties are an area of danger. If an unplanned event occurs it may be construed as an omission in the Designer's risk decision making process. The designer will need to be in an informed position to demonstrate he or she has knowledge of the uncertainty and how it had been taken into account in the project, and if not why not. If an event due to a safety failure does come to fruition and the unplanned event results in someone being killed, seriously injured or causes them to become seriously unwell, it would be detrimental to the defence of the Designer if he or she could not demonstrate that everything reasonable to combat the health and safety risk had been taken.

Risk–based decision making process

In summary, the risk management process discussed has provided a framework around which to work and has assisted in explaining how to achieve a planned proportionate approach to managing the construction risks and potential omissions. What has not been discussed is what is considered as a baseline of acceptable level of risk – an acceptable standard of health and safety practices. This will be considered next before we consider applying the process on projects.

Acceptable levels of risk

When accidents and incidents happen that involve legal proceedings, the courts will decide if the balance has been struck with regard to the working practices standards on site and the compliance standards emanating from legislation. The court in simplistic terms will be considering if the decisions made resulted in acceptable levels of risk on site. There are difficulties in defining what is an acceptable risk or tolerable risk under the legislative framework in place to regulate industry. What is suitable and sufficient, proportionate action?

This is because the law is 'goal setting' and the standard to be achieved will be continually changing to keep abreast of advances in technology. Therefore, there is a need to rely on professional judgement by competent people when making informed decisions based on the information available at the time on the foreseeable risks associated with the work activities certain to occur during construction, use, maintenance, cleaning, dismantling and eventual demolition. Reliance on current best practice and relevant industry standard designs in the risk management decision making process, within the context that the works are to be carried out will also form part of the decision reached by the courts.

The acceptable level of risk should be proportionate to the magnitude and complexity of the individual situation and be determined from hazard and risk assessment carried out in a coherent systematic manner. The construction risk management approach adopted must allow informed judgement on the degree of risk and provide confirmation that the balance has been struck between performance, whole life cost, timescale and the risk represents value for money.

An understanding of the relevant current best practice, an awareness of the limitations of areas of own experience and

knowledge and willingness and ability to supplement experience and knowledge when necessary by obtaining external help and assistance are all critical when trying to determine what is an acceptable level of risk.

The acceptable level is further complicated due to changes in societal pressure which are becoming less tolerable and are placing demands on government ministers to hold those making decisions more accountable for their actions, which in turn drives further changes to legislation and what is considered as acceptable. The risk based health and safety law under which we all operate in the EU is 'goal setting'. Therefore the never ending progress in technology will also drive change in the degree of mitigation to be applied to make the risk acceptable. What is interpreted as relevant good practice will be further complicated by the changes to legislation to raise the standard or set a new standard to tackle areas of the construction industry where it has become apparent that more stringent control levels are required to tackle a root cause of an unnecessary evil in the industry – injuries and ill health. A typical example is the recent enforcement of the *Work at Height Regulations*.

The decision making framework for Designers' duties relating to hazard elimination and risk reduction under the Construction (Design and Management) Regulations 2007 (CDM 2007) has a qualified standards of duty 'so far as is reasonably practicable'. To provide an insight to what this standard represents we will consider some recent HSE prosecutions (sourced from the HSE, 2009):

- Injured party falling through a fragile roof light while maintaining, repairing or extending an existing building.

- Discovery of an unsupported vertical excavated face approximately 4 m high due to lack of thorough ground investigation prior to commencement of work and lack of support material.

- Undertaking work to a building which involved coring out and boring through a structural concrete slab, the underside of which contained asbestos textured coating, without having identified the type of asbestos involved in the work. The outcome was that the works were carried out without effective measures to prevent the workforce being exposed to asbestos, nor taking measures to prevent the spread of asbestos into other parts of the premises.

- Failed to provide information relevant to asbestos to the contractor at an early stage to allow him to take precautions to protect against exposure to asbestos.

- Failed to make sure that everyone knew that the floors must not be overloaded. This resulted in the support under the floor collapsing due to the first floor joist being overloaded by lightweight blocks.

- Inadvertent exposure to asbestos due to type 3 survey not being carried out prior to refurbishment work. The duty holder had no asbestos register.

- Removal of ceilings in a three storey building constructed in asbestos insulating board. The property development company failed to undertake the survey prior to this work and the works were carried out adjacent to a pavement used by the members of the public with no control measures in place when the boards were being removed.

From review of the facts of the cases noted above it is evident what is not acceptable; what is acceptable is more difficult to define and this will be explored later.

One skill that we will not be able to develop is 20:20 hindsight. We will be measured on the quality of our professional judgement and the strength of our decision making process. In the event of an accident or near miss, under the legal reverse burden of proof in the Health and Safety at Work Act (HSW Act) the responsibility is placed on the decision maker to demonstrate that the actions taken were proportionate, suitable and sufficient to meet the standards set in law. This is the area we are now going to explore.

Qualified standards of duty

The risk based approach to health and safety legislation was introduced by the enforcement of the HSW Act in 1974 to replace the prescriptive legislation which was revoked following the Robens report (1972) (refer to Chapter 1 *Legal principles* for more details). The risk based law has introduced many concepts and set standards and liabilities that we must strive to meet or work beyond compliance to provide a safe and healthy work environment when carrying out the construction works. In addition, sometimes there is strict liability which must be achieved, e.g. obtaining a licence under a permission regime to operate a nuclear plant, licence for the removal of asbestos, etc.

The legislation in force places different standards on duty holders and when the risk assessments are being carried out it is important to establish the context for the assessment with due cognisance of liability to do something which is placed on the work activity and must be taken into account. The professional judgements in the risk–based decision making process will be judged against the differing legal tests, e.g. 'absolute', 'practicable' or 'reasonably practicable'.

The qualification to do something so far as is reasonably practicable is different from so far as is practicable. Practicable imposes stricter standards than reasonable practicable. Practicable is safeguarding measures that must be possible in the light of current knowledge. Reasonably practicable allows an assessment to be made in which the degree of risk is placed on one side of the scale and the sacrifice involved in any of the steps necessary for minimising or averting the risk, whether in money, time or trouble, is placed on the other. If there is a great disproportion between any of them – the risk being insignificant in comparison to the sacrifice involved – then the step was not reasonable. This introduces the term grossly disproportionate to the risk and the requirement for professional judgement to strike the balance on what is an acceptable level of risk – the action required in terms of safeguard measures so that in the event of something going wrong and creating an unsafe act or a unsafe condition it cannot be attributed to the accepted level of risk.

An ideal situation would be to agree acceptable level of risk with the enforcing authority but this may be construed as creating a situation that there would be deemed to satisfy or prescriptive control measures and remove responsibility from the designers to consider the context of the work activities are to be carried out. An example is the publishing of the 'traffic lights' – Red, Amber and Green CDM designers' list (see HSE, 2009a), capturing current good working practices within the design community. The HSE details design working practices – hazardous elements, materials or processes – which are unacceptable, working practices to be avoided if possible and working practices which were considered as acceptable. The designer should aim to move out of the red category and into the amber, and ultimately the green category.

Red category items are effectively prohibited and should be eliminated from the project except where no alternative can be found. The risks inherent in the design should therefore at worst fall into the amber category and generally be in the green category. A red category item would always require information to be included in the design.

Amber category items should only be used in the design with caution as they represent significant risks which may be acceptable on a risk management basis. These risks should be recorded on the project hazard elimination and risk reduction register and on the construction and as-built drawings as appropriate.

Green category items are the preferred working practices as they represent good working practices with significant risk reduction. Similar to the red and amber category items, green category item may require information to be included in the design if it is unusual or difficult to manage.

Examples of red, amber and green category items are provided below. These should not be treated as exhaustive lists or interpreted as approved by the HSE. The examples have been provided as suggestions to prompt the thought process. Most organisations will hold their own list of accepted practices and those to be avoided to aide their designers/CDM coordinators fulfil their duties.

Red

- Specification of fragile roof lights and roof assemblies using fragile materials.
- Substances identified as 'harmful' within the Control of Substances Hazardous to Health (COSHH) Regulations and current edition of EH40 (HSE list of approved workplace exposure limits) (HSE, 2007) used for on-site spraying or other application.
- Processes that give rise to the creation of large quantities of dust, e.g. dry cutting, chasing out of concrete/brick/block-work walls or floors for the installation of services.
- Locating heating and ventilation services on the roof of buildings without the provision for safe access and egress to carry out the planned preventative maintenance and repair works.
- Scrabbling of concrete.
- Designing piles that require to be cropped without the use of manual breakers.

Amber

- The specification of building blocks weighing greater than 20 kg.
- The need to include in the design a 'confined space'.

- Organic solvent based paints and thinners, or isocyanates, particularly for use in confined areas.

- Specifying manholes on heavily vehicular traffic routes/zones/marshalling yards.

Green

- Undertaking an asbestos survey to identify the nature and extent of asbestos-containing materials (ACMs) and provide brief, clear and precise information extracted from the asbestos analysis and report on the drawings.

- Limit the need to excavate and, when required, limit depth, e.g. foundations.

- Specify water based adhesives.

- Adapt the design to allow the use of less noisy solutions.

- Provision of a permanent access system and appropriate guarding of edges where there is a need to gain regular access/egress to/from a workplace at height.

- Specify materials located at height with longer life to first maintenance and reduced need for cleaning, repair and replacement.

- Design windows to enable them to be cleaned from the inside of the building.

- Providing information on overhead/underground high voltage electrical services and/or gas supplies to be isolated before construction works are carried out.

- Design machinery so that it is remotely operated and materials are fed automatically to separate the operator from the moving parts.

- Specifying slip resistant floor coverings.

Always remember reliance on PPE is the last resort and may be broadly divided as follows:

- hearing protection
- respiratory protection
- eye and face protection
- protective clothing.

PPE does nothing to stop the hazard at source, it simply has the potential to reduce the severity of the accident and only has the capacity to effectively control exposure, if worn, which the designer has no control over. The designers need to influence the contractors and promote the management arrangements to ensure the PPE will be worn.

Construction risk management – identification of main risk areas

The main areas to assist in identifying known significant risks in the construction industry are by reviewing the content of:

- Legislation enforced to control work activities involving:
  - significant hazards
  - levels of risk giving rise to wider public concern – complex/high hazard facilities;
- HSE Approved Codes of Practice (ACoPs);
- HSE guidance;
- Codes and formal Standards (e.g. BS, CEN, CENELEC, ISO, IEC);
- guidance produced by other government bodies;
- formal industry standards by an industry/occupational sector;
- relevant good practice;
- the most common causes of accidents and ill health in construction – accident and ill health statistics.

At this stage in the discussion it is useful to reflect and remind ourselves that HSE distinguishes between the terms 'good practice' – defined by HSE as satisfying the law – and 'best practice' which is interpreted as risk control above the legal minimum. This will be explored when we consider zero harm next (also see HSE, 2001).

The context in which guidance from the above sources is followed will need to remain under constant review, as the level of the risk and complexity of the nature of the works may result in the guidance not being sufficient to meet the standard set by the law in the circumstances that the works are to be undertaken. For example, general good practice road dressing repairs on a highway may significantly change due to a high voltage overhead power cable crossing the road necessitating introduction of further control measures to eliminate the risk of someone being electrocuted.

Always ask the question, has my professional judgement to create a safe place of work in the environment that the construction work activities are to be carried out, been suitable and sufficient – is there anything more that should be done to effectively reduce and control the risk specific to the project taking account of the context in which the works have to be carried out? Ultimately, it is for the courts to decide if the risks have been reduced 'as low as is reasonably practicable' – to an acceptable level of risk.

It is important to remember that good practice will change with time, particularly due to new technology, better engineering controls, environmental protection constraints, and health and safety legislation changes. There also may be external events or circumstances that will make the good practices no longer relevant.

Now review the legislation that the construction industry is operating under. By inspection of the titles of the legislation it will become apparent that the legislation has primarily been enforced to rid the industry of an evil that has resulted in many injuries, ill health or multiple fatalities. Further information on how to manage the hazards will be found in the ACoP, etc., which supports the regulation.

An example is the Control of Asbestos Regulations 2006, enforced to combat the single biggest killer in the industry – asbestos – with over 1000 fatalities per year. (For further information see (Chapter 8 *Assessing health issues in construction*) and HSE, 2009.)

Designers need to ensure that they have the knowledge and experience, or access to expert advice, to manage the risk of the presence of asbestos. So what part do designers have to play in tackling what is a significant hidden killer in the industry?

First the designers need to understand the extent of the problem. Asbestos was a popular building material since the 1950s and if the building was constructed or refurbished before 2000 it is most likely asbestos will be present in the building. Asbestos is found in many products used in buildings and was used particularly as an insulator, fire protection and to protect against corrosion. To demonstrate the variety of building products that contain asbestos a typical domestic dwelling house is considered:

- fuse wire asbestos flashing guard and fuse box panel;

- vinyl and thermoplastic floor tiles;

- insulation boards around domestic boilers;

- cisterns made of reinforced resin compound materials;

- insulation in wall and floor cavities and the loft;

- textured ceiling and wall finishes, e.g. artex;

- water tanks are generally made of asbestos cement in old houses (pre 1980s);

- asbestos cement on insulation boards used in facia and soffit boards;

- external wall cladding and internal wall panels particularly around windows;

- asbestos cement roof gutter and down water pipes.

By reviewing the list above it is evident that asbestos appears in many different guises, on a construction site, particularly on building refurbishment projects, and when carrying out maintenance and repairs to existing buildings.

The second step, which is essential and must be satisfactorily addressed, is that before design starts the designer needs to ask the people in charge of the workplace – buildings – to provide them with up to date information on the location and the condition of the asbestos present on the premises. The owner has a legal duty to provide this information, asbestos register, and the Design Team – CDM coordinator and designer – have legal duties to ensure that significant risks associated with the design are managed and controlled. This will include identifying and collecting the asbestos information for the project from the client and advising the client if asbestos surveys are required to infill gaps in the information needed for the design to be developed and to enable the contractor to manage the works safely on site.

The third step is to prepare the design such that the asbestos is not disturbed and if the asbestos needs to be removed design the works to reduce the risk from the remaining hazard as far as reasonably practicable and provide the information on the type, condition and location of the asbestos so that it can be removed safely.

Seems simple yet if you reflect on the number of prosecutions reviewed earlier, in many instances it just is not happening and that is not acceptable. It is essential that the information is provided on the drawing to make the workforce aware where the asbestos hazard exists.

One of the major problems is that the tradesmen are unknowingly working with asbestos despite the legal framework in place to safeguard their health. The workforce exposed to asbestos suffers the devastating illness known as mesothelioma: an incurable cancer. The number of deaths resulting from asbestos is significantly more than due to the single biggest cause of workplace deaths and the main causes of major injury, namely working at height (see Chapter 14 *Working at height and roofwork*) – between 30 to 50 fatalities occurring year on year – approximately 50% of the industries' annual fatalities figures. These statistics underpin the severity of the asbestos ill health problem. The effects of asbestos are not restricted to the workforce and it is known family members handling and cleaning the clothing worn while at work are also exposed to this killer and have died as a result. The HSE has estimated that every week 20 tradesmen die from asbestos related disease – eight joiners, six electricians and four plumbers.

As noted above, one of the main areas in identifying known significant hazards is the inspection of the accident and ill health statistics. In the other chapters of this book causes of occupational ill health and injuries caused by safety failures are discussed. (Refer to Chapter 8 *Assessing health issues in construction* and Chapter 9 *Assessing safety issues in construction* for more details.) By inspection of these chapters it will become apparent that the main causes are:

Injuries	Ill health
Falls – from height	Asbestos
Falls – fragile materials	Musculoskeletal
Entrapment or overturning of plant/equiment	Respiratory
Transport risk and falling objects	Skin
Confined spaces	Hearing
Electricity and water hazards	Vision

Similar to asbestos, it is of paramount importance that the Designer has a full knowledge of these known root causes of accidents and health risks, and the effects on the workforce who will construct or maintain a structure and those who may be affected by the works. The designers must be able to identify and assess the health and safety risks in the design, then take action to eliminate hazards, reduce risks and provide information. Again it is essential that the information is provided on the drawing to make the workforce aware where the hazard exists. If the designer does not have enough knowledge or experience to effectively manage the risk he will need to seek the assistance of someone with the expertise.

The next source of identifying the risk will be by reviewing the site information to identify the project specific significant and unusual hazards which may give rise to significant risk and the nature of the works which are inherently dangerous to build and will require careful and skillful planning. This review should extend to cover maintenance, use and, eventually, demolition of the structure.

Zero harm – beyond legal compliance

In industry there are several accident causation studies, the best known being the Heirich dominal model (1950), the Bird loss control model (1969), and Tye/Pearson (1974/75) (all cited in HSE, 1997). There is no universally accepted model. The output from these studies is illustrated by the 'accident triangle'. The ratio between major incidents – fatality or serious to near misses – non–injury or non–damage incidents differ due to the differences between how the accidents are defined, type of incidents and the industrial sector studied. However, there is an overall trend that for each major accident there are a large number of minor incidents with little potential to cause harm. The statistical relation that will be considered is that developed by Bird. The relationship was developed based on the research of approximately 2 million accidents in the USA that showed that for every major injury there can be as many as ten causing minor injury, 30 causing property damage and 600 that resulted in neither injury nor damage – near misses. This is illustrated by the Bird accident triangle shown in **Figure 2**. The triangle can be extended downwards to capture lesser events occurring in larger numbers – unsafe conditions and unsafe acts. This is illustrated in **Figure 3**.

To assist with the interpretation of **Figure 3**, reference to the Occupational Safety and Health Administration (OSHA) *Recordkeeping Handbook* (OSHA, 2005) is required. OSHA is the enforcing authority in the USA and undertakes a similar role to HSE in the UK.

First aid under OSHA is defined as 13 prescribed treatments including using non-prescriptive medication at non-prescriptive strength. Examples are using wound coverings such as butterfly bandages, steri strips (other than wound closing devices such as sutures, staples, etc., which are considered as medical treatment) and removing foreign bodies from the eye using only irrigation or a cotton swab.

For an injury or an illness to be recordable it must result in any of the following: death, days away from work, restricted

Figure 3 Level of harm and unsafe acts and unsafe conditions

work or transfer to another job, medical treatment beyond first aid, or loss of consciousness. The recordable criteria extend to serious injuries or illness diagnosed by a doctor even if it does not meet any of the above criteria. Restricted work or job transfer is when the work related injury or illness prevents the person from being able to perform routine functions – tasks regularly performed at least once per week.

Lost time injury involves one or more days away from work. The number of restricted or transferred days is also recorded.

To gain an understanding on how injuries and ill health occur, we need an understanding of the immediate, underlying and root causes behind these incidents. The immediate cause is the event that caused the event – e.g. unsupported excavation – the underlying causes are the failures that contributed to the event happening – lack of ground condition information – and the root causes are the fundamental, organisational and people factors – temporary works designer had not insisted on receiving the ground condition information relevant to the area of the works on site.

Unsafe acts that could be promoted by poor design could include:

■ Using plant, equipment and tools improperly, due to one heavy awkward lift necessitating special plant which may be very costly for the scale of the works which inadvertently encourages the contractor to devise a method of work that may unintentionally compromise safety or encourage the workforce into taking an improper work position.

■ Specifying the design incorrectly necessitating redesign.

■ Defeating safety devices – not making provision for isolating plant/equipment, not constructing the works in the planned sequence.

Unsafe conditions include:

■ excessive noise;

■ poor ventilation;

■ poor illumination;

■ working area without adequate space and inadequate guarding.

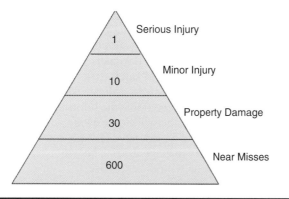

Figure 2 Bird accident triangle

ICE manual of health and safety in construction © 2010 Institution of Civil Engineers

Comparison of the two triangles allows an assessment to be carried out between the likelihood and consequence of an accident which is of assistance when prioritising the management of the hazards. For example, hazards with the potential to cause multiple fatalities/serious injuries will be ranked higher than those likely to cause a single fatality/serious injury, which will be ranked higher than those likely to result in persons being off work, which will be ranked higher than those likely to result in people needing first aid. This in turn will assist in implementing the health and safety in design risk management.

Investigating and targeting these lesser events provides the designer with the opportunity to maximise the learning experience available in the construction industry. This can extend to gaining an understanding on how others have dealt with similar issues. These events have low potential to harm and by investigating them we are able to build up a trend profile that will enable the designers to target their efforts to make the most significant safety improvements and prevent a major incident from occurring.

To realise incident free projects this is the area where the effort will need to be targeted, embracing the notion that every injury is preventable and no injury is acceptable. Once this is achieved the industry may realise step changes in performance opposed to the incremental changes currently being achieved year on year. By implementing the zero harm approach, health and safety in design will be taken beyond the baseline of acceptable risk and resolve the dilemma of only complying with the legal qualified standards of duty. The pledge to an injury free workplace also solves the schedule, cost or production versus, health and safety dilemma.

Opportunity to manage hazards and safety in design during the project process

Health and safety in design hierarchy

In essence, the objective of managing hazards and safety in design is to maximise the opportunity to eliminate hazards and reduce risk in the design, through the early consideration of how the structure will be built, used, cleaned, maintained, repaired and eventually demolished. If the opportunity management is effective it should reduce the threat of an early decision and assumptions, will have the potential of a disproportionate safety impact on the project. The process should underpin the principle of 'get it right first time' and avoid the need for redesign. If the need for redesign transpires during the construction works it will be significantly more costly than during the early stages of the project when the costs are more controllable and time is available to develop a solution that does not compromise safety due to time and cost constraints at the construction stage in the project process. If the opportunity management is effective it should reduce the threat that an early decision or assumptions will have the potential of resulting in a disproportionate safety impact on the project – the workforce are in

an informed position with regard to the hazards remaining in the design. This will enable them to plan appropriate safeguard measures with suitable and sufficient resources – people, plant, equipment and materials – to control the residual risks in the construction works emanating from the design and protect people that may be harmed.

For this to be achieved in a systematic, structured, coherent and timely manner, tailored to the context of the construction environment, it is important and beneficial for the early intervention of the opportunity management of the health and safety hazard inherent in the design. An integral part of this process will be the application of the health and safety in design hierarchy – avoid, reduce, protect. The designers can take actions that will avoid – eliminate – hazards which may give rise to risk and reduce risk from any remaining hazards. The designer's power to control the measures then adopted to protect those who may be affected by the works are most likely to be limited as it will generally be the contractor carrying out the construction works who will make these decisions. There are many instances when this will not be the case and the designer will have the opportunity to influence the method and sequence of construction, for example use of permanent shuttering to construct floor slabs in buildings, the use of prefabricated stairs to provide instant access between floors, the provision of the temporary and permanent means of structural stability. The application of the health and safety in design hierarchy on a construction project is demonstrated by inspection of **Figure 4**.

The profile of the opportunity risk avoidance graph will be influenced by the scale, nature and complexity of the project. On small projects with no special risks, minimal design input and involving routine construction the avoidance graph will most likely be very steep with very little opportunity to eliminate hazards once the design has been frozen at the concept design stage. For example, on a predominately one activity project – earthworks contract – the main hazard arising from the work activity is transport. Once a one-way transport system has been planned within the curtilage of the site or by temporarily taking more land adjacent to the site to avoid the need for vehicles to reverse and to segregate the transport and

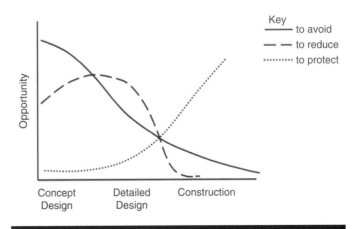

Figure 4 Opportunity to manage risk at different project stages

pedestrian movements to avoid people becoming entangled with the vehicles, the effort will focus on reducing risk, giving collective measures priority over individual measures.

On the larger or more complex projects or unusual or high risk projects constructed under a traditional procurement route the graph profile will more likely reflect that shown in **Figure 4**. The alternative forms of procurement Private Finance Initiatives projects, Design and Build, etc. most likely will influence the profile of the opportunity risk avoidance graph.

On the traditionally procured projects, at the concept design stage effort will be expended to avoid the site specific significant hazards associated with the physical obstructions and conditions, e.g. presence of adjacent structures, overhead and underground high voltage power cables, high pressure gas and oil pipelines, poor or contaminated ground conditions, presence of asbestos which is likely to be disturbed by the works, etc. and the operational constraints associated with the site or neighbouring premises, e.g. chemical process plant. At this stage the efforts will be also focusing on the inherently dangerous operations, e.g. large crane lifts and non routine in nature physical conditions and work activities that will be difficult to manage during the construction, use, maintenance, repair and eventual demolition of the structure. As the design develops into the detailed design stage and the opportunity has been exhausted to plan out the site specific significant hazards, the focus will in turn concentrate on the first level health and safety hierarchy associated with each of the significant hazards remaining in the design. These hazards, as previously discussed, are known to be the root cause of ill health and injuries in the construction industry, e.g. the working at height safety hierarchy. The first step will be to avoid working at height through pre-assembly, modularisation or the provision of leading edge protection during fabrication or manufacturing of the structural component to avoid the need to install protection at height.

It is important to routinely review the design to confirm that the elimination of one hazard or reducing the risk of a remaining hazard does not create a new and more significant hazard.

At the construction stage the opportunity to avoid the hazards will be significantly reduced and may only come to fruition by the contractor adopting alternative construction methods or materials or through advancements in technology and materials enabling them to be replaced or substituted by non-dangerous processes or materials.

With regard to the profile of the opportunity risk reduction graph it will be similarly influenced by the scale, nature and complexity of the project and the method of procurement. At this second level of the health and safety design hierarchy – reduce risk from any remaining hazards – the effort needs to be focused at combating the risks at source and the selection of collective protection measures over individual measures. Part of the decision making process will include making judgements on whether or not it is more appropriate to specify materials with higher risk during construction to take advantage of the lower risks of maintaining the material over its life span or the installation of large and heavy building components using mechanical handling methods to reduce the need for manual handling.

When considering reduction of risk, particularly at the concept stage in the design process, there will be broad assumptions on the method of construction with reliance on good industry working practices to control the risk inherent in the design including the personal protective measures to protect the workforce and those who may be affected by the construction works. The opportunity to reduce risk will be given more scrutiny during the detailed design, particularly on the larger or more complex projects or unusual or high risk projects. At this stage the design will most likely be frozen and the effort will be focused on developing the design with minor modifications or adjustments to develop safer solutions. Once the project enters the construction phase the collective protective measures and individual protective measures will become dominant as indicated by the profile of the opportunity risk protection graph – the last resort.

In summary, the graph indicates that at the concept design/front end engineering design stage there is maximum opportunity for the construction team to effectively contribute to the fundamental early design decisions to avoid – eliminate – hazards from the project. Once the options study is complete and the most appropriate engineering solution has been identified, the design development will progress into the detailed design stage where it will become more difficult to make fundamental changes to eliminate hazards associated with early design decisions. The focus will be on reducing risk with reliance on good working practices to control the risks inherent in the design. As the design develops into the construction phase the risk reduction will focus on the last resort, the personal protective measures to protect the workforce and those that may be affected by the construction works.

To enable the opportunity management process to be effective, an integral part of the design development strategy must be to clearly define a framework from the start of the project to secure the project specific information needed to identify hazards relating to the site and whole life cycle risks associated with the design and construction works. The information needs to cover the use, maintenance and eventual demolition of the structure. Part of the information management strategy will be to assess the adequacy of legacy asset knowledge, including the content of any existing health and safety files, and identify information gaps. In many instances this will include advising the client of any significant gaps or defects in the information and ensure these are filled by the client commissioning surveys or by making reasonable enquiries. This in turn will enable the uncertainties and unknowns to be removed from the construction works and will raise the confidence levels when planning and costing the works during design development and the construction phase.

The information to be collected will include the fixed hazards associated with the structures and associated infrastructure, the working operational hazards arising from the use of the site as a workplace or the client's undertakings, e.g. processing plant, operational constraints for the site or establishment and the risks arising from the neighbours' undertakings. Further, the availability and the management of the health and safety information to the right people, at the right time, and in an appropriate format to enable informed decisions to be made that will underpin the success of the project.

The effective opportunity management of the risk will enable informed decisions to be made with regard to the whole life cost and planning of the project. This will make delivery to time, cost and quality more likely. Unrealistic deadlines and failure to allocate sufficient funds are two of the largest contributors to poor control of risk on site.

Cost and planning

The objective during the cost estimation process will be to provide sound advice on the construction health and safety risk management to underpin the procurement strategy and investment decision and maximise the opportunity to plan the works so that they can be carried out safely and without risk to health. (Also see Chapter 10 *Procurement*)

To make the cost estimate process to be of value, once the scope of the works is formally frozen, no new design ideas should be introduced that change the scope of the works and this in turn will narrow down what is actually going to be delivered. Any scope creep will manifest into cost and planning issues which in turn will most likely impact on the safety of the construction works, provisions to undertake maintenance and use of the structure.

At the concept stage in the project processes, dependent on the client's procurement strategy, there will be a requirement to have engineered the design to enable the construction works to be estimated with a cost certainty accuracy of ±10–20% in the private sector and in the public sector a three-point valuation will most likely be adopted as recommended by UK Treasury guidance (HM Treasury, 2003). This type of estimate comprises three budget costs and is intended to quantify cost risk of the project so that a reasonable assessment can be made for budgeting purposes.

The three-point cost estimate comprises an anticipated tender return price, out-turn low budget price and out-turn high budget price.

The tender price is the best estimate obtained by procuring the works through a competitive tender procedure, and is calculated on the basis of quantities and a priced schedule at present day costs. The price must take into account the impact of the known project risk, including health and safety risk, e.g. constructability risks, presence of contamination in the ground or presence of asbestos within the structure.

The out-turn low budget price recognises an allowance for a reasonable level of risk and reasonable additions to the tender cost as the contract will typically tend to out-turn above the tender return cost. The out-turn low price is calculated by considering the tender cost as a base estimate and applying a contingency sum of 10%–20%, as detailed in the Treasury guidance, which is based on the uncertainty prevailing with the construction activities. The contingency sum percentage chosen will reflect the confidence levels on the construction activity quantities, the definition of the scope of the works, the experience and knowledge of the commercial

rates to execute the works, and understanding of mitigating the project risks.

The out-turn high budget price is aimed at providing an upper bound order of cost to reflect the potential financial impact of the identified construction risks, should all the identified risks prevail, and/or due to unforeseen conditions and scope creep. The out-turn high budget price is again a percentage increase on the tender price. The Treasury guidance statistically derived global factor of 60%–70% is the recommended increase for the feasibility stage best estimate. The contingency sum percentage chosen will reflect the confidence levels in the cost rates, information available on the site and ground conditions, and the perceived understanding of the project risks.

With proactive project management of the control measures to mitigate the impact of the project risks and strict financial management of change control, the out-turn cost of the project should be positioned between the out-turn low and out-turn high budget price. The key is to proactively manage the problems as they emerge, not deny that they exist.

In summary, the discussions underpin the principle that the effort expended in planning and managing the health and safety aspects of the project should be proportionate to the risk and complexity associated with the project.

The whole life approach to the application of the health and safety in design hierarchy enables informed decisions to eliminate hazards and reduce the risks inherent in the design, and in turn enable realistic deadlines and funds to be set. Once the structural solution has been selected with the monetary and time constraints imposed there will be less opportunity to eliminate hazards associated with early design decisions.

The integration of health and safety into the design life cycle of a construction project will now be considered in more detail.

Health and safety in design

The health and safety in design challenge is that the construction works emanating from a design can be built, maintained, used and demolished safely. The key components to be considered when preparing the Health and Safety in design are discussed below.

Obtain the information for the design
Legacy asset information

If the structures affected by the proposed works were built after 1994 a health and safety file (HSF) prepared under the CDM regulations should be available. By referring to it, health and safety issues specific to the structure constructed and the use of the site will be gained. If no HSF is available the designer will need to refer to the historic information available on the site.

Site visits

These are essential to confirm that the legacy assets information that is being used in the design is up to date with any site changes since the file was prepared, and to identify gaps in information that may not be apparent by carrying out a desk-top study. The visual inspection may also give a better understanding of the impact of known unusual or significant hazards on the site foot-print, any site operations by the client to undertake his business, and the land use of the neighbouring sites. It is very important to record the site observations in a manner that the information can be shared with the other designers on the project.

Site use

To obtain information about the use of the site the designer will need to communicate and consult with the client. This will include gaining information of the site specific significant and operational hazards associated with the materials, processes and equipment within the facility or plant. These consultations should also cover the use, hazards and operability reviews to provide a more in-depth understanding of the issues and uncertainties to be managed.

The output would include a site hazard plan with significant and unusual hazards clearly identified. The plan would indicate hazard management zones and underground and overhead construction hazards.

Design development

It is fundamental that, once the Designers have collected sufficient information to enable them to set the context of the environment, that the works are to be constructed, maintained, used and eventually demolished, they identify the hazards and establish how the structure can be built safely before leaping into the design.

The designer should produce a simple plan at the start of the project to confirm the method of construction, the envisaged build programme and the choice of construction materials. The risk management approach must take account of the activities that are certain to occur during the life cycle of the asset and those that are uncertain to occur but which may have a significant impact on the safety of the project.

One of the first steps will be to identify the construction, use, maintenance and demolition hazards and confirm that design interfaces have been properly addressed so that no areas have been overlooked. To record the output from the concept stage and the development of the design it is normal for a formally managed and documented process to be used to confirm that the health and safety project specific issues are captured in a Hazards Log or an elimination and risk reduction register. This provides clear evidence of how the legal duties have been fulfilled and will form part of the information to be included in the design. It should not replace the provision of information on the drawings as it is unlikely that the registers will be handed to the workforce being asked to build the project whereas the drawings will.

Part of these initial steps would be to assess if the design, or parts of the design, is likely to:

- use unproven design processes;

- involve technology/equipment unfamiliar to Designers or complex in nature;

- have critical schedule, budget, health and safety and/or quality impacts on the overall project.

If any two of above existed it would be advisable to set up a health and safety design integrity review as part of the design verification process.

Safety in design

To integrate safety into the design it is paramount that the context of the construction environment and the impact of the surrounding environment are fully taken into account and that the process is continuously repeated at each stage of the project life cycle – site selection, concept design, detail design, design and construction reviews, project construction, facility operation, maintenance, decommissioning and eventual demolition.

To benefit from the collective knowledge, experience, and lessons learned on previous projects by the Construction Team the Designers should consider engaging early contractor involvement in the design to draw on the contractors' construction expertise. Similarly the designer should engage with the asset manager to gain the knowledge and experience of operators and facility users to maximise on operational and maintenance safety opportunity that may be available. This would include asset and plant and process facility managers and operators.

The health and safety design process must extend down through each discipline and across the disciplines to ensure the process is seamless. The hazard elimination and risk reduction process can either be carried out and recorded on a discipline basis, or by area of the works. The hazard elimination and risk reduction registers by area secure designer cooperation, while by discipline requires resource input to secure coordination and cooperation in the design process. Irrespective of how the process is done at the end of each stage in the project process an interdisciplinary check should be completed.

The hazard and operability health and safety in design should focus on the cause and hazardous consequences of potential deviation, considering, identifying and resolving potential dangers. The safety in design should focus on altering the design to eliminate hazards; the risk reduction of the hazards remaining in the design should target the use of good working practices, special procedures to control the hazards, and features that have fail safe systems and devices. This may extend to specialist training of the workforce and operators. Personnel protection programmes and warning devices and signage will always be the last resort.

Box 2 outlines the basic health and safety areas that need to be covered in the health and safety design process, as discussed in other chapters. This list should not be treated as exhaustive; you should make your own project specific list as the design progresses.

Once the scope of the design has been frozen, the health and safety impact on the design can be assessed by considering the design in plan – two dimensions – to confirm that the traffic and pedestrian designated routes and space available around the structure, plant and equipment is sufficient to effectively construct, operate, use and maintain the structure. This would extend, where appropriate, to indicating the

Box 2 Site and Projects specific significant and unusual hazards and areas difficult to manage

- Site usage history.
- Site operational constraints (safety critical areas on/off site).
- Location and site security arrangements.
- Adjacent land use, e.g. schools, railways, roads.
- Location, nature and proximity of energy sources and operational constraints – gas, overhead underground electricity, process plant.
- Location and proximity of existing structures, e.g. foundations.
- Ground conditions – contaminated land, underground structures or water courses.
- Stability, structural form and fragile materials.
- Previous structural modifications, including weakening or strengthening of the structure.
- Fire damage, ground shrinkage, poor maintenance that may have damaged structure.
- Areas with difficult access – plant, equipment.
- Hazardous environments/workplaces.
- Hazardous operations and activities, e.g. live process plant and equipment.
- Hazardous substances, storage and use.
- Hazardous area clarifications, e.g. no go areas.
- Radiation and residual ionising radiation.
- Welfare arrangements.
- Emergency and contingency planning.
- Maintenance and storage areas.
- Fire prevention existing.
- Transportation on and off site.
- Resource availability (workforce/plant/equipment).
- Methods of transport – delivery modes and route, use of laydown and assembly areas.
- Modular construction on/off site.
- Sequence of construction.

- Utility services, water electricity and gas.
- Heating, ventilation, air conditioning plant – maintenance access, plant, equipment, valves and switches.
- Decommissioning and demolition.
- Anchorage points for fall arrest systems.
- Safety hazards.
- Stability of structures/extendibility.
- Working at elevation.
- Excavation requirements.
- Confined space.
- Transport and vehicular movements.
- Fire and explosion hazards.
- Working on, in, over and near water.
- Cofferdams and caissons.
- Crane requirements, location, ground stability and clearances to obstructions.
- Energy distribution systems.
- Fire detection and fire fighting.
- Fresh air, temperature and weather protection.
- Lighting and access for maintenance.
- Health issues in construction.
- Asbestos avoidance/management.
- Chemical hazards.
- Biological hazards.
- Physical hazards.
- Falsework.
- Personnel protection programme.
- Physical hazard prevention procedures, e.g. isolation, tag out.
- Operation and maintenance.
- Decommissioning and demolition.

proximity of adjacent structures, neighbouring properties, space required to accommodate the temporary works, welfare facilities and emergency evacuation arrangements from the site. Once this has been achieved by considering the structure in three dimensions, conflicts that may exist during construction, use, maintenance and demolition of the works can be identified and satisfactorily resolved. The overall objective of the review will be to confirm the design meets minimum project requirements for safety, functionality, operability, maintainability and constructability. To confirm that the design development is not introducing risks that are unacceptable, these reviews should be carried out to reflect the complexity of the project. In the petrochemical industry these would generally be carried out at 30%, 70% and 100% completion, and included as milestones within the programme. There should be a formal procedure to confirm the relevant hazard elimination and risk reduction issues have been properly assessed and that the relevant information has been included in the pre-construction information.

The risk assessments should be reviewed by the relevant discipline lead to evaluate the health and safety impact on the construction works emanating from the design and the outcome from the review coordinated with the other designers and contractors.

Providing the design information

The information about aspects of the design that could create significant risk or be difficult to manage during the construction works, that would not be obvious to a competent contractor or other designer, should be made available in a format suitable for use. This generally would relate to:

- site specific significant hazards (e.g. contamination, underground and overhead services, hazard area classifications, operation restrictions associated with working on/in plant or equipment, mechanical systems – boiler/pressure systems);
- constraints arising from site (e.g. maintaining the site critical infrastructure, effective emergency response arrangements, interfacing with routine, one-off and other activities of the clients' business);
- neighbouring structures (foundations, multi-site occupancy);
- the stability of the structure or a construction sequence.

The information would generally be provided on:

- general arrangement drawings;
- scope of works detailed in the contract;
- works specifications;
- health and safety file.

The use of notes on drawings, e.g. Safety, Health and Environment (SHE) boxes, provides a simple method of conveying the health and safety information in the design, although they should not be used to the exclusion of all other methods to convey vital health and safety information. The use of non text symbols, e.g. standard safety symbols (Highly Flammable Materials) to annotate the drawings provides a simple pictorial story of the significant risk data and overcomes the language barriers associated with the vulnerable workers, particularly those with reading difficulties and those unable to read the language in which the information has been provided. It also enables the location of the hazard to be clearly identified, readily shared with the other disciplines working on the project and overcomes the bureaucracy of information being hidden and overlooked during design development, costing and planning.

The three tests to confirm that the information is relevant, project specific and required to manage the remaining risk are, is the information:

- not likely to be obvious to a competent contractor or others who use the design?
- unusual?
- likely to be difficult to manage effectively?

An example of a SHE box is:

11 kV cable must be isolated prior to starting any excavation works. If for any reason the cables cannot be isolated for the duration of the excavation and protection works the Designer must be informed so that an alternative design can be produced.

The following would not be considered relevant or project specific:

appropriate measures to be taken to ensure safety when excavation adjacent to existing buried cable trenches.

The agreements with the client must confirm the health and safety file layout and format, arrangements for the collection and gathering of information, and method of storage of information.

Design change

The design can be altered at any stage during the design or indeed the construction process and it is important that arrangements are in place to capture these changes. It is important for the changes to be properly considered by all the relevant parties to enable them to be successfully closed out through a project change control procedure. The arrangements in place must be sufficiently robust for dealing with late changes to the design, and for securing cooperation with the contractors, so that the problems are resolved. Typical reasons for design changes include:

- Change to site conditions from those envisaged when developing the design.
- An instruction from the client which modifies the agreed scope of works.
- Alternative design to the 'definition design', particularly under a design and build form of procurement.
- Adapting the works to suit the contractor's preferred method of working or availability of plant, equipment and materials.
- Change to the project delivery strategy particularly sectional completion of the works to suit the client's business needs.
- Take advantage of technical progress to good industry practices.
- Design development arising from specialist contractor design elements of the permanent and/or temporary works, particularly those forming specialist design structural components.

The source of the design change includes:

- Written instruction from the Client, Minutes of meetings, confirmation of verbal instruction.
- Documentation received from suppliers and sub-contractors.
- Site instruction.
- Project specific change control system, e.g. NEC early warning and change control procedure.

Any redesign must be strictly controlled by using a variation note/change note instruction to secure cooperation from the other parties involved in the project and to coordinate the modifications to the agreed scope to secure the relevant changes to the safeguard provisions to deliver an incident free project.

Temporary works

The topic of health and safety in design of the temporary works is covered in detail in Chapter 17 *Falsework*, and will not be covered within this chapter. However it is important that these two topics are not considered in isolation and the interface of the designs of the permanent and temporary works is seamlessly managed. The permanent works Designer should ensure that sufficient information is provided in the design particularly relating to the temporary stability of the partly constructed structure.

The temporary works should be reviewed by the relevant discipline lead to evaluate the health and safety impact on the construction works, emanating from the design, and the outcome from the review, coordinated with the other designers and contractors.

Demolition and decommissioning

The topic of health and safety in design of the demolition and decommissioning works is covered in detail in Chapter 18 *Demolition, partial demolition, structural refurbishment and decommissioning* and will not be covered within this chapter. However it is important that it forms an integral part of the planning and design process from the inception of the project and sufficient information is provided in the design to enable load paths and stability to be assessed, prevent progressive collapse, services to be terminated, and fabric condition to be assessed.

Living the dream

Construction health and safety risk management throughout the project life cycle has challenged the industry for decades. Despite the development in good and best working practices and developments in understanding the risk management process, no solution has been identified to enable construction works to be carried out repeatedly in an incident and injury free environment. This is despite acknowledgement, within the construction industry, that good health and safety simply makes good business sense and a superior health and safety performance can drive superior business performance. This reinforces that the new goal setting legislation alone will not create the desired outcome of 'injury free projects', nor embed health and safety excellence into projects.

One of the key ingredients of carrying out the construction work so that no one gets hurt is to ensure that people are in an informed position with regard to the hazards that have to be controlled during the works, and are provided with the resources to enable the works to be carried out safely. The importance of early intervention of health and safety risk management in a project is demonstrated by inspection of **Figure 4**.

To live the dream of zero accidents on construction projects the industry will need to create a culture that lives and breathes health and safety with a moral commitment of caring for the workforce and stopping people getting hurt, by looking for the next potential incident and taking action before it happens. We need to create a construction environment health and safety lifestyle – fit for work, fit for life, fit for tomorrow, 24/7 approach – safety never stops in all we do. We need to work together to persuade, cajole and inspire all the industry stakeholders to come on board and make it happen. When this is achieved we may be in a position of sending the workforce home healthier than they arrive to work each and every day.

Commitment to the zero harm concepts solves the dilemma of health and safety versus production – schedule, cost and profit.

Summary of main points

In essence the objective of managing hazards and safety in design is to maximise the opportunity to eliminate hazards, and reduce risk in the design through the early consideration of how the structure will be built, used, cleaned, maintained, repaired and demolished. By effectively managing the opportunity it will reduce the threat of any early decision and assumptions leading to a disproportionate impact on safety at a later stage in the project. The process should underpin the principle of get it right first time and avoid the need for redesign.

The effective opportunity management of the risk will enable informed decisions to be made with regard to the whole life cost and planning of the project. It will also capture the knowledge and experience of the construction team up front when it can be most effective and avoid wasted effort during design development latter in the project life cycle. This will make delivery to time, cost and quality more likely – unrealistic deadlines and failure to allocate sufficient funds are two of the largest contributors to poor control of risk on site.

Health and safety in design is an integral part of the project design and decision making process and should be carried out in a systematic, structured and timely manner tailored to the context of the construction environment. To improve the effectiveness of the approach the overall framework needs to be dynamic, interactive and responsive to change.

The first step to health and safety in design is to identify the project specific and unusual health and safety hazards that are likely to be difficult to manage then apply the health and safety in design hierarchy.

To avoid work related injury the designers need to eliminate and reduce the risk from the design that will lead to unsafe acts or unsafe conditions which contribute to a person suffering from ill health and/or resulting in a person being injured.

An integral and fundamental part of the process is to routinely review the design to confirm that the elimination or reduction of a risk for a hazard remaining in the design does not create a new and more significant risk. The health and safety in design hierarchy reflects that designing out and the use of physical engineering controls and safeguard measures are more reliable than people.

The designer's 'tool kit' needs to needs to provide them with the knowledge and experience to understand:

- the most common causes of accidents and ill health in construction;
- the key construction health and safety legislation;
- a robust decision making framework;
- the information to be conveyed to the other parties involved in the project briefly, clearly, precisely and in a form suitable for the end user.

The project specific hazards need to be identified early in the design and assigned to the risk owner to enable them to be considered when appraising the usability, buildability, maintainability and eventual demolition of the project. The risk ownership should be assigned to the party best able to manage the risk.

There are a variety of mechanisms available to identify, assess, manage, inform and review construction risk and these have been detailed in this and other chapters of this book. The main challenge is to:

- gain a knowledge and understanding of how to manage the health and safety risk arising from the hazards;
- have an in-depth appreciation of what is considered as an acceptable level of risk in the industry;
- understand how this standard is interpreted into recognised good working practices with clear ownership of risk and clear delegation of authority.

The output from the design development should be brief, clear, precise information in a form suitable for the end user. It should be project specific and highlight significant risks that are not obvious to those who use the design and those that are unusual and likely to be difficult to manage.

The use of notes on drawings provides a simple non-exclusive method of conveying the health and safety information. The use of non-text symbols provides a picture and overcomes the barriers associated with vulnerable workers, particularly those with reading difficulties and unable to read the language in which the information has been provided. The use of SHE boxes provides a succinct means of conveying the information in design and overcomes the problems of information being hidden in documentation.

The challenge is that the construction works emanating from a design can be built safely with zero harm to the workforce and those affected by the works. The start is to change the concept that health and safety in design is tolerable of a set number of injuries. As an industry we must embrace and nurture a culture of caring for the construction workforce and those who may be affected by the works. Commitment to the zero harm concepts solves the dilemma of legal compliance and the dilemma of health and safety versus production – schedule, cost and profit.

Focus on zero harm and be a part of the step change in the construction industry health and safety performance.

References

Health and Safety Executive (HSE). *Five Steps to Risk Assessments* INDG163 (rev2), 1996, London: HSE Books. Available online at: http://www.hse.gov.uk/pubns/indg163.pdf

Health and Safety Executive (HSE). *Successful Health and Safety Management* HSG65, 1997, London: HSE Books. Available online at: http://www.hse.gov.uk/PUBNS/books/hsg65.htm

Health and Safety Executive (HSE). *EH40 – Table 1: List of Approved Workplace Exposure Limits (as consolidated with amendments October 2007)*, 2007, London: Health and Safety Executive. Available online at http://www.hse.gov.uk/COSHH/table1.pdf

Health and Safety Executive (HSE). *Designers – how can you reduce the health risks in construction?*, 2009a. Available online at: http://www.hse.gov.uk/construction/healthrisks/designers.htm

Health and Safety Executive. *HSE Public Register of Convictions*, 2009b. Available online at: www.hse.gov.uk./prosecutions

HM Treasury. *The Green Book. Appraisal and Evaluation in Central Government*. Treasury Guidance, 2003, London: The Stationery Office. Available online at: http://www.hm-treasury.gov.uk/data_greenbook_index.htm

Occupational Safety and Health Administration (2005) *OSHA Recordkeeping Handbook. The Regulation and Related Interpretation Regulations for Recording and Reporting Occupational Injuries and Illnesses*. OSHA 3245–09R 2005, Washington, DC: OSHA.

Robens of Woldingham Alfred Robens Baron chairman, Great Britain Committee on Health and Safety at Work, Great Britain Department of Employment. 1972. *Safety and Health at Work: Report of the Committee 1970–72. (Cmnd. 5034)*, 1972, London: HMSO.

Referenced legislation and standards

CIRIA report 166 'CDM Regulations – Work Sector Guidance for Designers.

Construction (Design and Management) Regulations 2007 Reprinted March 2007. Statutory instruments 320 2007, London: The Stationery Office.

Control of Asbestos Regulations 2006 Reprinted November 2006, January 2007 and March 2007. Statutory Instruments 2739 2006, London: The Stationery Office.

Control of Substances Hazardous to Health Regulations 1994. Statutory Instruments 1994 3246, London: HMSO.

Health and Safety at Work Act 1974 *Elizabeth II*. Chapter 37, London: HMSO.

ISO 31000: 2009 Risk Management – Principles and Guidelines, Geneva: International Organisation for Standardisation.

Work at Height Regulations. Statutory Instrument 2005, No. 735, 2005, London: The Stationery Office.

Further reading

Health and Safety Executive (HSE). *Designing for Health and Safety in Construction – a guide for designers on the Construction (Design and Management) Regulations 1994*, HSE guidance booklet C100, London: HSE Books.

Health and Safety Executive (HSE). *Principles and Guidelines to Assist HSE in its Judgments that Duty-holders Have Reduced Risk as Low as Reasonably Practicable*, 2001. Available online at: http://www.hse.gov.uk/risk/theory/alarp1.htm

Health and Safety Executive (HSE). *Managing Health and Safety in Construction. Construction (Design and Management) Regulations*

2007. (CDM) Approved Code of Practice. L144, 2007, London: HSE Books. Available online at: http://www.hse.gov.uk/PUBNS/books/l144.htm

Health and Safety Executive (HSE). *Asbestos is a Hidden Killer*, 2009. Available online at: http://www.hse.gov.uk/asbestos/hiddenkiller

Websites

Advisory Committee for Roofwork, publications http://www.roofworkadvice.info/html/publications.html

Behavioural Safety.com – a free website dedicated to promoting the use of behavioural safety techniques to reduce workplace accidents http://www.behavioural-safety.com/

ConstructionSkills, Guidance for CDM Regulations http://www.cskills.org/supportbusiness/healthsafety/cdmregs/guidance

Free Signage – UK statutory signs in pdf format for you to print and use http://www.freesignage.co.uk/index.php

Incident Investigation Systems www.kelvintopset.com

Institution of Civil Engineers, Health and Safety http://www.ice.org.uk/knowledge/specialist_health.asp

International Organisation for Standardisation (ISO) www.iso.org

Health and Safety Executive www.hse.gov.uk

Health and Safety Executive, Asbestos a hidden killer:

Real life stories http://www.hse.gov.uk/asbestos/hiddenkiller/real-life-stories.htm?ebul=cons/nov09&cr=4

Do you know the facts? http://www.hse.gov.uk/asbestos/hiddenkiller/facts-start.htm?ebul=cons/nov09&cr=3

Occupational Safety and Health Administration (OSHA) www.osha.gov

Office of Public Sector Information, CDM Regulations http://www.opsi.gov.uk/si/si2007/uksi_20070320_en_1

Safety in Design Ltd (SiD) – a not for profit company that exists to support designers in the built environment http://www.safetyindesign.org/

ice | manuals

Chapter 6

Establishing operational control processes

doi: 10:10.1680/mohs.40564.0071

Ciaran McAleenan Expert Ease International, Lurgan, Northern Ireland, UK

CONTENTS

Construction sites are hazardous environments and while some hazards can be eliminated or contained through good engineering design solutions, others, inherent in the process, have to be worked around. The risk management model starts off its analysis by looking at the hazards, trying to determine the likelihood of the hazards being realised (accidents), and then loses sight of what is important because its focus is often too narrow. Whereas the operational analysis and control model does not start with the identification of the hazard, rather it goes to the real starting point – the work operation – and determines from the outset what is needed to achieve a safe outcome. The manager's objective must be 'a safe outcome to a successful work operation' and in that there is a need to start to look at anything that will thwart the achievement of that aim.

Proper management of the entire operation requires that in consultation with the workers affected you define your operational outcome, provide the resources and review to consider the possibility of failure, prior to commencement, during the process and at various other stages. Operating to any less a standard will only guarantee a negative outcome and ensure that accidents continue.

Box 1   Key learning points

In this chapter you will be introduced to:

- Approaches to risk assessment, including the myths and pitfalls to avoid.
- The operational analysis and control model.
- Workforce involvement.
- Method statements, permits to work and safe systems of work.
- Accident analysis.

Introduction

Construction workers have a right to a working environment that protects and promotes their health. As professional engineers and designers we must fully support the need to improve the health of workers within the construction industry. For far too long this industry has concentrated more on the safety than the health issues. We are all very familiar with the main causes of accidents, but we probably spend less time considering the main health issues facing construction workers. Workers in this industry have high rates of ill health related to work; specifically musculoskeletal ill health, noise and vibration damage, exposure to respirable silica as well as asbestos related diseases. The time to raise health issues facing workers to the same level as their safety has arrived.

The core message, that designs are to be capable of being built, used, maintained and demolished in a manner that will not cause harm to the worker, includes the harm related to the workers' health just as much as it relates to their safety.

However, the link between construction site accidents and the designer is not always so obvious at the point and place of accident. Recent analyses of accidents, however, have shown the connection to be greater than would first be apparent to most investigators. Consequently, in the drive to reduce construction site accidents, the role of the designer must be very much to the fore.

In this chapter we will discuss workplace operations and the necessary controls. The reason for having operational control processes is to ensure that work operations are carried out in strict accordance with all relevant 'healthy and safe working' procedures. In this way we can make sure that people, plant and property are protected from harm prior to, during and after the work operation, regardless of the nature of the hazards faced. Therefore, operational control processes should have a work objective that states: '... activities are to be completed on time and in a manner that does not cause harm to the employees, non-employees, and the company'.

This chapter will examine:

1 How to analyse work activities to establish what can cause harm and what needs to be done to prevent harm from being realised.
2 How to manage operations, knowing what needs to be done, what resources are needed and how and when to review effectiveness.
3 In reviewing effectiveness considering how the activity has progressed, whether anything has changed since the previous review and what effect the changes have had/will have on operational management.

We will be discussing the range of control options from engineering solutions to personal protection and discussing how engineers' and designers' decisions can affect the final choice of controls. Remember that as professionals we can play a significant role by working out how our designs can influence the immediate health and safety of construction workers, and the longer term health of the workers and those impacted by the structures and buildings we design. In addition we will address health surveillance programmes that can and should be delivered within excellent construction projects. Now it is important to note, before proceeding any further, that health surveillance is not a control measure, nor is it a substitute for controls to deal with health hazards. Rather, health surveillance is a means of methodically watching out for early signs of work related ill health in employees. It is one measure of the effectiveness of the controls you have put in place to protect workers from the effects of exposure to health hazards.

Finally, remembering that accidents are control failures, the chapter will examine the thought processes and key questions needed whenever an accident has occurred, both in the context of emergency responses and incident investigation.

Dispelling some risk assessment myths

Some hazards can be eliminated or contained through good engineering design solutions while others, inherent in the process, have to be worked around. Risk assessment is the process used to lead to the appropriate solution. It is not the solution in itself! And that is what we are going to look at now. But first I would like to dispel a few myths. I will open by discussing two concepts: 'written risk assessments', and 'significant risks'. Much has been discussed lately about 'over interpretation' of regulations and this discussion went as high as to be discussed at a UK Parliamentary Committee in 2008, where there was a degree of finger pointing, with no one willing to take responsibility for the situation. However, at this point I wish to single out and query the suggestion that there is in fact an 'over interpretation of regulation' in the first place.

There are two key concepts that have no basis in regulation, but which many, including some practitioners, educators and enforcers, suggest are requirements in law. They are written risk assessments, and significant risks (including its corollary, the trivial risk). First, no one can point to the health and safety regulation that requires all of these so called excessive risk assessments. I would suggest that rather than there being an over interpretation of regulations, what has happened has been an incorrect interpretation of the regulations in the first instance which has led to the creation of the monster that is the written risk assessment. Let's examine regulation 3 of the Management of Health and Safety at Work Regulations 1999 (MHSW, 1999), which states:

> (1) Every employer shall make a suitable and sufficient assessment of – (a) the risks to the health and safety of his employees to which they are exposed whilst they are at work; and (b) the risks to the health and safety of persons not in his employment arising

out of or in connection with the conduct by him of his undertaking, for the purpose of identifying the measures he needs to take to comply with the requirements and prohibitions imposed upon him by or under the relevant statutory provisions ...

Just to note that in Northern Ireland these regulations are known as Management of Health and Safety at Work (Northern Ireland) Regulations 2000.

However, the key thing to keep in your mind is the requirement for employers to make suitable and sufficient assessments of the hazards associated with their work operations and that risk assessment is the process preferred by most to identify the potential sources of harm to ascertain the actions (controls) needed to prevent that harm from being realised. The process is so simple that children are taught to do it at an early age and can apply it quite effectively to, for example, crossing the road. Remember the childhood mantra that went: 'Look left, look right, look left again ...' or 'Stop, look and listen ...' (depending on your age)?

I'll bet you can still hear that mantra in your heads occasionally. Now as we grow older we become more skilled at this process, conducting risk assessments practically all of the time, mostly subconsciously and certainly without the need to write anything down or to have a written risk assessment available to be read before we act. For example, move away from crossing the road and now consider the act of driving a car. Can you think of any hazards you might be faced with during this activity? And yet you may be doing this on a daily basis and making safety decisions virtually every second to consider what could cause you, or your passengers, harm and acting appropriately to avoid that harm being realised. No written risk assessment there, because you are a competent driver, adequately resourced and authorised to act, which allows you to analyse the situation, you know what controls are at your disposal and you know how and when to operate the controls and lastly you are the decision maker. So what difference occurs between this activity and the act of driving a large earth moving machine, or operating a pile driver or tower crane? OK the activity is different but surely the operator is:

■ competent

■ resourced, and

■ authorised to act,

which allows him to analyse the situation, know what controls are at his disposal, know how and when to operate the controls and lastly make the decision. As workers become competent, that same learned ability to assess their work operation grows with their competence to undertake the tasks required of them, in a regular and intuitive manner.

What regulation 3 does is make a duty of this process as it relates to the employer and his duties to those affected by his undertakings. The duty does not extend to a requirement that all risk assessments are written risk assessments; indeed if you read regulation 3 (6) there is no requirement to produce any written risk assessments at all:

(6) Where the employer employs five or more employees, he shall record – (a) the significant findings of the assessment; and (b) any group of his employees identified by it as being especially at risk.

What an employer is required to do is to record any *significant findings* of his assessment. And I would contest that it is by combining the requirement to conduct risk assessments with the requirement to record significant findings that we have the source of the error that is the belief in a 'written risk assessment' for every activity, every situation, every disabled person, every pregnant woman, etc. in the workplace. I'll say it again as it is time to set the record straight:

Risk assessment is a process; a means to an end. Not the end in itself!

Whilst it may be prudent in more complex operations to have the process of assessment written up, it is not necessary to do so (later in the chapter we will discuss various written safety procedures, including: method statements, permits to work, lifting plans and emergency response arrangements). There are those who would argue that it is only by having the written assessments that we know that the process has been carried out. This is a logical fallacy. The safe work practices that have been developed by operatives, employers and industry in general point to the same conclusion. The better safe work practices/procedures (or whatever other name is given to them) are the result of full operation assessments (and not just risk assessments) and therefore point to the risk assessment process having been carried out suitably and sufficiently. Others have argued that the written risk assessment should and must be carried out only where there are significant risks. And this is the second of the interpretation errors. The regulations make no mention of 'significant risks' – anywhere.

The error stems from a misinterpretation of regulation 3 (6), which requires that *significant findings* be recorded. Now, in the context of risk assessment, a significant finding is a major or substantial conclusion about the risk that still exists and which renders the work operation unsafe to proceed. Therefore something further is required in order to control the work operation or the environment. A suitable method statement would be mindful of any significant findings and a competent worker would be aware of and able to implement appropriate controls that will eliminate or protect him from the effects of the hazard, taking care of any significant findings. Once such controls have been ascertained and put in place, there should no longer be any significant findings; the work operation should be safe to proceed.

In the light of this you can be assured that written risk assessments, of whatever length, are not an over-interpretation but a misinterpretation of the regulations and with the correct approach to the analysis of any operations, appropriate controls can be developed and implemented with the option to write up the risk assessment component of the process being left up to the employer and his view on the necessity to do so and based on the competent advice of his employees.

Remember that the point of the exercise is to consider and develop where necessary prevention and loss control programmes. But what's in a name? The outworkings of the risk assessment, if written down, are likely to be referred to by many different names, the more common ones being: method statements, operational safety control sheets, safety procedures and safe systems of work. The critical feature being that prevention is fundamental and that work activities must be safe to start, safe to execute and safe to finish.

Safe to start

I'd like you to consider for a moment what it means when I say 'safe to start'. Now as I discuss this with you I will be considering often misunderstood terms such as, risk matrices and hierarchy of controls.

In the Health and Safety Executive's (HSE) document *Successful Health and Safety Management* (HSE, 1997a) there is a section dealing with prioritising health and safety activities and this is where I believe the misunderstanding with regard to risk matrices has developed. The HSE document states:

Systems of assessing relative hazard and risk can contribute to decisions about priorities. They are also a useful aid to answering questions of importance and urgency arising at other stages in planning and implementing a health and safety management system …

The book goes on to explain that, 'While there is no general formula for rating hazards and risks, several techniques can help in decision-making. These differ from the detailed risk assessments needed to establish workplace precautions to satisfy legal standards … The techniques involve a means of ranking hazards and risks …' HSE suggests a simple enough formula for numerically calculating a risk rating based on numbers being assigned to the degree of expected harm, should a worker be exposed to a particular hazard multiplied by a figure relating to the likelihood of this occurring. The output from this exercise is to establish, among others, the priority for meeting different health and safety objectives, establishing training and competence development priorities. But one thing is clear: this suggested approach is predicated on the fact that a work activity has to be safe to start, safe to execute and safe to finish.

I shall explain this further by referring to what has become known as the hierarchy of controls. The popular view expounded by the various codes of practice is that there is a hierarchy of controls which begins with elimination of the hazard and ends with the use of personal protective equipment. You are likely to often hear the statement 'personal protective equipment (PPE) should only be used as a last resort' and that is clearly not right. Let me elaborate. Take, for example, a work activity where the best option would be to eliminate the task altogether such as entry into a confined space to clear blockages. Now, in design, perhaps entry to clear blockages could be designed with automatic flushing, camera systems (CCTV) and fixed water jetting facilities, but there are countless drains, sewers and tanks across the country where such items do not exist and even to install a CCTV or portable jetting hose requires entry into the confined space for a short period then, after venting or purging the space of hazardous gases, wearing the right respiratory and personal protective equipment and having appropriate warning devices the entry worker will have the correct controls in place to ensure that the job is safe from the start. These controls are further down the so called hierarchy of controls but are nevertheless effective. At a later stage, finances and time permitting, the automatic systems described above could be retro-fitted thereby negating the need for any future entry into the space. Please note that neither solution is any safer than the other, but there is a convenience about the second option which offers better working conditions so all things being equal that should be the preferred solution. The following are typical examples of the hierarchy of controls (range of control options):

1 Eliminate the hazards, i.e. use a different product or construction method or substitute it for an inherently safer product.
2 Combat at source, i.e. use engineering controls (e.g. machine guards) and give priority to collective protective measures.
3 Minimise, i.e. design suitable and sufficient safe methods of working. and/or use PPE.

So what I am contending is that rather than having a hierarchy of controls to consider you have in fact a range of control options to choose from and, depending on your circumstances at the time of the work operation, you should select the most appropriate option, always bearing in mind the core objective that the job has to be safe to start.

Workforce involvement

In this chapter I have been discussing with you an approach to health and safety that ensures that work operations are carried out in strict accordance with all relevant 'safe working' procedures advocating that those who are involved in the project, at whatever stage, have a contribution to make to the elimination or control of hazards. This is not a unique precept; indeed it is a fundamental aspect of the International Labour Organisation's (ILO, 2001) health and safety management standards (the Seoul Declaration is discussed in more detail in Chapter 4 *Managing Workers' Conditions*). Discussing his experience of turning around and making his company profitable during Brazil's recession in the 1980s Dr Ricardo Semler (2001) stated that:

> Accepting there is no such thing as a 'special worker' perfectly suited for one company means accepting worker individuality. And once you do that, you set the stage for making the most of that individuality by encouraging workers to tap their inner reservoir and find a balance between their aspirations and the company's.

In the context of this chapter I will develop the theme that the best people to involve when developing operational controls that take full account of health, safety and welfare are the workers themselves. Recognising that a competent,

resourceful worker can make the right choice to effect a successful outcome will remove some of the barriers between those who make the decisions and those whose task it is to carry them out. In discussing organisational maturity and the need for workers and their representatives to be involved in health and safety decision making, Ayres (2009) recognised that such an approach is consistent with the cultural concept of occupational health and safety, which supports the hypothesis of social and cultural relations at the workplace. Organisations that are good at managing health and safety create an authoritative, multi-directional, leadership structure to maximise the contribution of competent individuals and groups in the delivery of successful prevention and loss control programmes.

Operational analysis and control

There are no fixed rules about how a risk assessment should be carried out. The purpose is to help the employers to determine what measures should be taken to comply with their statutory duties. Now the core legal requirement for employers is to provide workplaces and environments that are free from recognised hazards that cause or are likely to cause death or serious physical harm to employees. The parent law is then supplemented by regulations, approved codes of practice and guides that provide more specific details on how the fundamental duty may be met with regard to specific work operations. For many years now legislators have been intent upon creating the conditions whereby risk in the workplace would be eliminated. In 2000, the HSE (Great Britain) issued a discussion document on regulating higher hazards in the workplace. Principal 2 stated:

> *Permissioning regimes require operators to describe how they plan to* achieve and maintain control, and to demonstrate active commitment to the effective management of risk. The overall objective is to secure an integrated and coherent approach to eliminating hazards and managing residual risks that would work without the intervention of the safety regulator.

However despite the existence of our present laws, the introduction of new laws, and the continuing implementation of supplementary regulations and codes of practice, workplace accidents continue to occur unabated, albeit with a decreasing rate of incidence in many instances. The state aims to eliminate workplace hazards and many safety professionals understand that intent behind risk management. Yet the mistaken notion has grown up around safety that risk management means the reduction of risk to acceptable levels. This is far from being the intent of legislation. If the fundamental legal requirement is to provide safe working environments and products, then we do not need additional laws to control industry, rather we need to reappraise, work with what we have and manage it better.

Many risk assessment approaches start with the hazards and lose sight of what is important because the focus is too narrow. Your objective has to be 'a safe outcome to a successful work operation' (McAleenan and McAleenan, 2002) and in that you start to look at anything that will prevent you from achieving that aim. Consequently your starting point is the work operation.

Within the construction industry you will have a wide range of work operations and many employees involved in the different work activities associated with each operation. Start by listing all of the work operations and categorising them into general areas (that is those with common activities). Work with the staff in each of the areas to ensure that all operations have been properly listed and categorised. From that list it will be possible to identify the hazards by asking, 'What can cause harm?' Remember to think of this in the widest sense:

■ What can cause the workers harm? (Plant and machinery, equipment, etc.)

■ What can cause harm to the operation? (Workers/visitors/lack of resource, etc.)

■ What harm can the environment cause to the workers? (Noise, vibration, substances, biological hazards, asbestos, etc.)

■ What harm can be caused to the environment? (Emissions, spillage. etc.)

Once you know what can cause harm you can look for the ways to:

■ Eliminate the hazard (look for alternative ways of doing the work).

■ Contain the hazard (look for ways of isolating the worker from the hazard or isolating the hazard from the worker).

■ Control the hazard (proper management procedures such as safe working procedures, permits to work, training and assessment, use of personal protective equipment, health surveillance*, monitoring and auditing).

It will depend on the nature of the work operation and the skills within your company whether you can do all of this from within. Your own expertise and that of your colleagues will be central, as will the skills of the workforce. Remember those who carry out the work task often have a very good idea of what will make the operation safe. Do not be tempted to dismiss this valuable source of information and advice. If you tell them what you are trying to achieve, ask for their assistance where it is needed and tell them why it is important that their company keeps them safe and healthy throughout their working life and beyond. If you think it is necessary, you can look for outside assistance but this should only be after you are clear what it is you require.

This is an iterative process, which you can work at until you feel that you have exhausted all of the possibilities. Then you ask 'is that it?' In that way your assessment approach works on the basis of continual improvement. You remain focused on the desire to ensure that all that could be known is known and that all that can be done is being done.

Now there are some points that are peculiar to occupational health that are worthy of note, namely exposure assessment, required to measure the quantities or concentrations of a particular hazard, and they may be necessary:

■ where there are airborne particulates or fume that may be toxic or harmful in nature;

■ where asbestos is present;

■ where noise or vibration is a possible problem;

■ where heat levels are likely to be high.

Having established the likely exposure to the various hazards, you may need to carry out a dose response assessment to determine the dose received by individual or groups of workers. The dose response assessment seeks to establish the relationship between the exposure concentration and the adverse health effects. Both of these techniques are complex in nature and normally the reserve of the occupational health specialist. The degree of exposure or dose response assessment will depend on the preliminary information gathered in the hazard identification stage of the process. Many tools can be used to establish the extent of the problem such as sickness absence records, health surveillance programmes, and epidemiological studies within the industry.

Assessment is only one aspect of operational management and when viewed in isolation often fails to achieve the desired effect; namely the safe completion of the work operation or activity. The purpose behind operational analysis and control (**Figure 1**) is to ensure that work operations are carried out in strict accordance with all relevant safe working procedures. In this way you can make sure that your people, plant and property are protected from harm prior to, during and after the work operation, regardless of the nature of the hazards faced.

Having carried out the analysis you must list what needs to be done to manage the work operation to ensure a safe outcome (e.g. have employees been made aware of what can cause them harm and what they must do?, do you know what training is needed?, are there written safety instructions? does everyone know who is responsible and for what? etc.)? You will need to know what resources are needed and make sure that they are available. Note that some resources will be needed well in advance of any work operation so you may have to build your controls into your budget and business plan. This is a simple but not simplistic approach that relies on workers and their managers being competent and appropriately authorised to make decisions. Appropriate authority to make safe operating decisions must extend to everybody in your business. To do anything less fails to recognise the important part that each person plays in delivering safe outcomes.

*Remember that health surveillance is not a control measure; rather it is a measure of the effectiveness of the controls that have been put in place to protect your employees from the effects of exposure to health hazards.

Operational analysis and control (OAC) model

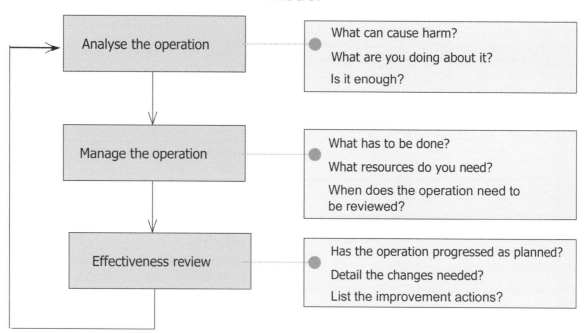

Analyse the operation	What can cause harm? What are you doing about it? Is it enough?
Manage the operation	What has to be done? What resources do you need? When does the operation need to be reviewed?
Effectiveness review	Has the operation progressed as planned? Detail the changes needed? List the improvement actions?

Figure 1 Operation analysis and control
Reproduced with permission from Expert Ease International. © McAleenan and McAleenan

Finally, believing that you have a safe workplace is the route to ensuring that you have not. Like every aspect of your work, safety needs to be continually managed and improved, as necessary. Things can go wrong and you must be able to anticipate and act in advance to ensure that they don't happen. So don't despair if you do not always get it right: there is no reason to give up or accept a lesser standard. Accepting accidents as inevitable is fatalistic. Your objective of integrating the highest standards of health and safety with improved business performance means that your end product must be achieved in a manner that protects your employees and the public from harm.

Method statements

Method statements are operational controls, specific to the task at hand, and are the outworking of your risk assessment process. They can also be referred to as safe working procedures, standard operating procedures, plans of work or safety statements; however, within the construction industry, they are most commonly known as method statements. Method statements need only specify site specific details, providing that generic company health and safety information (standard arrangements) are readily available. Don't forget what was said earlier about significant findings when assessing, monitoring compliance with or indeed preparing a method statement. It is normally only necessary to develop a method statement where

controls to address significant findings or work processes differ from the standard arrangements.

Method statements should identify the sources of harm for the work activity and set down the operational safety controls needed to prevent harm. The typical information required should include but not be limited to the following:

- A description of the work to be undertaken.
- The location of the work.
- The roles and responsibilities of key people, including the supervision arrangements.
- The sequence of work activities, including specific site rules and restrictions.
- A list of necessary resources, including work equipment and personal protective equipment.
- The emergency response arrangements.
- The competence level for workers and their managers, with names provided, where possible.

It is becoming more common for Clients to ask contractors to produce method statements as part of their health and safety documentation or within their construction phase plan for projects that are notifiable under the Construction (Design and Management) (CDM) Regulations.

Permits and safe systems of work

A permit to work system is a precise sequence of documented steps with strict checking and supervision built in. It is, in essence, an operational control process or method statement but for a particularly hazardous work activity or working environment, and, for that reason it is of necessity inflexible in its execution.

HSE in their *Confined Spaces* leaflet (HSE, 1997b) said:

A permit-to-work ensures a formal check is undertaken to ensure all the elements of a safe system of work are in place before people are allowed to enter or work in the confined space. It is also a means of communication between site management, supervisors, and those carrying out the hazardous work. Essential features of a permit-to-work are:

■ clear identification of who may authorise particular jobs (and any limits to their authority) and who is responsible for specifying the necessary precautions (e.g. isolation, air testing, emergency arrangements etc.);

■ provision for ensuring that contractors engaged to carry out work are included;

■ training and instruction in the issue of permits;

■ monitoring and auditing to ensure that the system works as intended.

Permits are generally used for operations involving:

1 Work in or near to confined spaces.
2 Hot work, or work where there is the potential to cause a fire or explosion.
3 Breaking into pressure pipelines.
4 Working on roofs or at height.
5 Repair and maintenance work on electrical or mechanical systems.

The permit is a declaration, which states that the work location has been isolated from the danger and is safe to work in, for a limited period. Once issued the permit becomes the principal work instruction, which will override all other instructions issued in connection with specific operations until it has been cancelled or it expires. Cancelling a permit can only be done by an authorised person (Issuer). The person accepting (Acceptor) the permit is the person whose responsibility it is to ensure safe conduct of the work and precise adherence to all of the specified controls. Only the Issuer of the permit has the authority to alter or cancel it.

If, for example, there is to be work carried out on an electrical system, one control aspect will be locking out the electrical system (and perhaps tagging the locks) in order to ensure that the system is not inadvertently re-energised by something or someone not associated with the work operation. The lock-out will be done, checked and signed off by both the Issuer and the Acceptor of the permit before work commences and will be reversed by the same two people before the permit can be cancelled. Remember it isn't the existence of the permit that makes the work safe, rather it is strict observance of all of its requirements, by all parties that will ensure a safe and healthy outcome. Without going into excessive detail, a permit procedure will follow the following steps:

1 Operations coming within the parameters of 'permit required work' are to be identified and the list periodically reviewed for continued validity.
2 Authorised persons (i.e. Issuers, Acceptors) are to be identified and confirmed as competent.
3 A permit request is to be submitted to an authorised Issuer.
4 The Issuer and an authorised Acceptor are to assess the hazards associated with the task and agree the permit boundaries and controls.
5 The Acceptor is to ensure that all controls are in place before accepting the permit.
6 The Acceptor is to ensure that all workers associated with the permit-required work operation are fully conversant with it restrictions and competent to carry out their requisite tasks.
7 The work operation is to be carried in strict accordance with the permit and within the allocated time limit.
8 Compliance is to be monitored throughout.
9 On completion of the work operation, the Acceptor is to inspect the work, the plant and equipment to ensure that normal operational safeguards are back in place, before handing back of the work area.
10 The Issuers is to confirm that the work has been satisfactorily completed and that the plant and equipment are safe to return to normal service before accepting the hand back.
11 The permit procedure is to specify what is to happen in the event of an emergency arising.

You should note that a permit to work system is based on comprehensive, up to date information in order that operations are carried out safely and it defines exactly what work is to be carried out. It applies equally to direct employees and to any contractors.

Accidents are control failures

It would be remiss of me not to mention incidents, accidents and near misses (also known as unsafe acts or omissions) in a chapter concerned with operational control processes, for there is a strong link between controls and accident prevention. In construction, as in many other industrial operations, ignorance of key elements of the operation/process reduces the control and introduces uncertainty into the final outcome. The National Institution for Occupational Safety and Health (NIOSH) refers to accidents as 'preventable injuries' (Becker, 2001), a useful definition to bear in mind when considering dynamic safety management.

To begin this discussion, I would like you to consider that accidents are control failures and not some unforeseen event beyond anyone's control. I believe that an incident that leads to a worker being injured is in fact foreseeable, given the right amount of attention to detail in the development of operational controls. Earlier we discussed how OAC took you away from the 'haven't I done enough?' mentality to an 'is there anything more I could do?' attitude. And in these circumstances you are continually evaluating and re-evaluating the work activities to determine what is likely to cause harm so somewhere in that ongoing process you will foresee the events that could lead to harm and put the appropriate controls in place.

There is a phrase you will come across, namely 'we have the intellectual and technological capability to prevent accidents'. Always bear that in mind, since some safety practitioners will argue that, 'there are always going to be injuries and deaths in the workplace'. However, accepting this fatalistic approach to safety as an 'inevitable outcome' sets both the expectations and limitations of operational control. It is only possible to get closer to an acceptable standard of safety if it is defined at the outset and the operation is properly managed to allow it to be achieved. There can only ever be one standard, no matter how you dress it up, namely that the product or service is produced in a manner that will not injure workers or others who come into contact with it. That straightforward position, extended, includes no damage to the environment or to profitability. In other words, it is not unreasonable to consider all the potential losses and put the proper controls in place prior to commencement. Proper management of the entire operation requires that you define your operational outcome, provide the resources and review to consider the possibility of failure, prior to commencement, during the process and at various other stages. Operating to any less a standard will only guarantee a negative outcome and ensure that accidents continue.

The purpose behind any 'accident' prevention strategy is to prevent accidents from occurring. Bearing in mind the view that accidents are control failures, this drive to eliminate accidents can only happen when control failures, whether or not they lead to injury incidents, are fully investigated, lessons learned and the appropriate remedial actions addressed. Investigations need to follow a rigorous and robust pattern, and can equally be applied to injury event and observed unsafe acts or omissions.

Table 1 gives a checklist of key questions and actions to assist in any investigation.

Final comment on this topic; as well as workplace fatalities there are many types of injury accidents, ill health occurrences and dangerous occurrences that you will have to report to the HSE. You should make yourself familiar with them. The information sources at the end of each chapter of this book will give you some useful information in this regard.

Duty of care

Duty of care derives from the common law obligation to act towards another in a manner that is reasonable in all circumstances, so as to avoid injury to him or his property. It requires that there is a sufficiently proximate relationship between the parties such that obligates them to behave towards each other in way that will not lead to loss or injury either through a reckless

act, an unintentionally careless act or an omission. (In the UK the authority for duty of care is the leading Scottish case of *Donoghue* v. *Stevenson* 1932 SC (HL) 31.) There is a long established history of the courts upholding the duty of care principle and in the modern world the duty is often incorporated into contracts and statutes, which define the nature of the relationship between parties, e.g. employers/employees, client/contractor, the specifics of any duty owed by one to the other and the remedies that may be sought for breaching the duty. Nevertheless, the general duty of care, as outlined, continues to exist through the nature of the relationship between the parties, regardless of any contractual obligations. 'Duty' was first put forward as a unifying concept in the law of tort in Buller's *Nisi Prius* published in 1768.

In this regard it may be argued that statute law, from the primary legislation through its supporting regulations and codes of practice, adds substance to (makes specific) the fundamental duties and rights necessary for the proper functioning of social relationships. Statute law and the courts' interpretations, although often the final arbiter, do not present the final word. The qualification of duty of care by the legislatures of many states is indicative of the universality of the principle.

For those countries that are party to the intergovernmental Organisation for Economic Cooperation and Development (OECD, 2004), the development of and adherence to the corporate governance principles will ensure that the interests and rights of stakeholders are taken into account and respected by the Boards of corporate bodies. (The OECD has 30 member countries including Ireland and the UK.) The principle is further made manifest through the United Nations (UN) wherein the International Labour Office promotes social justice and internationally recognised human and labour rights in a tripartite structure that involves employers, employees and governments through the development of standards and conventions applicable to the world of work. With 178 member countries, there is little doubt as to the universality of the principle and the advent of the Decent Work Agenda (2002), which advocates the necessity for safe work, guarantees that the prevention is an international (if not yet a universal) responsibility. The level of participation from the member countries will determine the success of the prevention principle.

Remember an understanding or awareness of the universality of the principle of duty of care does not automatically translate into knowledge of how we are obliged to act. The issue often only arises when there has been a failure of that duty and individuals or organisations are faced with defending their actions.

Summary of main points

Any good safety management system relies upon quality information and, while there are always many competing factors that make it easy to be swayed by a crisis or the issue of the day, the managers' role is to work to demonstrate how well they are geared up to deal with it. The danger is that this will not always achieve the desirable position of properly managing the safety of the operation on an equal basis with all the other important

1. Establishing the facts

   - What work activity was going on at the time of the incident?
   - Where was the work activity taking place (address of location and description of the environment if appropriate)?
   - Were any other activities going on in the vicinity?
   - Who was involved in the work activity?
   - Who were the key decision makers and for which elements of the activity?*
   - Was there a safe working method identified for the work activity?
   - What controls were in place prior to and during the activity (include competences, training, equipment, PPE, etc.)?
   - What resources were available for carrying out the activity (material, human, financial)?
   - Describe the sequence of events that led up to the incident (based on witness statements).

2. Collecting the evidence

   - Has the investigation team got a copy of the safe working method?
   - Is there a history of compliance checks being carried out for this work activity?
   - Is there a history of compliance checks being carried out for this group of workers?
   - Is there a history of similar incidents?
   - Have witness statements been taken from all those involved in the work activity? (Note this is to include a statement from those involved in the work activity who did not witness the incident, stating as much.)
   - Have witness statements been taken from others (non-employees) who witnessed the incident?
   - How have the witness statements been verified for accuracy?
   - List material evidence gathered from the scene of the incident (e.g. photographs, broken or damaged components, measurements, etc.)?
   - Have medical reports been obtained in respect of any injuries sustained?

3. Analysis

   - Was the hazard or hazards associated with the incident identified in the safe working method?
   - Were competent persons involved all the appropriate stages in the planning and implementation of the work activity?
   - Were all of the necessary controls identified in the safe working method?
   - If yes were the controls being properly applied?
   - If the controls were not being properly applied was this a common practice?
   - Which controls were not being implemented?
   - Could additional controls have been easily identified and put into place?

4. Conclusions

   - Immediate cause of the incident.
   - Underlying cause of the incident.

5. Action plan

   - Address any immediate actions to be implemented *before* the work re-commences or similar work commences elsewhere.
   - Address intermediate and longer term actions required at operational, supervisory and manager levels.

* Chain of authority question.

Table 1 Incident investigation checklist
Reproduced with permission from Expert Ease International. © McAleenan and McAleenan

business areas. Consequently, an organisation's approach to establishing control processes needs to reflect the nature, scale and impact of its activities across the entire spectrum of its activities. Therefore not only is it correct to customise these controls to fit the needs of the business, it is expected of any quality organisation.

Finally there are 24 words that sum up all that is important in health and safety:

Health and Safety fundamentals; identify sources of, set controls for and get resources to prevent harm. Monitor compliance and success. Adjust if need be. Involve workers. (McAleenan, 2009)

Remember these words or better still text them to a colleague, friend or acquaintance. Let's get this message circulating.

References

Ayres, G. *Consultation, Organisational Maturity and Influence Decision Making at the Workplace: Has the Construction Industry the Maturity to Allow OHS Representatives Real Influence in the OHS Decision Making Process?* CIB W099 Conference, Working Together: Planning, Designing and Building a Healthy and Safe Construction Industry, 2009.

Becker, P. Lecturer West Virginia University. Private conversation with the author (ISSA Paris December 2001).

Health and Safety Executive (HSE). *Successful Health and Safety Management*, 2nd edition, HSG65, 1997a, London: HSE Books. Available online at: http://books.hse.gov.uk/hse/public/saleproduct. jsf?catalogueCode=9780717612765

Health and Safety Executive. *Safe Work in Confined Spaces* INDG258, 1997b, London: HSE Books. Available online at: http://books.hse. gov.uk/hse/public/saleproduct.jsf?catalogueCode=INDG258

Health and Safety Executive (HSE). *Regulating Higher Hazards: Exploring the Issues* (discussion document DDE15), 2000, London: HSE. Available online at: http://www.hse.gov.uk/consult/disdocs/ dde15.htm

International Labour Organisation (ILO). *Guidelines on Occupational Safety and Health Management Systems, ILO OSH 2001*, 2001, Geneva: International Labour Office. Available online at: http://www.ilo.org/asia/whatwedo/publications/lang--en/ docName--WCMS_099129/index.htm

McAleenan, C. *Health and Safety Case Study: Construction Incident – An Inspector Calls ...*, 2009, Institution of Civil Engineers Graduates and Students Professional Development Masterclass presentation.

McAleenan, P. and McAleenan, C. *A Different Approach – Operational Analysis and Control*, 2002, National Safety Council (USA), Congress proceedings.

Organisation for Economic Cooperation and Development, Principles of Corporate Governance, endorsed by Ministers at the OECD Council meeting at Ministerial level on 26–27 May 1999, last revised 2004.

Semler, R. *Maverick!, The success story behind the world's most unusual workplace* 2001, London: Random House Business Books.

Takala, J. *Decent Work – Safe Work, Introductory Report to XVIth World Congress on Safety and Health at Work, Vienna 26–31 May 2002*, Geneva: International Labour Office.

Referenced legislation

Management of Health and Safety at Work Regulations 1999 Reprinted February 2005. Statutory Instruments 1999 3242, London: The Stationery Office.

Management of Health and Safety at Work Regulations (Northern Ireland) 2000. Statutory Rule 2000 388, London: The Stationery Office.

Further reading

Benjamin, K. and White, J., *Occupational Health in the Supply Chain: A Literature Review. HSL/2003/06*, 2003, Sheffield: Health and Safety Laboratory. Available online at: http://www.hse.gov.uk/ research/hsl/ochealth.htm

Construction (Design and Management) Regulations 2007. Reprinted March 2007. Statutory Instruments 320 2007, London: The Stationery Office.

Construction (Design and Management) Regulations (Northern Ireland) 2007. Statutory rules of Northern Ireland 291 2007, London: The Stationery Office.

European Agency for Safety and Health at Work. *Assessment, Elimination and Substantial Reduction of Occupational Risks*, 2009, Luxembourg: Office for Official Publications of the European Communities. Available online at: http://osha.europa. eu/en/publications/reports/TEWE09001ENC/view

International Labour Organisation. *Seoul Declaration on Safety and Health at Work*, 2008, Seoul, Korea: International Labour Organisation.

McAleenan, C. and McAleenan, P. *Dynamic Safety Management in the Construction Industry*, 2001, Presentation to International Safety and Security Association (ISSA), Paris 2001.

Websites

European Agency for Safety and Health at Work http://osha. europa.eu/

Health and Safety Executive (HSE) http://www.hse.gov.uk

Institution of Civil Engineers (ICE), Health and safety http://www. ice.org.uk/knowledge/specialist_health.asp

International Labour Organisation (ILO) http://www.ilo.org

National Institution for Occupational Safety and Health (NIOSH) http://www.cdc.gov/niosh/

Organisation for Economic Cooperation and Development (OECD) http://www.oecd.org

SafeWork Bookshelf http://www.ilo.org/safework_bookshelf/english

Chapter 7

Occupational health and safety management systems

Philip McAleenan Expert Ease International, Downpatrick, Northern Ireland, UK

doi: 10:10.1680/mohs.40564.0081

CONTENTS

The competent company is one that recognises and maintains the distinction between the governance of the company and the daily management requirements to meet the strategic objectives of the company/project. An effective management programme commences with the establishment of the strategic objectives followed by the allocation of sufficient human, financial, material and time resources to meet those objectives.

The appointment of competent persons at all levels in the project, empowered with the appropriate authority to make critical decisions within their area of competence and sphere of influence, is a key element in building the management structures. Channels of communication between individuals, teams and departments, and the provision of appropriate and sufficient information is a necessary prerequisite for effective decision making to ensure quality and safety in the achievement of project objectives. An effective management programme will integrate quality, safety, environment, resources, etc. into a unified structure where the key decision makers act in unison to meet the project objectives.

Box 1	Key learning points

Readers will acquire information on the interrelationship between strategic and managerial roles of the directors and senior management team in a company, the requirements for the development of a competent company and the organisational tasks for the development and implementation of prevention strategies and practice.

Introduction

The commencement, progress and successful conclusion of any project happen only to the degree to which it is effectively managed. Even a poorly managed project will eventually arrive at an end point, but the price paid in terms of quality, safety and health may well render the project ineffectual and without merit. A successful project is based on good management which in turn derives its effectiveness from a number of core principles, all of which are outlined in a variety of specifications and guides which inform the development of safety management systems.

These principles cover the structure of management, the roles and responsibilities of the company, its managers and workers to assess and control hazards, timely and relevant communication of information to all who require it, monitoring and review of the effectiveness of policies and procedures, and the anticipation and preparation of any unintended event or outcome that would jeopardise workers, the community or the company.

The key to developing a safety management system is an understanding of these principles and their implementation in a way that meets the requirements and ethos of the company. It is appropriate to begin any consideration of management programmes, whether occupational safety or otherwise, by examining their position within the governance structures of the company, and in that respect it is worthwhile spending a little time on looking at what governance means and how it is to be exercised.

Governance

The Institution of Civil Engineers (ICE) 2008 review on its internal governance structure describes it as being the art of leading and directing a company. The current Organisation for Economic Cooperation and Development (OECD) *Principles of Corporate Governance* are based on the concept that good governance is a continual process of self-appraisal in order to ensure that the interests of shareholders and stakeholders are met (OECD, 2004). We can combine both these ideas and say that those charged with leading the company are competent to develop strategies in accordance with the interests and requirements of the shareholders and those affected by the business of the company and to direct their management team to establish appropriate operational programmes that will achieve their objectives.

Governance is also concerned with the professional and ethical behaviour of the company, more so since the financial scandals of the early 2000s (cf. Enron in 2001, Arthur Anderson LLP in 2002, and WorldCom in 2002) and the financial crises of 2008/2009. The OECD principles require openness and transparency on the part of the boards of companies combined with the highest standards of behaviour that will meet the member states' objectives of sustainable economic growth and continual improvement of the living standards of all members of society.

In this respect there is a clear distinction between strategic and operational tasks of a company. The strategic objectives must look towards the long term success of the company exercised in a manner that benefits not just the shareholders and owners of the business but also the communities in which the company operates and impacts upon. Conversely the management tasks are to translate the strategic objectives into short term, even day-by-day activities that will inexorably meet those objectives.

These objectives find resonance in the Bilbao Declaration (European Agency for Health and Safety at Work, 2004) (signatories to this declaration include the European Council for Civil Engineers, European Federation of Engineering Consultancy Associations, Architects Council of Europe, construction industry and building federations, and the Dutch presidency of the EU at that time), which calls for all partners in any construction project to ensure that the pursuit of high calibre buildings and civil projects are quality driven. Quality contributes to the success and sustainability of the company and thus is in the interests of the shareholders. It is in the interest of all stakeholders as it represents value for money for the client, it will impact positively of the community in which it will exist and, in the context on this manual, it results in and from good health and safety practice throughout the entire life of the project, from design through construction and maintenance of the finished project to its eventual demolition: again a matter that Bilbao considers integral to the construction project process.

Individual and organisational competence

Sir John Egan (1998), reporting to the UK's Deputy Prime Minister on the state of the construction industry, stated that:

> If the industry is to achieve its full potential, substantial changes in its culture and structure are also required to support improvement. The industry must provide decent and safe working conditions and improve management and supervisory skills at all levels.

What Egan was talking about was the establishment of a competent and sustainable industry, achieved through commitment to safety at all levels in the industry, which means that:

- Companies and their clients work in partnership within the industry, the health and safety enforcement agencies, fellow client bodies and workers and their representatives.

- Board members commit to making safety work for all their staff and for all those affected by their work.

- Business strategies and objectives are prefaced with a commitment that goals will be achieved in a manner that does not cause harm to workers or end-users.

- Companies go beyond compliance where occupational health and safety is critical.

- Individuals, workers and employers will act as they would expect others to act, i.e. competently.

Organisational competence requires that individuals in all positions in the company acquire knowledge of and develop and exercise competence in the safety aspects of their position. The Quebec Protocol (International Social Security Association, 2003) adopted by the International Social Security Association (ISSA) International Section for Education and Training for Prevention established a set of principles to integrate occupational health and safety competencies into the education processes for all occupations and for the key players in the education and prevention communities to pool their resources to achieve this.

Recognising the importance for companies to have access to professional and competent health and safety advice and assistance, the Institution of Civil Engineers (ICE) maintains a health and safety register for those professionally qualified members of relevant professional bodies (membership is not limited to ICE members) who can demonstrate at least 10 years construction experience and have attended appropriate health and safety courses.

It is incumbent upon those working in the industry to acquire and maintain their competences in health and safety, and upon the company to recognise, support and utilise these competences to ensure inherently safer designs and structures.

Prevention

The International Labour Organisation (ILO) Occupational Safety and Health Convention of 2006 (ILO, 2006) (this was ratified by the UK on 29 May 2008) puts prevention at the heart of national strategies which are to be developed in conjunction with the most representative employer and worker organisations. It follows that health and safety must be a key consideration both in the development of a company's corporate strategy and in the way that it manages its activities.

Strategically, the Board (in the context of this chapter, the term 'Board' is used generically to refer to the owner(s) as well as the representative body of the owners of a business whether a limited company or not) must decide not to become but to be (in the present tense) a safe company and to establish policies that reflect this reality. This doesn't mean a health and safety strategy that is distinct from other policies, such as fiscal, procurement, human resources, operational, etc. A competent company will ensure that quality, health and safety are integral elements to all other policies. Procurement policies will include a requirement that companies in the supply chain respect and exercise their legal duties to ensure good employment practices and safe working conditions; human resource strategies will aim to recruit competent persons to all positions and jobs, or to train successful recruits up to the required standard of competence; financial policies will ensure that priority is given to resourcing all measures necessary for the health, safety and welfare of the workforce, and the protection of the communities affected by the work of the company, and so on.

In this way any dichotomy between health and safety and what the company does is removed. No longer an adjunct, health and safety becomes the responsibility for all managers and employees to exercise continuously as an integral aspect of their work.

This strategic orientation is the commencement point for the development of an effective safety management system. In keeping with the requirements of legislation and the counsel of the Health and Safety Executive (HSE)/Health and Safety Executive Northern Ireland (HSENI), leadership of health and safety is a corporate responsibility and how the company is to exercise its responsibilities is determined by the strategies and policies that are adopted by the Board and implemented by the management team and employees.

Owners, shareholders and stakeholders

Before we look at the management structures and systems let us look at the question of shareholders and stakeholders.

Shareholders are those who own shares in a limited company and collectively are the owners of a company. It is from their ranks that the Board is elected with majority shareholders having the greater number of votes, but irrespective of the numbers of shares, all have equal rights as described in the OECD principles.

Not all companies are publically traded and they come in many forms and structures, but in essence they will all be owned either individually, in partnerships, as limited companies with at least two directors, and as collectives and cooperatives. The duties and responsibilities for health and safety begin and end with the owners of a company and this is incorporated into the Health and Safety at Work Act 1974 (HSW Act 1974) and the Health and Safety at Work (Northern Ireland) Order 1978 (HSW Order 1978). It is the employer or the self-employed owner of a company that owes the duty to ensure that no one is harmed by the activities of their company, and under the recent corporate manslaughter legislation a company may be successfully prosecuted for a fatality if it is shown that the way in which they organise and manage their business contributed significantly to the fatality.

In most projects there are a number of companies involved and the law designates specific duties to different players, from the client, through the designers to the contractors and Construction Design and Management (CDM) coordinators (these duties are explained in Chapter 3 *Responsibilities of the key duty holders in construction design and management*).

Stakeholders are all those other persons who have an interest in the business and what it does: employees, whose investment is time and expectation is fair remuneration, the client whose expectation is a quality end product, the community in which the finished project exists and who may be served by the function of the project, the wider community whose interest relates to the social, environmental and economic impacts of the project (especially in respect of civil projects), and the state whose interests lie in sustainability, economic development and improvements in living standards. As with shareholders, the interests of stakeholders vary according to their investments and their justifiable expectations of the outcome. And their duties and responsibilities will be greater or lesser in proportion to their relationship to the project and the companies involved, with employees having legal duties associated with cooperation with the companies' safety policies and professional obligations stemming from codes of conduct. The government will have duties in respect of legislating for safety and health of the citizens of the state and appointing appropriate enforcement bodies and judicial bodies to ensure compliance.

On this note we now return to the issue and principles of safety management, and the roles and responsibilities of managers.

Management

The task of management is to translate the policies of the company into workable activities that will ensure the successful achievement of the company's strategic objectives. For senior management that role will encompass the broadest perspective of the company and its position within the market, local, national or global; gathering and transmitting information to the Board in a timely manner to enable it to monitor the correctness and the effectiveness of its policies and to make such strategic adjustments as are necessary for the success of the business and the continuity of the company.

The relationship between senior management and Board is critical to the effectiveness of the company. It is not the task of the Board to manage the company (with the exception of small companies where the owners of necessity are undertaking key management functions) but rather to appoint competent professionals, respect their professionalism, accept the information proffered and give due consideration to guidance they offer. Likewise it is not the task of senior management to develop strategy or overarching policies that determine the nature and direction of the company, but to objectively and professionally offer guidance to the Board as well as direction, based on the policies of the Board, to their management teams (OECD, 2004).

Communication and information

The nature and effectiveness of the interface between the Board and the senior management will impact on what the company ethos will be like and, in this context, whether it will be viewed as having a positive or negative safety culture. Part of this is due to the approach it takes on communication, whether it is an open two-way channel, how much or how little is communicated, what is being communicated and what is restricted, and, crucially, how the information obtained is to be used.

What the Board asks for and the information that it provides will transfer into the whole company. Having corporate policies that integrate safety and health requires that it receive reports from the different sections on how the safety and health issues are being met; financial reporting should detail the input to and benefits from good safety practice; procurement should detail the impact of safety strategies on the quality of products and services that company obtains and the effects this has on the company's output.

When the Board directs management to make changes in the interests of the company and which impact upon the way people work (or indeed whether they work or not), information to support their rationale is vital to ensure that there is an understanding of the decision. It is not necessary that this results in an acceptance, indeed the provision of as much relevant information across all divisions of the company and up and down the hierarchy will elicit greater participation in effectiveness, efficiency, creative solutions, and improved and expanded productivity.

The key to effective communication, openness and transparency in the disclosure of information is not simply a requirement of good governance, but it stems from a respect for the competence of the workforce and a trust that, whatever position of job they are doing, workers and managers are capable of receiving information, assessing it and applying the knowledge gained to what they do. I have often used the analogy of a company that does not give purchasing responsibility on a quarter million pounds budget to a cleaner or labourer because as cleaners 'they are not competent', yet in their private lives these two people are capable of making decisions to spend that amount on a purchasing a home, to deal with solicitors, mortgage companies, estate agents, and to go on to take responsibility for the welfare of children, decide on educational needs and so on. Ricardo Semler grasped this bull firmly by the horns and transformed his company, Semco SA, into one of the most successful companies in Brazil in the 1980s. The key to the success was participatory industrial democracy.

Planning and resources

Having made the decision to incorporate health and safety into all aspects of what the company does, and communicated this down the line, the next task is for management, particularly second tier management, to plan how the objective is to be translated into practice.

Although distinct and with differing functions, the various departments are interconnected and the decisions of one will impact upon the ability of another to conduct its activities successfully or within the goals of the company. Procurement may be driven by a need to ensure that those in the supply chain comply with the requirements for health and safety yet are aware of the cost implications, while the site managers comply with the requirements for quality products, and finances to maintain the profitability of the company. There is ample scope in this scenario for conflicts between the various managers as they compete for their own position and needs. Traditionally managers exercised authority within their own domain without sideways accountability. Such an approach is not conducive to good health and safety, and is a negation of the integrated approach to health and safety here described.

The contradiction is resolved when the management team approaches the task of achieving the company's goals in a safe and sustainable manner as an integrated team rather than as a collection of individuals. All the members contribute to the planning, exercising their basic functions as managers of a particular aspect of the company and interacting with their colleagues to the extent necessary to coordinate the activities and communicate essential information, and all done within each manager's sphere of influence (McAleenan and McAleenan, 2005).

The achievement of a safe and healthful outcome to any project commences at the concept and design stage. We have this in the CDM Regulations 2007, so I won't labour the point here, but it is worth stressing that the planning of the project and of safety is not a one-off action but something that is continuous throughout the design, tendering, construction and commissioning phases of the project. How is harm to be prevented, how is the health and welfare of the workforce and the community to be maintained?

Scheduling, procurement, recruitment, engagement, notifications, contracts, funding, commencement, none occur without efficient management systems; neither will safety. In a coordinated manner the hazards are identified, the controls agreed upon, the timing of the different phases of the construction (at macro and micro level) is established, as is the arrival of supplies, and the engagement of contractors and sub-contractors. Everything is explored by the management team with safe and effective solutions agreed upon and implemented. Thus effective safety management is integrated fully with the management of all aspects of the project.

When it comes to resources, the ability of all managers and employees to exercise their competences and carry out their work activities in a safe and healthful manner is dependent upon having sufficient and necessary resources to do so. The withholding of resources, for whatever reason, compromises safety. Thus, in budget planning, the financial resources for safety must be determined at the outset, and this means obtaining information on what will be required; material resources such as plant and equipment, tools and personal protective equipment (PPE), notices and signs (multilingual as required), and sufficient and adequate welfare facilities are to be included in the budget.

Scheduling must include the time necessary to improve workers' levels of competence, particularly where new methods of engineering, construction techniques or particular environmental conditions pertain. Where workers are engaged whose first language is not English or suppliers are arriving from abroad, the schedule must allocate time to effective inductions, communications, re-training, etc. Qualified translators must be engaged, especially on large scale projects, to translate essential literature (e.g. safe systems of work, control sheets, etc.).

Human resources go to the engagement of competent workers at all levels, and to the provision of essential training to bring workers up to the desired levels of competence. The nature and size of the project will determine the size of the workforce, which must be sufficient to meet the time constraints established by the client without unduly overextending their abilities to carry out the work in a safe and effective manner.

Remember that a competent workforce must also be recompensed at a level commensurate with the company's expectations that they produce a quality outcome in a safe manner.

Emergency planning and business continuity

When the OECD was established it had as a key objective the attainment of the highest sustainable economic growth and employment and the raising of the standard of living in member states. This objective is satisfied if there is sustainable growth and improvement in the standard of living across a state, irrespective of the success or failure of individual companies, (though the current worldwide financial crisis [2008–2009] stems from a few large companies in key influential financial positions whose failure impacted on all sectors of the economy rendering people homeless and unemployed, and business bankrupt and shareholders losing vast sums). The subprime mortgage crisis in the USA impacted directly on the housing market and the economy, which in turn impacted on the economies abroad leading to severe financial difficulties and the collapse of many businesses including the construction sector.

However, things are not so callous and in relation to corporate governance the principles require that companies take cognisance of the rights of stakeholders and to act in their interests as much as in the interests of shareholders.

This means that the company must be aware of the potential for catastrophic failures over which it has control, that will result in the collapse of the company should they occur. This covers the wide spectrum of business decisions from risky ventures to the application of poor design to major projects, to unpreparedness for safety and quality failures of whatever magnitude they may occur.

The degree of control is critical. In May 2004 a portion of Terminal 2E at Charles deGaule airport collapsed. The inquiry found that the design had little margin for safety, the concrete roof was not resilient and that it had been weakened in places by the pierced openings for the pillars. The inquiry also found that the whole building chain had worked as close to the limits as was possible in order to keep costs down. The architect, Paul Andreu, laid the blame at the feet of the builder for not preparing the concrete correctly.

This analysis suggests that the collapse was preventable and I have no hesitation in stating that by definition all accidents are preventable. But being preventable doesn't mean that they won't continue to happen, and because of that the planning team must give consideration to the range of possible accidents and risks that could occur and have a similar range of plans for responding as quickly and as effectively as possible.

Prioritise the planning on the basis of the potential extent of the emergency envisaged. A fire in a chemical plant, for example, will have graver consequences than someone falling off the scaffolding, and even though the controls to prevent the former are likely to be greater than the latter, logic and recent history demonstrate all too clearly that a clear, unambiguous and speedily implemented emergency plan will effectively reduce the human and environmental damage.

Remaining with the fire example, an outbreak occurring at the beginning of a construction project is probably less consequential than a fire occurring at the end stage of the project. In this respect, emergency plans should not be fixed but rather be developed in relation to what is pertaining at the time and reviewed and updated as the project progresses.

Planning should consider what is to be done in response to the emergency and to delimit it, what personnel are required and what their respective roles in the emergency are, what authorisations do they have and how is this authority communicated to others, and indeed how are the plans to be communicated to all those who are likely to be affected (workers as well as public if the situation calls for it).

What resources will be required, will these be acquired by the company (just in case) or will external resources be used, and who will provide them?

Are there any potential emergencies that will require the active participation with external bodies, not solely the fire and rescue services but, for example, the local authorities, utilities or any civil organisations? How is this to be coordinated, will practice drills be required?

Some of the same questions can be applied to lower level internal emergencies, such as fire drills, emergency evacuations, confined space rescue, etc.

Business continuity looks towards ensuring that, whatever emergency arises, the business is able to continue, vital documentation, designs, contracts, etc. are backed up and stored separately from the originals. If premises are lost, can alternative premises and equipment be obtained quickly and work brought back up to speed in the shortest time practicable? If plant and equipment are damaged, how quickly can they be repaired or replaced?

Consider too that emergencies may be brought about by prohibition notices that themselves are issued to prevent other human emergencies. These notices result from an oversight (in quality companies) in something critical to the safety and welfare of their workers and a strategy for responding to prohibition must be in place to ensure that the requirements of the notice are carried out and arrangements in place to recommence work on the project as soon as possible. The enforcement authorities are not in the business of halting business but in preventing harm. The manner in which a company responds to any requirement from the enforcement body will determine whether and how quickly work resumes.

In the end good emergency planning contributes greatly to the process of identifying and controlling hazards. Planning for emergencies, although no guarantee that it will prevent them, raises issues and demand solutions that will reduce the likelihood of the emergency occurring.

Monitoring and reporting

Effective governance and management need a feedback loop that will allow the principals to ascertain the appropriateness of their decisions and the effectiveness of all activities to achieve the objectives of the company.

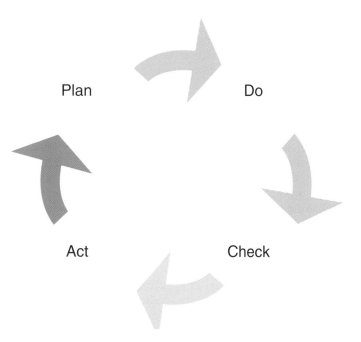

Figure 1   The Deming Cycle

The Deming Cycle (Plan – Do – Check – Act) (PDCA) is commonly used and familiar to many in quality control circles (see **Figure 1**). It has the advantage in being an improvement cycle in that it specifies what is necessary to determine whether the stated outcome has been achieved, and allows for the modification of activities to keep the process on track.

It is an outcome focused method for monitoring, where the end product or result is defined at the outset. Contained within the definition are the requirements for time, budget, quality, safety and health, customer satisfaction and environmental impact. For efficacy, such statements, when worded positively, favour pro-active methods of working. A negative target, such as 'reduce accidents by x%' requires a re-active approach; the accidents must happen before a plan of action to deal with them

can be put in place. But requiring that a project be carried out in a manner that is safe and healthy requires a knowledge and understanding of what would harm the operatives and implementing activities that add to the hazard controls. For example, set a target for the adoption and use of alternative methods to working at heights. This will generate activities from the design rooms to the site itself with all parties involved in the activities. Monitoring will be looking at the efficacy of the new methods in respect of quality of construction, impact of the schedules, the budget and on the safety and welfare of the site operatives. It is a holistic approach wherein the safety of workers is monitored within the overall context of monitoring the project, and adjustments are made to improve the whole process.

Box 2   Monitoring case study

Laing O'Rourke used pre-constructed concrete walls for the 2008 Project Omega Wastewater Treatment plant in NI and raised them to the vertical position, thus reducing the need for a considerable amount of working at heights.

Good PDCA can be viewed as a spiral of improvement, the aim of each cycle being to advance quality, safety and sustainability, rather than maintaining the status quo. A variation of the PDCA utilises a dialogical, sometimes known as praxic approach. This is a feedback loop that involves a degree of self-reflection and two-way influencing on what and how a project develops.

Traditionally the rationale of the project determines what the outcome will be and how it is to be achieved. It is a mono-directional journey from point A to point B where the end point, and to a degree the methods, once decided upon, remain the target, although the journey may be modified to achieve an improvement. Conversely, a praxic approach (see **Figure 2**) will reflect upon the end point and the methods to achieve it, considering not just the efficacy of each but also the appropriateness of both. It allows for adaption to changing circumstances, internal to the project as well as external factors. It contains within it a greater degree of industrial democracy in that all the stakeholders, from

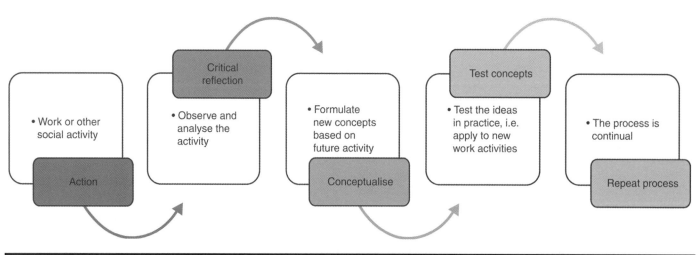

Figure 2   A praxic approach

client, designer and contractor to the managers and workers, whose competence is recognised and respected, contribute to the continuous review of the project and methods adopted.

Its advantage for monitoring safety management, indeed for any management system, is that it permits the monitor to ask not just the question of whether things are being done the right way but also whether the right things are being done. (It is possible for the wrong thing to be done the right way and, if problems arise, there will be a less than complete understanding of why and thus a less than effective remedy developed.)

The reporting mechanism in an effective management system must take into account the purpose of the report and who is to receive it. Generally reports move up the line from the monitor/auditor through a specific manager to senior management and thus to the Board, and in quantity from full report, to summary report and possibly to a one or two paragraph summary inside a broader company report. It is not unreasonable that this be the case. Those charged with responsibility for understanding the practicalities of the project as it progress and making decisions about the allocation of resources or modifying practice, schedules etc. need the detailed reports. At senior management level, the CEO and directors are less concerned with the minutiae of daily activities, but nonetheless require essential information on how the company is meeting the corporate strategy and policies in order to inform and offer guidance to the Board.

Less often are reports filtered down the line and sometimes the 'need to know' argument is used to rationalise this practice; workers only need to know what has been decided and what is required of them – anything else doesn't need to be explained.

This paternalistic approach to the workforce, that the company knows best and if you do what is asked of you, you will come to no harm, is not acceptable, and indeed is negated historically by accident and fatality statistics. When you consider that currently 2.3 million workers die annually from workplace accidents and illness, there are 270 million >three days lost time injuries (LTIs) and 160 million new cases of industrial disease annually, companies are failing badly in their health and safety. The construction industry is one of the major contributors to this statistic. (ILO, 2009). Current thought from the OECD principles that require openness and transparency of companies, to the various international conventions and declarations that advocate worker/employer partnerships and effective two-way communications within companies, would suggest that such monitoring and review reports are freely available and open to discussion by all those who are affected by the recommendations contained within them. And, after due consideration, is this not one of the key elements of respect for the professionalism and competence of those employed by the company?

Occupational health and safety management systems standards, accreditation and certification

We have looked at the principles of an effective management system that integrates health and safety into all aspects of what a company does. Where, then, do such specifications and guides as the British Standard OHSAS 18001 *Occupational Health and Safety* (BS OHSAS 18001), ILO Guidelines on Occupational Safety and Health Management Systems (ILO OSH 2001), American National Standard for Occupational Health and Safety Management Systems (ANSIZ10–2005), and Successful Health and Safety Management HSG65 (HSE, 1997) fit in?

None of those listed require mandatory compliance. All recognise that it is the duty of employers to effectively manage their undertakings in such a way that they do not cause harm to their employees or to those affected by what they do. How the employer manages the business is less important than the results of that management as it pertains to health and safety; however there are levels of managerial competence which at one end of the scale will not result in good health and safety and at the other there is a high degree of assurance that the health and safety outcomes are positive.

Rather than considering safety management on a linear scale that goes from 'bad' to an ideal system I would suggest viewing it as a three-dimensional matrix that encompasses many and varied effective systems, some very similar and others strikingly different meeting unique and particular circumstances. Such is the case is recognised by all of the above, although Adele Abrams suggests that the ANSI standard may at some time in the future be incorporated into US law by the Occupation Safety and Health Administration (OSHA) (Abrams, 2006).

What these documents provide is recognition that there are a number of principles underpinning effective safety management whatever systems an employer follows and they provide guidance and support for employers to develop or improve upon their systems.

We are all familiar with the HSG65 *Successful Health and Safety Management* (HSE, 1997) publication. This is a guidance document aimed at assisting employers in improving health and safety. It is not mandatory but, in keeping with other similar documents produced by HSE, employers who follow the guidance contained are going some way to meeting their statutory obligations.

BS OHSAS *18001 Occupational Health and Safety*, an assessment specification used by many on the route to certification, states that it is a framework to assist employers to identify and control safety risks in, and reduce the potential for, accidents and comply with legislation. It is not mandatory and its popularity comes from the fact that it has been developed by the British Standards Institution, has a degree of international recognition and is supported by an audit and certification scheme.

The ILO-OSH 2001 *Guidelines on Occupational Safety and Health Management Systems* also contain a voluntary programme designed according to internationally recognised principles for safety management, which aims to assist employers develop a sustainable safety culture.

Thus employers are free to develop their own safety management systems with the assistance of all (or none) of the guidelines above. What is essential is that whatever system the Company has in place meets the fundamental requirement to safeguard their workers and others affected by their undertakings. But in saying that, there is a growing requirement, from Clients in the construction industry, for employers to demonstrate that they have an effective safety (and quality and environmental) management system in place. This brings about the matter of accreditation and certification.

Accreditation relates to an organisation's competence to manage the process of certification and applies to organisations that are recognised by an accreditation body as having the capability to assess and certify third parties against specified standards. In the UK, the Government body that accredits certifying organisations is the United Kingdom Accreditation Service (UKAS); this is the sole national accreditation body recognised by government to assess, against internationally agreed standards, organisations that provide certification, testing, inspection and calibration services. An accredited company is one that has successfully demonstrated to the accreditation body that it is capable of exercising good certification practices as defined in the requisite standards. It is not essential for a company that is good at assessment to have accreditation, though it does ease the process of recognition.

Certification is the process of having your system assessed against a standard and being awarded a certificate on the basis of the results of the assessment. It is not necessary for the certifying body to be accredited by UKAS to do so unless they are certifying against specific standards or specifications such as BS OHSAS 18001 that require any certifying company to be accredited. Nor is it a requirement for the certifying body to be a third party body; second party (i.e. by the client) or first party (i.e. internal) certification is permissible and the only provision that I would make is that these parties follow good assessment/audit practices, for which ample standards and guidelines are available.

Summary of main points

In summary, the social, moral and legal requirement for any company is to appreciate the need to be aware of the impact their activities will have on workers, communities and the environment and to conduct their activities in such a manner that they cause no harm. Managing what they do with a wholehearted commitment to continuous improvement will naturally draw them toward good and more effective management practices.

Whether or not a company chooses to submit its management system to the scrutiny of a certification body is immaterial and is based on the benefits it brings to the company. What matters most is the efficacy of the system that has been adopted to achieve its safety, health, quality and environmental targets.

References

Abrams, A. *Legal Perspectives – ANSI Z10–2005 Standard Occupational Health and Safety Management Systems,* ASSE 2006. Available online at: http://www.asse.org/practicespecialties/international/docs/92ArticleaboutZ10LegalPerspectives.pdf

Egan, Sir John. *Rethinking Construction, The Report of the Construction Task Force,* Department of Trade and Industry, London, July 1998.

European Agency for Health and Safety at Work. *Bilbao Declaration,* 2004. Available online at: http://osha.europa.eu/en/publications/other/20041122/view

Health and Safety Executive (HSE). *Successful Health and Safety Management (HSG65),* 1997, London: HSE Books. Available online at: http://books.hse.gov.uk/hse/public/saleproduct.jsf?catalogueCode=9780717612765

International Labour Organisation (ILO). *Convention Concerning the Promotional Framework for Occupational Safety and Health,* 2006, Geneva: ILO.

International Labour Organisation (ILO). *World Day for Safety and Health at Work 2009.* Facts on *Safety and Health at Work.* Available online at: http://www.ilo.org/wcmsp5/groups/public/---dgreports/---dcomm/documents/publication/wcms_105146.pdf

International Social Security Association (ISSA). *Quebec Protocol for the Integration of Occupational Health and Safety Competence into Vocational and Technical Education,* 2003, Quebec, Canada: ISSA, International Section for Education and Training for Prevention.

McAleenan, C. and McAleenan, P. *Prevention – a Universal Responsibility,* 2005. World Congress on Occupational Safety and Health, Florida.

Organisation for Economic Cooperation and Development. *Methodology for Assessing the Implementation of the OECD Principles on Corporate Governance,* 2006, Paris: OECD.

Organisation for Economic Cooperation and Development. *Principles of Corporate Governance,* 2004, Paris: OECD.

Referenced legislation and standards

ANSI Z10–2005 American National Standard for Occupational Helath and Safety Management Systems, Washington, DC: American National Standards Institute.

BS OHSAS 18001 2007 Occupational Health and Safety Management Systems. Requirements, London: British Standards Institution.

Construction (Design and Management) Regulations 2007, London: The Stationery Office.

Health and Safety at Work, etc. Act 1974, London: Her Majesty's Stationery Office.

ILO-OSH 2001 ILO Guidelines on Occupational Safety and Health Management Systems, Geneva: International Labour Office.

The Health and Safety at Work (Northern Ireland) Order 1978, London: Her Majesty's Stationery Office.

Further reading

Financial Reporting Council. *The Combined Code on Corporate Governance,* 2008. Available online at: http://www.frcpublications.com

Freire, P. *Pedagogy of the Oppressed,* 1970, Harmondsworth: Penguin Books.

Institution of Civil Engineers (ICE). *Health and Safety Register.* Available online at: http://www.ice.org.uk

Joyce, R. *CDM Regulations 2007 Explained,* 2007, London: Thomas Telford.

Lakha, R. and Moore, T. (Eds) *Tolley's Handbook of Disaster and Emergency Management: Principles and Practice,* 2002, London: LexisNexis Butterworths Tolley.

McAleenan, C. and McAleenan, P. *Development of the Competent Company in the Context of the Seoul Declaration*, 2009, Calgary: CSSE PDC.

Semler, R. *Maverick! The Success Story Behind the World's Most Unusual Workplace*, Random House Business Books, Reissue edn 2001.

Websites

American National Standards Institute (ANSI) http://www.ansi.org

British Standards Institution (BSI) http://www.bsigroup.com

European Agency for Health and Safety at Work (EU-OSHA) http://osha.europa.eu

Health and Safety Executive (HSE) http://www.hse.gov.uk

Health and Safety Executive Northern Ireland (HSENI) http://*www.hseni.gov.uk*

Institution of Civil Engineers (ICE), Health and safety http://www.ice.org.uk/knowledge/specialist_health.asp

International Social Security Association (ISSA), International Prevention Sections http://www.issa.int/aiss/About-ISSA/International-Prevention-Sections

International Labour Organisation (ILO) http://www.ilo.org

Occupation Safety and Health Administration (OSHA) http://www.osha.gov

Organisation for Economic Cooperation and Development (OECD) http://www.oecd.org

Chapter 8

Assessing health issues in construction

Alistair G. F. Gibb Loughborough University, UK

doi: 10:10.1680/mohs.40564.0091

Occupational health is a major problem for construction. The most effective action is to assess, then eliminate or significantly reduce any health risk early in the project, at design or pre-construction planning stage. Residual risks must then be communicated to the contractor who should further assess, reduce, control and manage the risk to reduce its effect on the construction workers, having informed and trained them in the hazards and appropriate control measures. Health surveillance may be required. The main construction occupational health problems include: asbestos related conditions; musculoskeletal disorders; hand–arm vibration syndrome; problems caused by materials hazardous to health, including dermatitis and asthma; deafness and stress. Many occupational health problems become more acute for older workers.

CONTENTS

Box 1 Key learning points
■ Early action is best.
■ Effective assessment of health hazards and risks is essential, but useless unless effective action is taken.
■ Designers and pre-construction planners can eliminate and reduce health risks.
■ Construction managers and supervisors can reduce and control risks through effective management.
■ Health problems are cumulative and older workers are likely to suffer more than younger workers.
■ 'Zero harm' and not just 'zero accidents' should be the target and vision for the construction industry.

Significance of occupational health

'Health is a state of complete physical, mental, and social well-being and not merely the absence of disease or infirmity.' (World Health Organisation, 1992)

Occupational ill health is a major problem for construction

Occupational health (OH) is the poor relation of occupational safety. It is often helpful to consider OH as a slow accident. Ill health continues to kill and disable significant numbers of construction workers, and the delay in the outworking of the effects is one of the main reasons why the subject should be taken seriously. Construction managers understand their role in safety and the prevention of injury, but are less clear regarding the prevention of ill health (Gyi et al., 1998). In her inquiry into the underlying causes of construction fatal accidents, Rita Donaghy stressed that 'for too long, health has had minimal attention when compared with safety. Thus, while significant progress has been made by the industry on safety issues during the last decade, it has failed to achieve the same for ill health' (Donaghy, 2009).

The costs of sickness absence to government and industry are substantial. In the United Kingdom more than 370 million working days are lost each year due to certified incapacity, with a cost to British business estimated at £13 billion (Statistical Office, 1985 and Confederation of British Industry, 1993, both cited in Marmot et al., 1995). As for accidents, the real cost to industry due to ill health is considerably greater and includes: lost production; compensation, insurance, re-training and recruitment. In addition, society incurs the costs for social security, health care and early retirement.

Each year more than 2000 people die of mesothelioma, and thousands more from other occupational cancers and lung diseases (COHME, 2009). There are estimated to be around 4000 deaths per year from exposure to asbestos; many of these deaths are associated with workers from the construction industry. More than 500 people died from silica related lung cancer in 2004 (COHME, 2009). In 2007–2008, 1.7 million days were lost due to work related ill health, which equates to 0.77 days per worker. This is almost twice the number of days lost due to accident related workplace injury (HSE, 2009). **Figure 1** compares construction's health record with that of all industries.

Table 1 provides data of construction's main ill health conditions compared with the all industry average. This provides data from The Health and Occupation Reporting network (THOR) and the compensatable prescribed diseases under the Industrial Injuries Disablement Benefit (IIDB). It can be seen that there are some differences between these data sets and some data are missing. Nevertheless, the overall picture remains the same:

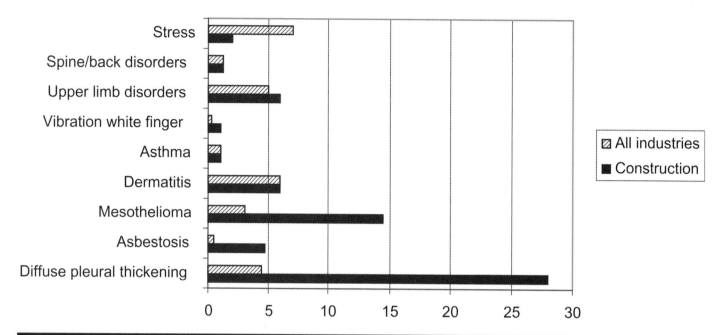

Figure 1 Incidence rates per 100 000 workers/employees (reproduced from HSE, 2009. © Crown Copyright reproduced with permission of Her Majesty's Stationery Office)
From annual average incidence rates of occupational diseases seen by disease specialist doctors in the THOR surveillance schemes; 2005–2007.

construction does not perform well against the all industry average in many of these conditions.

But health management is a complex issue. Health management is associated with large set-up costs and the benefits are not immediate and often difficult to demonstrate. It is hard to categorically apportion responsibility or blame for the cause of ill health. Typically, the macho culture leads to workers being even less interested in OH than in the more immediate and obvious risks of accidents. This 'reality-gap' must

be acknowledged and eradicated. Notwithstanding, effective, strategic management action early in the process will reduce accidents and incidents leading to injury and ill health (Gibb et al., 1999).

This chapter introduces the topic of occupational health management in construction. Many ill health conditions, or the hazards and risks associated with them, are covered by specific legislation. This chapter does not attempt to cover the detailed legislative aspects – readers should ensure that they are aware

Condition	THOR hospital specialist cases				IIDB prescribed disease cases	
	Construction		All industries		Construction	All industries
	Average annual cases	Average annual rate per 100 000	Average annual cases	Average annual rate per 100 000	Average annual rate per 100 000	
Diffuse pleural thickening	624	28.5	1263	4.6	12.8	5.8
Mesothelioma	317	14.5	761	2.7	44.7	2.8
Asbestosis	54	2.4	160	0.6	21.7	1.4
All MSDs	168	8	1874	7	0.8	n/a
Upper limb disorders	128	6	1358	5	n/a	n/a
Spine/back disorders	26	1	379	1	n/a	0.5
Vibration white finger	22	1	110	0.4	6.5	0.8
Dermatitis	124	6	1624	6	1.2	0.7
Stress	43	2	1873	7	n/a	1.5
Asthma	14	1	358	1	0.9	0
Occupational deafness	n/a	n/a	n/a	n/a	2.4	0.9

Table 1 Construction and 'all industries' ill health data, 2005–2007
Reproduced from HSE, 2009 © Crown Copyright reproduced with permission of Her Majesty's Stationery Office

ICE manual of health and safety in construction © 2010 Institution of Civil Engineers

of these requirements. More information on good practice is provided at the end of this chapter.

Managing occupational health

What can designers and pre-construction planners do?

As for accident related safety, the most effective action for any health risk is taken early in the project, at design or pre-construction planning stage. It is only at this early stage that risk can be eliminated or very significantly reduced. Effective assessment of health hazards and risks is essential, but useless unless effective action is taken. Non-conventional residual risks must then be appropriately communicated to the construction team. Specific duties on designers and pre-construction planners exist in common law and legislation, such as the Construction (Design and Management) Regulations (CDM 2007). Advice related to each condition is provided in the following sections but these fundamental actions are taken as a 'given'.

ASSESS → ELIMINATE/REDUCE → INFORM constructors

It is essential that designers don't just 'do' a design risk assessment, but rather they assess the risks and then take action to eliminate and reduce them wherever possible. Residual risks that would not be obvious to a competent contractor should be identified and communicated to the contractor in the form of pre-construction information.

What can managers and supervisors do?

Providing early action to eliminate and reduce risk by designers and pre-construction planners, as well as residual risks, have been identified and appropriately communicated, a competent contractor, employing competent managers and supervisors, should be able to further reduce and control the residual risks. Managers and supervisors must first understand the risk and its effect on their workers. They must also understand the competence and pre-existent health conditions of their workers. They must then be aware of, and apply appropriate reduction and control measures, informing and training the workforce as appropriate, with all of these actions being incorporated into the construction phase plan.

ASSESS → MANAGE INFORM TRAIN → REFER to medical experts

This approach is described in more detail in **Figure 2**. Since OH problems tend to have a lengthy and complex history, the use of pre-employment or pre-deployment health assessments for workers is of significant benefit. However, currently, this is rarely done in construction. Furthermore, ongoing health surveillance will increase the likelihood that conditions will be identified and action taken to reduce the consequences at the earliest opportunity.

There are many companies in the UK that offer OH screening. Access to an OH service offers the opportunity to explore workers' concerns about the effect of the workplace on their health. By advising the company on how to control hazards at work, the OH service should be able to make the workplace a safer place.

As many of these issues are less familiar to many construction site staff, OH professionals should be involved in assessing risks, training and awareness. Specific duties for many of these situations exist in common law and legislation (see Section 1, Chapter 1 'Legal principles' for further information on health and safety legislation). Increasing the understanding of causes and early symptoms of ill health conditions and encouraging open reporting of hazards and risks are important parts of addressing the challenges.

These actions are taken as a 'given' in the following sections, although specific advice for each ill health condition is provided.

Many construction staff are less familiar with the need to reduce OH risks than they are for occupational safety. The move from several major contractors to talk of 'zero harm' initiatives rather than just 'zero accidents' is to be applauded. However, continual vigilance is required to ensure that OH is treated seriously.

Construction occupational health problems

This section covers the most common 'construction' OH problems, to assist in the development of company or site policy and procedures. However, medical advice must always be sought in suspected cases. In each case, advice for designers, pre-construction planners, managers and supervisors is given. More advice can be found in specific health related publications, cited in 'References' and 'Further reading' sections at the end of the chapter.

Asbestos related conditions

Asbestos related diseases are covered under the generic term pneumoconiosis (literally 'dusty lungs'), referring to conditions that can cause permanent damage to the lungs. The diseases and ill health conditions include diffuse pleural thickening, mesothelioma and asbestosis, and are, by a long way, the largest cause of death of construction workers. One asbestos spore can result in the development of the diseases. One major challenge with these conditions is the significant time delay in the onset (and hence diagnosis) of the disease. Since its ubiquitous use in the middle of the last century, the use of asbestos has been banned in the UK, but workers are still beginning to be affected by these conditions caused by previous exposure. Current exposure is limited in new-build projects, but, as the majority of construction projects involve repair or refurbishment, further exposure is still a considerable risk.

Action	By whom
Recognise the health problem	Worker, health and safety advisor, line managers, nurse, doctor
Diagnose the problem	Nurse, doctor
Treatment	Nurse, doctor
Discover the cause	Worker, health and safety advisor, hygienist, nurse, doctor
Monitor and control the cause	Health and safety advisor, line managers, hygienist, ergonomist
Monitor the workers health	Nurse, doctor, epidemiologist, toxicologist

Figure 2 Action and responsibilities for control of health hazards
Reproduced, with permission, from Gibb et al., 1999.

Silicosis is a similar condition caused by silica exposure, often through silica dust. Whilst this is not currently as prevalent as asbestos related conditions it may become much more significant in the future. High risk groups are those involved with drilling, cutting, grinding or scabbling concrete or other silica-containing materials (mortars, rock, stone, etc.). Operations such as demolition or refurbishment of brick, stone or concrete products are high risk. Specialist advice should be sought.

The COSHH Regulations state the maximum exposure limit (MEL) for respirable crystalline silica as 0.3 mg/m^3. However, it is accepted that there is a much higher risk of lung damage than had previously been thought. Therefore exposures should be controlled to 0.1 mg/m^3 (eight hour time-weighted average (TWA)) or below.

What can designers and pre-construction planners do?

Asbestos-containing products should not be specified or allowed on any projects irrespective of the specific national legislation (some countries still do not have a complete ban).

All projects involving work in or around existing buildings or facilities should require an asbestos survey during the design and planning phase.

Designers have a particular duty when working on refurbishment, decommissioning or demolition projects where the presence of asbestos is very likely. A full asbestos survey is required

and a detailed consideration of the likely construction methods necessary to identify areas of the project or particular operations and tasks where workers will be at risk. There are specific legal duties on clients and building/facility users, to identify and notify of the existence of asbestos in their properties. However, designers' responsibility is heightened by the fact that many owners, despite the regulations, do not know the extent of asbestos within their buildings or facilities. Other owners take the 'easy option' by making a blanket declaration that asbestos is present but without any detailed information. Designers and pre-construction planners must take early action to address these challenges to prevent workers being put at risk or the project programme being compromised by the discovery of unexpected asbestos.

Any operations that disturb asbestos-containing material (including asbestos cement), such as breaking, drilling or cutting, release fibres into the air. The following are situations where asbestos is most likely to be found in buildings:

■ Sprayed or loose packing as fire breaks in ceiling voids.

■ Moulded or pre-formed sprayed coatings and lagging – generally for thermal insulation of pipes and boilers.

■ Sprayed asbestos mixed with hydrated asbestos cement – generally as fire protection in ducts, firebreaks, panels, partitions, soffit boards, ceiling panels and around structural work.

■ Insulating boards used for fire protection, thermal insulation, partitioning and ducts.

- Some ceiling tiles.

- Vinyl flooring backed with asbestos paper.

- Paper and paper products used for insulation of electrical equipment; asbestos paper has been used as fireproof facing on wood fibreboard.

- Asbestos cement products – usually flat or corrugated sheets, often used as roofing and wall cladding. Other products include gutters, rainwater pipes and water tanks.

- Certain textured coatings.

- Bituminised products (e.g. flashing, floor tiles). (Adapted from Gibb et al., 1999, with permission.)

What can managers and supervisors do?

- Be aware of the likelihood of hidden asbestos in older buildings.

- Licensed specialist contractors must be used for both removal and disposal of asbestos related products.

- Carry out risk assessment and identify, assess and manage the risk of asbestos exposure to workers. It may be necessary to use specialist expertise.

- Take steps to eliminate or at least minimise the risks.

- Ensure efficient and rigorous control methods for removing asbestos dust in the workplace.

- Inform supervisors and workers of the presence of asbestos, its location and the type of health risks if it is disturbed.

- If supervisors and workers must work with asbestos, ensure that they are aware, fully trained in working with asbestos and are implementing control measures, e.g. personal protective equipment (PPE) (especially masks/respirators), minimising asbestos dust, keeping the material wet, and disposal is in sealed labelled containers whilst still damp.

- Set up a workable system for the open reporting of any material suspected of containing asbestos.

- Refer at-risk workers to occupational health professionals for health surveillance, who can also be proactive in educating and training supervisors and workers.

- Ensure adequate washing facilities and cleaning procedures for protective clothing (unless disposable). Protective clothing should not be taken home to wash. Disposable clothing, respiratory masks and filters should be disposed of as asbestos waste.

- Where there is a high degree of hazard, surveillance should include a more detailed pre-employment assessment, a detailed history and lung function test – statutory requirements apply for workers exposed to asbestos. (Adapted from Gibb et al., 1999, with permission.)

Musculoskeletal disorders

Musculoskeletal disorders, as the name suggests, affect the muscles or the skeleton, particularly joints between bones. These include: upper limb disorders; back pain; hand–arm vibration syndrome (HAVS – also called vibration white finger); repetitive strain injury (RSI); and a number of other ill health conditions. This section concentrates on the first three, as those most prevalent in construction workers.

As with many ill health conditions, it can be difficult to differentiate between work related musculoskeletal disorder (MSD) conditions and those brought about or exacerbated by out of work events, particularly sporting activities. Notwithstanding, this should not be used as an excuse to ignore this important area.

Since 1946, France has required regular medical examinations of all workers in all sectors and this has resulted in an extensive database of occupational ill health conditions. According to French occupational health physicians, the most significant condition for construction workers by some considerable margin is damage to the shoulders, followed by the lower back and then the knees.

Upper limb disorders

Upper limb disorders are also known as RSIs or MSDs in the neck, shoulders and upper limb. These include a number of medical conditions including: carpel tunnel syndrome; tendonitis; tenosynovitis (e.g. trigger finger); bursitis; and tennis elbow.

Workers most at risk include those exposed to repetitive tasks involving the fingers, hands or whole arm, for instance tight gripping, hammering, twisting, pushing or pulling. Problems can be exacerbated by awkward postures and working on the same task for long periods. Of all the occupational health (OH) conditions, this is the group that can also readily apply to all employees rather than just those involved in site operations. Symptoms can be acute, recurrent or chronic and may cause permanent damage. Early symptoms include localised pain, fatigue, discomfort and numbness. Upon inspection, swelling, tenderness and inflammation can be observed. Severe cases can cause loss of function, limited movement and loss of muscle power. Pre-employment assessment can sometimes identify individuals who are at risk, e.g. those with previous injuries, but is otherwise of limited benefit.

What can designers and pre-construction planners do?

- Remove or minimise manual handling; for example, through the increase in the use of offsite solutions, where more mechanisation can be used in the factory and in the installation of the completed units.

  - For example, modern factory facilities typically use more craneage and other lifting devices than most sites and can deliver materials and components direct to the workface without manual handling.
  - Furthermore, the work can be arranged such that it is done at a convenient height to reduce overstretching and awkward postures.

What can managers and supervisors do?

- Assess risks from tasks/activities that could result in back pain, involving workers and outside experts – eliminating or minimising manual handling wherever possible.

- Involve specialists in design and development of work activities to minimise strain.
- Build in flexibility in task planning, encouraging workers to take rest breaks.
- Train all staff:
  - to recognise and report early signs and symptoms of back pain; and
  - in preventative measures relevant to the workplace.

Spine/back disorders

Work related spine and back disorders are typically caused by a series of repeated, cumulative injuries rather than by a single accident or injury and can lead to permanent disability if preventative action is not taken. Most at risk are workers handling loads, particularly heavy or bulky loads; repetitive handling work, static and awkward postures or those at the extremes of the range of movements. The nature of construction operations can lead to back injury in most activities. Symptoms can be acute, recurrent or chronic. Early symptoms include localised pain, discomfort, fatigue and numbness of the muscles, tendons or soft tissues. Swelling, tenderness and inflammation typically follow. In severe cases loss of function, limited movement and loss of muscle power can occur, leading to permanent disability if preventative action is not taken. Pre-employment assessment can sometimes identify individuals who are at risk, e.g. those with previous injuries, but is otherwise of limited benefit.

What can designers, pre-construction planners, managers and supervisors do?

- Specify lighter products.
  - However, designers should be aware that workers may just carry more items, or have to make more operations to complete the same amount of work.
  - For example, using smaller concrete blocks will reduce the individual load per unit installation but require more units to be installed (presumably in the same time period).
  - Another example is that 1.2 x 2.4 sheets of plasterboard can create a considerable manual handling risk, especially when used for ceiling work. Using smaller sheets reduces the individual load, but considering alternative methods of forming ceilings or wall finishes can eliminate the risk completely.
  - Actually, making things heavier so that they can't be lifted manually can also be an effective solution providing effective mechanical handling solutions are available – this highlights the need for designers to discuss such aspects with installation experts.
- Other preventative and remedial action is very similar to that for upper limb disorders.

Hand–arm vibration syndrome – HAVS (e.g. vibration white finger)

HAVS results from prolonged use of high-vibration tools, causing damage to hands and arms. In the early stages, improvement may occur by stopping tasks associated with vibration. In the later stages the condition is more likely to become permanent. Early signs include tingling or pins and needles in the fingers and whiteness at the tips of the fingers when cold. With continued exposure, attacks are longer and occur more frequently causing pain and reduced dexterity and also occur when the hands are warm. In very severe cases, blood supply is permanently impaired causing a blue-black appearance of the hand. Over recent years many manufacturers have had their equipment tested (see HAVTEC on OPERC website) and there has been significant improvement in tool design. Nevertheless, it is important to note that so-called 'normal' use of vibrating tools in construction is very likely to result in HAVS – in other words, custom and practice is unlikely to be acceptable in this case. Pre-employment assessment will establish a baseline to evaluate results of routine assessment and may also help identify at-risk workers. Annual health surveillance is recommended for workers in high risk activities with more frequent checks for newer workers.

What can designers and pre-construction planners do?

- Eliminate or reduce the risk of HAVS through pre-construction action by removing or significantly reducing the need for workers to use hand-held vibrating tools or machinery.

Examples include:

- The breaking down of the tops of in situ concrete bored piles (Gibb et al., 2007). Alternatives include: designing the piles to avoid the need to 'over-cast'; the removal of the excess concrete using techniques such as the Elliot hydraulic splitter or an expanding grout technique. However, despite these techniques being available for a number of years and having been demonstrated to be cost and time effective, there are still many projects where they are not employed.
- The specification of self-compacting concrete which removes the need for vibration during placing and tamping for floor slabs.
- The specification of cast-in fixings for in situ concrete in lieu of post-drilled fixings.
- The avoidance of the need for scabbling concrete by redesign of joints or specification of alternative methods of removing concrete laitance.

What can managers and supervisors do?

- Eliminate or minimise tasks that require the use of hand-held vibrating tools.
- Minimise periods for each worker using vibrating tools and machinery.
- Know the vibration data and implications from all equipment and tools.
- Vibration measurement is a complex task requiring a competent person.
- British Standards describe the normalisation of vibration measurements to an eight hour working period, A(8). Workers regularly

exposed to an A(8) of 2.5 m/s^2 are particularly at risk, but it should be noted that this action level would not eliminate risk.

■ Ensure latest, well designed tools are used and remove old and faulty equipment.

■ Train all staff to recognise and report early signs and symptoms of HAVS and introduce preventative measures.

Problems caused by materials hazardous to health

The use of materials hazardous to health is prevalent in all industry sectors and specific legislation covers their use (e.g. COSHH: Control of Substances Hazardous to Health). Registration, Evaluation and Authorisation of Chemicals (REACH) regulation places greater responsibility on industry to manage the risk of chemicals and provide appropriate safety information to professional users and consumers. There are new standards for symbols for such materials (see HSE globally harmonised systems website) and a globally harmonised system (GHS) is now available where any substance considered to be toxic or environmentally hazardous will be labelled all over the world using the same symbol. However, currently, there are still various systems in use worldwide.

Additional challenges for construction include the temporary and changing nature of project sites which can limit the effectiveness of technical and organisational systems to control such substances. Most construction operations are likely to use some materials that are hazardous to health.

The hierarchy of action is the same for all materials that may be hazardous to health. First the designers and specifiers must identify and then avoid the use of hazardous substances wherever possible. A simple example might be the specification of mechanical fixings instead of solvent adhesives. The second action should be to substitute the hazardous substance with a less hazardous alternative. In some cases the contractor may be left to decide the use of certain substances. In this case the contractor has the designers' responsibility mentioned above.

If either of these actions is not possible and the use of a hazardous substance is necessary, then this should be advised to the contractor as part of the pre-construction information. The contractor must manage the residual risk by:

■ Examining health data sheets, text books, guidance documents, websites or research papers, to identify the hazards and the means by which they can cause health problems (e.g. inhalation, skin absorption and ingestion – often called exposure routes).

■ Evaluate how and where the substances will be used and whether the exposure routes may occur (e.g. confined spaces). Change methods where possible to minimise the escape and spread of hazardous substances.

■ Evaluate the risk against existing control strategies.

■ Consider new or upgraded control measures.

   ■ First, mechanical measures (e.g. dedicated extract systems).

   ■ Second, and only if mechanical measures are impractical or insufficient to adequately control the risk, specify and control the use of PPE (e.g. respirators, eye protection).

■ Implement these control measures and monitor their effectiveness.

■ Consider the effectiveness of the control measures for emergency situations (e.g. spills).

■ Implement health surveillance where appropriate.

■ Disseminate information to users and provide suitable training on the hazards and risks from the substances with which they work and the use of control measures.

■ Check that the control measures themselves do not increase the overall risk to health and safety.

Designers and preconstruction planners should also consider the effect of the use of such substances on members of the public or others, not directly involved in the construction works. For example, Rey and Berriel-Cass (2009) say that 'Hospital construction and renovation projects have been associated with outbreaks of infection in immuno-compromised patients. Successful implementation of dust control measures requires a partnership between infection control, project managers and contractors throughout the project from the planning and design phase to final clean-up.'

The two most significant health conditions affecting construction workers are dermatitis and asthma.

Dermatitis

Dermatitis is a skin condition caused by exposure to certain substances, activities or environments. Cement related dermatitis is a particular problem in construction. Correct diagnosis, leading to timely corrective action, can usually lead to improvement which may allow workers to continue their vocation. Since substances hazardous to health are common in construction, all workers are at risk. However, workers whose skin is exposed to windy, wet and cold weather are more susceptible to the penetration of irritants such as cement, mortar or plaster. Workers with pre-employment conditions such as eczema should be encouraged to take particular care and may even be unsuitable for certain tasks. All workers should be trained to identify the early signs of work related dermatitis and preventative measures.

What can designers and pre-construction planners do?

■ Eliminate substances hazardous to health wherever possible.

■ Specify cement without chromium VI to reduce the dermatitis risk from concrete, grout and mortar.

■ Designers should be aware of and use the numerous web-based guidance documents to eliminate the use of problematic substances wherever possible, replacing them with more benign alternatives (see 'References' and 'Further reading' at the end of this chapter).

What can managers and supervisors do?

■ Understand and control the COSHH risk and also the risk to certain workers from more innocuous substances.

■ Eliminate where possible and at least minimise the risks by avoiding the use of harmful substances or work operations in exposed conditions wherever possible.

■ Use pre-employment and periodic health surveillance to identify workers sensitised to certain common materials (e.g. cement).

■ Inform supervisors and workers about any substances they work with that could cause dermatitis.

■ Train supervisors and workers in control measures and to recognise and report early signs and symptoms.

■ Provide adequate washing facilities and cleaning procedures for protective clothing.

Asthma

Asthma, which can be a non-work related condition, can be caused or exacerbated by the inhalation of respiratory sensitisers. Work related asthma occurs when a respiratory sensitiser triggers an irreversible allergic reaction in the respiratory system. Once sensitised, the body's response may occur within minutes or take several hours. Sensitisation can take months or years of breathing in the sensitiser. Therefore, those working in areas where such sensitisers exist, or doing work that generates them, are particularly at risk. Pre-employment assessment can identify workers particularly at risk, for example those with other allergies. Most workers will be able to develop strategies to deal with asthmatic symptoms but a few people who develop occupational asthma will be forced to discontinue work as their health deteriorates.

Some materials may be fairly benign when in solid or liquid form but create a considerable health risk as a dust or vapour. Therefore, the reduction and control of dust- and vapour-creating activities is crucial.

What can designers and pre-construction planners do?

■ Eliminate substances that contain respiratory sensitisers wherever possible.

■ Designers should be aware of and use the numerous web-based guidance documents to eliminate the use of problematic substances wherever possible, replacing them with more benign alternatives.

■ Designers should avoid specifying solutions that require site works that create dust; for example, chasing walls for cables or pipes. Where possible, conduits or preformed ducts should be used.

What can managers and supervisors do?

■ Preventative and remedial action is very similar to that for dermatitis. However, there are additional requirements to reduce dust-creating activities where possible and ensure adequate dust extraction at source where the hazard can not be eliminated. Personal protective equipment (PPE) should be a last resort.

■ Dust suppressants other than water include calcium chloride flake, calcium chloride solution, magnesium chloride flake, and magnesium chloride solution – their use should be carefully monitored.

■ Where the use of respiratory sensitisers is unavoidable, the appropriate use of PPE must be enforced and ongoing health surveillance of workers should be carried out.

Occupational deafness

Noise at high levels, whether intermittent or continuous, can lead to permanent hearing damage with the volume, intensity and period of exposure being the key risk factors. All construction workers are potentially at risk and not only those actually operating the noisy equipment. Hearing loss tends to begin at the upper level of speech appreciation (4 kHz) before extending over time to 3–6 kHz, considerably reducing the individual's ability to distinguish speech. Other symptoms include tinnitus, stress and disturbed sleep. Hearing aids amplify all sounds but do not improve hearing distortions. Pre-employment assessment including audiometry can provide a baseline for further surveillance. The UK has a number of exposure action levels based on a time-weighted average day of eight hours (see HSE Noise website):

■ Lower exposure action level   Daily or weekly: 80 dB(A)
  Peak: 135 dB(A)

■ Upper exposure action level   Daily or weekly: 85 dB(A)
  Peak: 137 dB(A)

What can designers and pre-construction planners do?

■ It is acknowledged that designers are unlikely to be aware of the specific noise levels generated from particular construction activities. However, they should reduce the need for noisy operations wherever possible, particularly in enclosed spaces where the effects can be amplified and more workers exposed. This may include activities such as scabbling concrete, chasing concrete for cables or work within containers or vessels.

■ Discuss alternatives with construction experts.

■ Many actions are similar to those to reduce HAV, as noise is often linked to tasks involving vibrating machinery.

What can managers and supervisors do?

■ Assess risks from tasks/activities that could result in noise-induced hearing loss, involving workers and outside experts – eliminating or minimising activities wherever possible.

  ■ Assessing noise risks requires competent 'experts'.
  ■ If daily noise exposure remains above the lower action level, ear protection must be provided and affected zones clearly marked.
  ■ Noise levels should be reduced to as low as reasonably practicable, prioritising methods of noise control.
  ■ Ensure suitable ear protectors are worn by exposed workers.

■ Inform and train workers and supervisors regarding the risk of exposure, the damage caused, minimising risks, personal responsibilities and obtaining ear protection.

■ Refer workers for health surveillance or examination where necessary.

ICE manual of health and safety in construction © 2010 Institution of Civil Engineers

Stress

Stress is the one occupational health condition where the statistics suggest that it is less of a problem in construction than in other industry sectors, and yet the sector is claimed to have the highest rate of suicide is in the UK (Broughton and Pearson, 2003). The low rate of reported stress may be affected by the relative freedom of construction workers but may also be an over-positive picture influenced by construction's macho-culture, where acknowledging stress problems can be seen as an admission of weakness and a fear that an employer may consider the individual as not suitable for such a high pressured industry. Furthermore, there seems to be a relatively fine line between constructive pressure leading to increased performance and stress, leading to poorer performance and related health problems.

A survey by the Chartered Institute of Building (CIOB) (Campbell, 2006) showed the 68% of the 847 construction participants claimed to have suffered from stress, anxiety or depression, and 154 had taken medical advice. However, only 6% had taken any time off work as a result of their stress. The highest recorded 'causes' in the CIOB survey were 'too much work'; 'pressure'; 'ambitious deadlines'; and 'hours worked'. 'Poor planning', 'poor communications' and 'lack of feedback' featured more highly than 'interpersonal conflicts'. The long working hours typical within construction, poor working environments, high job demands, working away from family, exposure to traumatic events (e.g. witnessing accidents and fatalities) are all recognised stress hazards (Loosemore et al., 2003). Statistical linkages have been found between these agents and levels of stress in other workforces (Haslam and Mallon, 2003).

Stress has its effect on business through poor performance; inefficiency; poorer quality work; increased health and safety risks; customer and colleague complaints; and higher staff turnover. The affected employee often suffers through worry and depression; low self esteem; emotional outbursts; tiredness; lower interest in work; irregular attendance and increased health and safety risk.

Stress is one of the more complicated health conditions to deal with, both within and outside of the work environment and both for the individual employee and the employer. Stress is a process that builds into a downward spiral. The earlier in the process that stress is recognised and action taken, the more likely the action is to be effective. Therefore, it is important to recognise the onset of stress although clearly there is a danger in employers trying to play the role of amateur medics. This may be indicated by factors such as: general irritability; elevated heart rate; increased blood pressure; anxiety-anxious feeling for no specific reason; trembling; insomnia; headaches; indigestion; pain in neck and/or lower back; and changes in appetite or sleep pattern.

What can designers, employers and managers do?

■ Since stress can affect all employees, action is required by line managers at all levels in both design and construction organisations.

■ Stress is often exacerbated by excessive work pressure. Those involved in setting and monitoring deadlines throughout the construction process should be aware of the effect that unrealistic time deadlines can have on the stress of those involved in achieving them.

Other health issues

The conditions covered so far are the most prevalent in construction, based on the number of cases reported. However, there are a number of other important conditions that may be more significant depending on the type of project or the nature of the task. These include:

■ heat stress and strain
  ■ physiological or pathological change resulting from heat stress, e.g. increase in heart rate and body temperature, sweating and salt imbalance
■ ultraviolet radiation
  ■ by working in direct or reflected sunlight; significant exposure can lead to skin cancer
■ radiation diseases
  ■ through exposure to ionising or non-ionising processes such as weld testing
■ hyperbaric conditions
  ■ through working in an increased pressure environment, which could lead to decompression illness, burst eardrum and ear infections, burst lung, oxygen poisoning and carbon monoxide poisoning
■ leptospirosis (Weil's disease)
  ■ a bacterial infection carried in rat's urine, which contaminates water and soil.

Further information on these and other conditions is available from the references cited in this chapter and other sections of this manual (see Section 4 Health hazards and Section 5 Safety hazards).

Implications for ageing workers

Worldwide demographic trends show an ageing population. As a consequence, the working population is growing older. Retaining the older worker in the workforce will not only become a necessity, it is also beneficial to industry. Older workers accumulate invaluable job-related knowledge and experience, and it should be desirable for employers to retain this skill base.

The issues covered so far affect all construction workers. However, it should be noted that occupational health stressors have a cumulative affect. This means that the challenges faced by younger workers are magnified for older workers. While ageing is an individual process, it can be accelerated by arduous working conditions such as manual handling of heavy loads, excessive noise exposure or atypical working hours. This is particularly the case for MSDs. This also leads to a further complicating factor where younger workers, often spurred on by a macho-culture, use their greater physical ability to overstress their bodies, not realising the long term problems that they are creating. Often, by the time the effects are noticed, the main damage has already been done. This requires all construction stakeholders to work together to re-educate the workforce to highlight such problems. Designers can play a particular role by not accepting the 'we can do anything' culture that pervades much of the construction sector and specifying health enhancing options covered in this publication.

Summary

Occupational health is a major problem for construction. The most effective action is to assess, then eliminate or significantly reduce any health risk early in the project, at design or pre-construction planning stage. This needs to be done by designers and pre-construction planners – by the time the project gets to the site it is too late to eliminate these risks.

Any residual risks that designers and planners have not been able to eliminate must then be communicated to the contractor who should further assess, reduce, control and manage the risk to reduce its effect on the construction workers, having informed and trained them in the hazards and appropriate control measures. Pre-employment and ongoing health surveillance may be required.

The main construction occupational health problems include: asbestos related conditions; musculoskeletal disorders; hand–arm vibration syndrome (HAVS); problems caused by materials hazardous to health, including dermatitis and asthma; deafness and stress. Many OH problems become more acute for older workers.

References

Broughton, T. and Pearson, A. The Dark Side of Construction, *Building* 2003, 27 June, No. 25, 40–43.

Campbell, F. *Occupational Stress in the Construction Industry*, 2006, London: The Chartered Institute of Building. Available online at: http://www.ciob.org/filegrab/stress.pdf

COHME. *Managing Occupational Health Risks in Construction*, 2009, London: Health and Safety Executive. Available online at: http://www.hse.gov.uk/construction/healthrisks/

Confederation of British Industry. *Confederation of British Industry, 2005 Statistics, Reported in HSE Sickness Absence*, 2005. Available online at: http://www.hse.gov.uk/sicknessabsence/

Donaghy, R. *Report to the Secretary of State for Work and Pensions: One Death is too Many – Inquiry into the Underlying Causes of Construction Fatal Accidents*, 2009, July. Available online at: http://www.dwp.gov.uk/docs/one-death-is-too-many.pdf

Gibb, A. G. F., Gyi, D. E. and Thompson, T. (Eds). *The ECI Guide to Managing Health in Construction*, 1999, London: Thomas Telford.

Gibb, A. G. F, Haslam, R. A., Pavitt, T. C., Horne, K. A. (2007) Designing for Health – Reducing Occupational Health Risks in Bored Pile Operations. *Construction Information Quarterly* Special Issue: Health and Safety 2007, **9**(3), 113–123.

Gyi, D. E., Haslam, R. A. and Gibb, A. G. F. Case Studies of Occupational Health Management in the Engineering Construction Industry. *Occupational Medicine* 1998, **48**(4), 263–271.

Haslam, C. and Mallon, K. Post-traumatic Stress Symptoms Among Firefighters. *Work & Stress* 2003, **17**, 277–285.

Health and Safety Executive (HSE) *Work-related Injuries and Ill Health in Construction*, 2009. Available online at: http://www.hse.gov.uk/statistics/industry/construction/index.htm

Loosemore, M., Dainty, A. R. J. and Lingard, H. *HRM in Construction Projects: Strategic and Operational Approaches*, 2003, London: Spon Press.

Marmot, M., Feeney, A., Shipley, M., North, F. and Syme, S. L. Sickness Absence as a Measure of Health Status and Functioning: From the UK Whitehall II Study. *Journal of Epidemiology and Community Health*, 1995, **49**, 124–130.

Pendlebury, M. C., Brace, C. L., Gibb, A. G. F. and Gyi, D. E. *CIRIA Site Health*, 2004, London: Construction Industry Research and Information Association.

Rey, J. and Berriel-Cass, D. Construction Planning and Design: Beyond the Infection Control Risk Asessment. *American Journal of Infection Control* 2009, **33**(5), e85–e86.

World Health Organisation. *Basic Documents*, 39th edn., 1992. Geneva: WHO.

Referenced legislation

Construction (Design and Management) Regulations 2007. Reprinted March 2007. Statutory Instruments 320 2007, London: The Stationery Office.

Control of Substances Hazardous to Health Regulations 2002 Reprinted April 2004 and March 2007. Statutory instruments 2677 2002, London: The Stationery Office.

REACH Enforcement Regulations 2008. Statutory instruments 2852 2008, London: The Stationery Office.

Further reading

Health and Safety Executive. *Working with Substances Hazardous to Health: What You Need to Know About COSH* (INDG136(rev4) 06/09, free leaflet), 2009, London: HSE. Available online at: http://www.hse.gov.uk/pubns/indg136.pdf

United States Department of Labor. *A Guide to the Globally Harmonized System of Classification and Labeling of Chemicals (GHS)*. Available online at: http://www.osha.gov/dsg/hazcom/ghs.html

Websites

Further information on good practice

British Standards Institution (BSI) http://www.bsigroup.com

Chartered Institute of Building (CIOB) http://www.ciob.org.uk

CIRIA Site Health http://www.ciria.org

Constructing Better Health http://www.constructingbetterhealth.com/

Construction Occupational Health Management Essentials (COHME) http://www.hse.gov.uk/construction/healthrisks/

Control of Substances Hazardous to Health (COSHH) Essentials (from the HSE) http://www.coshh-essentials.org.uk/

ECI Managing Health in Construction http://www.eci-online.org

Health and Safety Executive (HSE) http://www.hse.gov.uk

HSE, Globally Harmonised Systems – Implications and guidance http://www.hse.gov.uk/ghs/implications.htm

HSE, Noise – Advice for employers http://www.hse.gov.uk/noise/advice.htm

Institution of Civil Engineers (ICE) Health and safety http://www.ice.org.uk/knowledge/specialist_health.asp

Off-highway Plant and Equipment Research Centre (OPERC) http://www.operc.com

OPERC Hand Arm Vibration Test Centre (HAVTEC) http://www.operc.com/pages/havtecwelcome.asp

Chapter 9

Assessing safety issues in construction

Philip McAleenan Expert Ease International, Downpatrick, Northern Ireland, UK

doi: 10:10.1680/mohs.40564.0101

CONTENTS

Safety in construction is an integral aspect of the whole process from the time that the initial thoughts are being scribbled down and the concept developed through to the final stages of a structure's life when the demolition company is taking down and removing the debris to suitable recycling and waste sites. Responsibility for safety begins with the client and continues through the design team who are responsible for ensuring that their designs present no unnecessary hazards to personnel and users at any stage of the life of their structure, through to the contractor who must ensure that any hazards associated with the construction are controlled by all appropriate means and that any residual risks associated with use of the structure are notified to the client at the end of the construction phase.

The core strategic objective must be to achieve inherently safe designs that are, as far as is practicable, hazard free. Focusing on safety at the beginning of the process will contribute substantially to the quality of the structure and the safety of those who will build and use it. The goal is to achieve a structure that stands acknowledged by all as a symbol of excellence in the built environment.

Box 1   Key learning points

Readers will be introduced to a range of hazard assessment methodologies and concepts for the elimination of hazards at the earliest stages of the design process. They will be provided with information designed to raise awareness of the hazards presented at the different stages of the construction and use of the structure and of the opportunities for eliminating them or reducing exposure to those that necessarily remain.

Introduction

It has been said that the million-pound mistake can be traced back to the early stages in the design process and to the decisions made at the first scribbling of the design.

What the Client wants and what he provides to achieve it and how the Designers, Engineers and Architects respond to his requirements will determine whether the project outcomes will be achieved in a manner that is non-injurious to the workers (who will construct it), those who'll use and maintain it and, at some future point, those who'll be charged with the decommissioning and demolition of the structure. It is at this critical stage when the feasibility of the project is being considered, when the design and engineering solutions are being sketched out, and land, budgets, resources and schedule to completion are being discussed that the impact of the project on the environment, and on the safety and health of those who will be affected by it throughout its whole life, must be given equal

consideration and the prevention of harm prioritised within the solutions agreed upon.

A useful concept, common in process engineering design, is that of 'inherently safe design'. Underpinning inherently safe design is that hazards are controlled through the design of the plant, and, in the case of construction, the structure and the means for erection. Hazards are eliminated as a priority, followed by the reduction in consequences where they cannot be eliminated, e.g. by construction methods that reduce exposure to hazards such as working at heights, and finally by reducing the likelihood of an event occurring through simplification techniques and clarity of information (Moore, 2000).

The success of the process commences with the client and his willingness to commission a project that will bear the test of time and stand acknowledged by the present and future generations as a symbol of excellence in the built environment. The governance of the project begins and ends with the Client. It is he who will establish the strategic objectives of the project, who will select and appoint a competent team of Designers, Engineers and Architects, and Contractors to translate his objectives into the drawings and plans, to construct and complete the project and who will advise on and support him in meeting the core strategic objectives. Those core objectives must consider the requirements of all the stakeholders – owners, workers, users, community, government, etc. – and lead to a design solution that incorporates these in relation to construction, maintenance, management, flexibility, health and safety, sustainability and environmental impact.

If they do not receive due consideration at this stage, and, with all the best intentions in the world, the project proceeds leaving such matters to be decided upon as and when they arise, invariably there will be mistakes that will result in loss of life, injury and costly redesign and remediation. These issues lie with the client and the top-tier team and, on the basis of top-down leadership, appropriate consideration and direction from this team will ensure that those who follow – contractors, employees, specialist advisors and consultants – will be brought on board with those same values and commitment to the quality of the project and the safety of those who will be affected by it.

Hazards assessment

Risk assessment has been fundamental to health and safety since the introduction of the Health and Safety at Work Act 1974 (HSW Act 1974), the Health and Safety at Work (Northern Ireland) Order 1978 (HSW Order 1978) and the subsequent Management of Health and Safety at Work Regulations 1999. It has proved to be beneficial and has contributed enormously to the successful implementation of safe working practices and the reduction of accident rates in workplaces throughout all industries. It is not without controversy, though, and in 2008 the parliamentary Work and Pensions Committee undertook an inquiry into the role of the Health and Safety Commission (HSC) and Health and Safety Executive (HSE) in regulating health and safety in Great Britain (Work and Pensions Committee, 2008), from which 'a good deal of evidence' emerged that duty holders can over-interpret legislation and produce voluminous risk assessments that are not required by law. Not unrelated to this is the concept of 'significant risk', again not an aspect of the primary legislation but is a concept that is used often by advisers and consultants and which contributes to the problem of written risk assessments. For the designer it is important to have a clear understanding of the assessment processes and the options available for conducting and, where necessary, recording the process.

Risk assessment is first and foremost a process of thinking about the activity that you are undertaking with a view to deciding the next step. In that respect it is a continuous process and, in familiar circumstances and situations it tends to be largely a subconscious but no less effective activity. This fact about risk assessment differentiates it from the idea of risk assessment as an output, a finished product taking the form of a written (sometimes verbose) report carried out prior to an activity taking place. It also negates that idea of generic risk assessments done once and applied every time the same activity or hazard is carried out or encountered. Indeed risk assessment as a process is holistic in that it takes cognisance of the activity, the objective, the hazards and the likelihood of success. Good risk assessment eschews the notion that a single hazard or condition can be effectively abstracted from the overall activity and controlled before moving on and being applied to the next element.

In this respect the requirement that the Engineer or the Architect assesses the project (often referred to as designers'

risk assessments or design safety analysis) from the outset from the perspective of developing a design and structure that is quality built and safe to construct, use and maintain is one that assesses the hazards posed by his design or by the environment in which the structure is to be sited and in as far as it can be practically achieved, develops solutions that will prevent the realisation of the harm that would result. It is of necessity more than a simple risk assessment, but nonetheless may incorporate the assessment of risk posed by any hazard found. Such assessments will seek to identify the hazards with the assessor then moving on to making an appropriate determination of the controls that will be required to eliminate the hazard or its potential to cause harm.

Box 2 Probability

A problematic point with such assessments is attempting to firstly assess the risk, i.e. the probability of an unwanted event occurring and, second, how to make use of the score attained. This arises from the matrices that have been developed on the heels of a suggestion in the HSE/Health and Safety Executive Northern Ireland's (HSENI's) risk assessment guides. These matrices are typically qualitative (where the probability of injury is graded high/low/medium), or quantitative (probability is assigned a number based on likelihood) and are two dimensional (likelihood versus severity) or three dimensional with the added parameter of frequency of exposure.

The probability of an event occurring lies between 0 and 1; 0 meaning that it will not occur and 1 that it will. In simple systems with a small known number of outcomes the probability of a particular outcome can be calculated with ease. In complex systems with ever-increasing numbers of possible outcomes, the probability cannot be calculated with ease, and therefore reference to historical data is used, i.e. in so many operations over so many years how often has such and such an outcome occurred. Rigorous mathematical discipline will be applied to the development of safety critical elements in, for example, the development of a nuclear power plant, but not necessarily so with something like working methods, and therefore rather crude estimates are applied, hence matrices with high/medium/low or 1/2/3 probabilities. (Liz Bennett, former Chair of ICE's Health and Safety Board, asks a similar question on the Safety in Design forum, http://www.safetyindesign.org/)

Knowing the probability of an event occurring does not allow for a prediction about when it will occur; a highly probable event may theoretically never occur just as a remotely probable event may occur today and again tomorrow.

The one thing that we can say about risk, and this is something not often considered by assessors, is that where it exists, even at the lower probabilities (see Box 2) resulting from the engineering and human controls introduced, its existence implies that the structure, system or process is not controlled. This may result from either the application of the 'so far as is reasonably practicable' principle or because there is a gap in our information about that hazard that prevents us developing an appropriate design control, or it may be that it is so remote that it is worth taking the risk. None of this is problematic where the consequence is negligible but as you move up the scale of severity towards catastrophic, greater diligence is required of the assessor to determine how much needs to be done to prevent the harm occurring. And whilst it may not be possible to know absolutely all the variables that will impact upon the project, the absence of certainty should not translate into less controls than is necessary for safety but with severe consequences erring on the side of caution does not equate with going overboard but makes good practical sense (Malone, 2007).

Models for assessing hazards

'Eliminate, reduce, inform and control', going by the acronym ERIC, is promoted as a simple qualitative model that can be used in all but the rarest of situations (e.g. design of nuclear facilities) (Construction Skills, 2007). It describes a sequence of steps to be followed by designers at the earliest stages of the design process after having identified the hazards associated with the project and the proposed structure.

The first step is to consider eliminating the hazard by an alternative design solution. This may be a mandatory requirement or obligation for which the Eurocodes and national statutes will provide details; otherwise the principle of 'so far as is reasonably practicable' will be relevant. There is no doubt that seeking to eliminate the hazard has led to many new and innovative solutions to both common and unique hazards. Sometimes this may be led by a strategic objective that is very specific, e.g. to reduce working at height during construction as far as practicable has led to prefabrication of structural components both on and off site with erection on site being speedy and requiring a fraction of the time spent working at height by the workers involved. Concrete tilt-slab walls can be constructed flat at waist height and raised into position when ready in a matter of hours (this method is dominant in North America and Australia and New Zealand but as yet is not commonly used in the UK or Ireland – see **Figure 1**). Scaffolding can be eliminated, again by policy decision and replaced by the use of appropriate elevating platforms, eliminating the need for fragile structures that require frequent inspections and adjustments and are vulnerable to weather conditions (as well as providing a wonderful opportunity for youngsters to play on when the site is closed for the night). Queen's University Belfast (Centre for Built Environment and Research, School of Planning Architecture and Civil Engineering) has developed the FlexiArch bridge, a 'flat-pack' precast concrete solution that is delivered to site and hoisted into position by crane. As it is not reinforced it is more durable, quick and easy to install, less disruptive to local communities and cheaper to transport.

The model reminds the user that alternative design solutions that eliminate one hazard may introduce another. ERIC is a model that requires you know the hazards before they come into play. It does not itself facilitate primary hazard identification. Nor is it an iterative process that is to be applied continuously until all hazards have been identified, considered and controlled. However, in reminding the designer that secondary hazards may arise with hazard controls, it follows that the process should continue to be applied until all new hazards are identified and dealt with.

Not all hazards can be eliminated and the second step in ERIC requires that the designer reduce the remaining risks associated with the hazard, so far as is reasonably practicable. This step moves the hazard identification and control procedure into the risk assessment model and the concomitant problems that have been identified above. Although in practice many designers will modify their designs and put in place a range of controls for such hazards that remain, rather than reducing the risk, what is happening in these measures is that the severity of the hazard is varying from high to low. It must be remembered that the existence of a hazard does not necessarily equate with the existence of a risk; this generally comes about when, in relation to the structure, it is not engineered to the requirements necessary for the environment or the use to which it is put, or, in relation to human interaction, when workers and users interact with the hazard.

The efficacy of this element of ERIC is greatly enhanced if it is viewed as pertaining to hazard reduction via a range of alternative solutions coupled with suggested controls that will eliminate the potential to cause harm. This can be achieved by utilising appropriate combinations and competent use of engineering, mechanical, safe systems, and personal protective equipment (PPE) protective measures, prioritising collective measures above individual measures.

Step three is to provide the contractor with hazard information that they need to be aware of that has not been controlled in the design solutions and, through the health and safety file, inform those who will maintain and those who will demolish the structure. It is not for the designer to specify what controls are to be implemented on site – that arises in step four and is for the Contractor to consider and implement unless there are modifications to the design of the structure or other factors come to bear.

It is worth keeping in mind that the pre-construction information provided should be that which a competent contractor could not reasonably be expected to know.

ERIC is a useful tool provided the limitations are recognised and the fundamental requirement to design a structure that is inherently safe is to the fore in the thinking of the designers. ERIC is not a requirement of Construction (Design and Management) (CDM) Regulations 2007 or the HES's Approved Codes of Practice (ACoP) but has emerged in subsequent guidance to good practice from models that have been in existence since the early days of risk assessment in health and safety.

Another method used by some Designers to comply with the duty to inform the Contractor and others down the line is the risk register. Depending on who is producing the register, this can be a useful tool provided it is used for noting hazards that the Contractor needs to be aware of with a few comments from designers. It should not become a cumbersome document full of 'risk assessments' that detail the process used and information on the control of all risks, minor and major. Some would suggest that the register should contain this information, as well as details of the applicable legislation, regulations, codes, industry guidance and standards despite this being contrary to the intent of the lawmakers, the HSE/HSENI and Institution of

Figure 1 Tilt-slab construction

Civil Engineers (ICE), their case being that without this there is no proof that the assessment was done should anything happen and the matter end up in court.

> ICE believes that the process of design risk assessment remains an essential designer's tool but supports the emphasis in CDM 2007, the ACOP and the Notes for Guidance on producing an effective end product, not on the process itself. In [their] view this end product should be the application of *engineering judgement* to eliminate and where it is not reasonably practicable, to reduce hazards and risks within the construction and subsequent phases. This should be recorded in a simple, short Residual Significant Risk Register to be passed to the principal contractor and others. (ICE, 2007)

Remember, Designer risk assessment is a process, the output of which is reflected in the drawings and in any note that the Designer makes about hazards that the Contractor or the in-use maintenance team needs to control on site. A competent person following the plans will recognise from this output that the thought processes have been gone through as well as the degree of diligence that has been applied. An effective register would therefore be one that records any significant or unusual hazards that the contractor and end user need to take into their consideration when planning the on-site/in-use activities and controls.

This register could be improved with additional columns that identify the actions required from designers and those from contractors (possibly with the inclusion of action dates). Additionally, rather than delete risks from the register, a suggestion would be to change their colour and drop them down the list as the project progresses.

The elimination or containment of hazards in the structure is achieved through good engineering design whilst process hazards have to be worked around by the contractor and subcontractors during the construction phase. Identification and analysis of the hazards at the earliest opportunity is essential, as is the development of appropriate controls to ensure that people and property are protected prior to, during and after the work operation. Design Safety Analysis and Control is a model that addresses the issue of safety in design through the elimination of risk through the absolute control of all stages of the process. This model suits the requirements of the European Council Directive 92/57/EEC on the implementation of minimum safety and health requirements at temporary or mobile construction sites (EEC, 1992) with particular reference to project preparation and assists designers in delivery of inherently safer designs.

Hazard identification begins with an initial proposal by the Client and the preliminary assessment measures his requirements against the resources that are assigned to the project. At this point the Designer will be able to identify any hazards that may render the project non-viable or that will need to be adjusted to make the project feasible. This means looking at the design, location, environment and materials to be used for factors that can cause harm and modifying the design (or the resources) accordingly. This process is continuous throughout the design stage and into the construction stage; identifying sources of harm and eliminating or containing the hazards through barriers, safe systems and competent work practices.

Critically this model is not concerned with reducing the risk posed by hazards, but rather it is focused on the elimination of the risk via robust and appropriate controls. Whether it is the Designer, the Contractor or the competent worker, each player must at the relevant stage consider what he has to do in order to prevent harm occurring. And since elimination of a hazard is the primary objective, it is often necessary to return to and review the designs. Where elimination is not an option, consideration must be given to other control measures that will prevent harm; often this will be through the application of a matrix of measures that in combination will meet the requirements to safeguard workers.

At the design stage and before embarking upon the construction of the project, it is important to ensure that enough has been done to prevent harm, or inform the relevant persons of hazards that continue to exist.

There are limitations to the knowledge that any one individual possesses and it is imperative that whenever a hazard assessment is being undertaken specialist advice and cross-checking are sought (see below). Sources of specialist advice include Engineers, Architects and other Designers, the project's CDM Coordinator, Contractors, Manufacturers and suppliers of materials, plant and equipment, HSE/HSENI, trade and professional associations or other relevant safety professionals. In this way the Client and the Designer can be assured that national and/or international design standards are being applied.

Things can go wrong and it is important to try to anticipate what they may be as early as possible in order to ensure that the most effective controls are employed. Asking questions about what could happen focuses the mind and ensures that all the foreseeable incidents have been considered and planned for at the preparation stage. These questions also prompt consideration of emergency preparedness and what should be in place before any particular aspect of the project is commenced.

Stages in the assessment process

Concept and feasibility

Assessment begins when the Client introduces his concept to the lead designers. How much of an idea the Client has in regard to how he envisions the project will vary, but what is certain is that his project relates to a functional outcome. Before pencil is put to paper there will be questions about the location of the project, the estimated usage, size, lifespan, and the social and economic need that the project aims to address. At this stage the Designer must consider the objective technical feasibility of the project taking into consideration such matters as the geology of the location, the general environmental impact of the project, current levels of engineering capability and regulatory requirements. These and other factors establish the parameters within which a design is feasible. How wide those parameters are is dependent upon what is found in this preliminary investigation and the effectiveness of this investigation requires that appropriate specialists are engaged to address the various issues that arise. If the project is not feasible under any circumstances,

1.	References/Location/Phases etc.	
2.	Hazard or issue	Brief description (if helpful, these can be wider project hazards, e.g. planning issues).
3.	Elimination? If No	If 'yes' – describe action to be taken, by whom and by when.
4.	Associated risks and reduction methods taken	Note: action to be taken in accordance with the hierarchy and principles of prevention and protection (see 2.3.3).
5.	Information to be conveyed to others	Is the method chosen to communicate this information adequate? (see section 1.7)
6.	Date of assessment	
7.	Action by: Required by:	
8.	Action cleared	

Figure 2 An example of a Risk Register (headings)
Source: CDM 2007 – Industry Guidance for Designers 2007

Sample Designers Checklist (Apply appropriate International Standards)				
Phase	**Activity**	**embodying Hazards / Harm**	**Design Solution** (Note whether it eliminates, contains or controls the hazard)	**Information required for** (Insert: construction phase plan and/or client safety file)[1]
Construction	• Bridge construction,	• working at heights	• Pre-assemble spans and raise to position (eliminates a substantial amount of work at heights)	• Construction safety plan
Occupation & maintenance	• Painting exposed steelwork (e.g. bridges) • Pipe and cable maintenance or replacement •	• Fragile roof/surface	• Use weathering steel (eliminate need for painting) • Lay pipes and cables along accessible routes (eliminate need for heights) • Construct and mark permanent walkways, erection of barriers, warning signs (contains)	• Construction safety plan • Construction safety plan and client safety file • Construction safety plan and client safety file
Demolition	• Pre-stressed concrete supports	•	•	• Construction safety plan and demolition sequence in the client safety file
Note: This table is simplified, incomplete and is included as an example to illustrate the DSAC process				

[1] Construction phase plans are dynamic documents that describe the measures necessary for ensuring safety at all stages of construction. Client health and safety files are produced at the end of the construction phase and are intended for use by the client during the occupation/maintenance and demolition phases of the project.

Figure 3 – Sample Designers Checklist
(This table is simplified, incomplete and is included as an example to illustrate the DSAC process)

this must be immediately conveyed to the Client, but in other respects the parameters within which it is feasible are established for the next stage of assessment.

At this point, if not provided initially, the Designer must obtain information from the Client on the resources that will be available for the project, in particular financial and time constraints, as well as any notion of how the Client envisages the project, its appearance, materials to be used, its footprint of the available land, etc. This information will narrow the parameters further and, where any of the require-

ments of the Client render the project unfeasible because, for example, the vision involves a technically poor design or unsafe solution, or because the constraints limit what he can obtain, the design team must at this point inform him of such limitations.

Design

With this information the Designer has enough to develop preliminary solutions. Initially a range of solutions may be developed to provide the Client with options before one is decided upon for further development (presenting the Client with a range of options from which he makes a choice does not make him a Designer). And while these may be essential concepts, there are a number of issues relating to safety and environment that must be considered if the finally selected solution is to be developed into detailed drawings. The designs must reflect the following key engineering principles:

■ fitness for purpose

■ buildability/constructability

■ maintainability, including emergency preparedness and security

■ demolition and disposability.

Fitness for purpose

All structures have a number of different elements and a number of ways of interpreting those elements that will depend upon the nature of the structure and functions to be carried within. Initially it may be regarded by the Client as having a single purpose that will persist throughout the whole life of the structure, but from the Designer's point of view he must regard it as having the potential to be put to alternative uses at some time in the future and prepare his hazards information accordingly. He does not have to foresee what uses it will be put to but he will know the limitations of the final structure and will be able to detail these limitations in the information that will eventually go into the health and safety file, for example the strength and suitability of various floors for increased point loading, or the safety issues surrounding the removal of internal walls.

The design should permit the maximum freedom of movement of people, vehicles and product without unnecessary obstruction or meandering.

Different functional areas should be clearly defined, and facilities for reception, visitors, work, rest, welfare, hygiene, meals, etc. clearly marked and easily accessed. Regulations pertaining to disability and other equality measures must be considered in the design layout. It is important to note that new-build structures or major improvements to existing buildings must not disadvantage or put at risk people (worker, visitor or other user) who have disabilities and physical impairments.

Internal light, heat, air and general environment should be of good quality and individually controllable where possible. In deciding how the structure will look, internally and externally, the Designer must consider the impact of the spaces created, materials used and aspect of the building on these factors. Consideration too must be given on how the aesthetics of the structure will impact upon workplace stress. Factors such as colours, textures, architectural features may enliven the environment and enhance the welcoming effect of the structure or alternatively reduce it.

Security and safety of all those who will use the finished structure must be considered at this stage. This will entail applying a range of scenarios and exploring the impact of

each one on the safety of all who are likely to be affected by them. The guiding principle is prevention of harm to people and then to property. This means that in the design, measures to prevent structural failure will take precedence over, but not replace, other measures that safeguard people in the event of structural failure. Further information on these issues is provided in a series of booklets developed by the Commission for Architecture and the Built Environment (CABE); for information on prevention as a global responsibility see the Seoul Declaration on Safety and Health at Work, 2008 (ILO, 2008).

Buildability

The design solution that is eventually chosen must be technically feasible, safe to construct and capable of being carried through to completion within the resources allocated. When considering this element of the design process the knowledge and expertise of the principal contractor (and engineers if they are not part of the initial design team) will prove beneficial. This element involves consideration of the sequences necessary for effective and safe construction with engineering factors being of primary importance. These will include assessments of the technical knowledge required to bring the structure into reality, the capacity of the materials to be employed in the construction, the sequencing and pre-ordering of fabricated components and units, pre-construction preparation of the site, or technologies required to facilitate the construction, e.g. custom built tunnelling machines.

The logistics of the construction will be considered, as the footprint of the structure on the site will impact on on-site storage, vehicular manoeuvrability, welfare facilities, neighbouring buildings and passing public. The greater the footprint relative to the size of the site the tighter the project becomes with an increase in the hazards to workers and public alike. In these circumstances, technical solutions may require an increased awareness of how the project is managed, and relevant expertise will be needed. For example, a simple technical solution to limited space may be that large components are fabricated off site, but this throws into relief problems associated with the transport of those components to the site, the suitability of the roads to take the width, weight and size of the units, and at the site the similar considerations to the suitability of the road to take the weight etc. of mobile cranes if tower cranes are not in use.

The impact of blasting, noise, dust, increased traffic flow to the site and decreased public pedestrian space on neighbours must be considered, and satisfactory solutions built into the process. The solution may be the remit of the principal contractor but the hazard identification will come within the remit of the design team who must be aware of and inform the contractor that such matters will arise.

The safety of the end structure is without a doubt a function of good design. But equally important is the safety of the structure at the different stages of construction. Additional interim measures including temporary support structures (see Chapter 17 *Falsework*) need to be considered and included in the detailed plans or referred to information provided to the contractor. The more complex the interim structure the greater the necessity to include the design details in the plans.

In the previous section it was noted that the designer had to give some consideration to the future of the project in terms of change of use and possible emergencies that would impact upon the users of the finished structure. An additional element of foreseeability will require an assessment of the possible effects of further development in the surrounding area on the finished structure. Just as the design will impact upon neighbouring structures and thus due consideration given to preventing harm, the effects of future structures in the immediate environment must be reflected in the robustness of the design to withstand, for example, changing wind patterns that may result. And, as always, not every future event is capable of being predicted, or designed for, but information about the limitations of the structure can be inserted into the health and safety file and conveyed to future developers as appropriate.

Maintainability

It has been estimated that the whole life use and maintenance costs by far exceed the design and construction costs of most substantial structures and buildings. These costs begin immediately the structure is handed over to the Client and include the arrangements for safeguarding maintenance workers and the public who will work in or use the finished structure.

Maintenance requirements can give rise to substantial hazards that have the potential to injure workers and the public, to damage the primary structure as well as neighbouring structures, to interrupt industry and commerce, disrupt the functioning of urban and rural transport systems and infrastructure and cause public and political discontent. For example, in May 2004 a crane used to change light-bulbs on the external face of the Taipei Tower, Taiwan, shifted during the operation, damaging a connecting bridge between two buildings and causing it to be closed (Huang, 2003).

The aesthetics of the structure must not sacrifice safety for visual impact and, where it is unavoidable, the design must contain solutions to the maintenance issues or provide information to future users on hazards that they will face. External maintenance work will be required, but the design should ensure, by the use of materials chosen, higher standards of robustness, self-cleaning and weathering steel, that the need for workers to access external parts of the structure are infrequent.

Common problems encountered by maintenance workers include climbing to unsafe heights, traversing fragile surfaces or entry into restricted crawlways in order to access pipes, cables and switch points. Routing cables, and pipelines alongside fixed walkways, elimination of fragile surfaces in the vicinity of plant, and the siting of plant on lower levels or at places that can be accessed by elevating platforms can and must be considered as integral to and an essential requirement of the design process.

It is important to recognise too that the structure must be built within a high safety margin or factor of safety determined by the rate of deterioration of the materials used. All materials deteriorate and as a consequence the safety of the structure will be compromised. Constructing to the limit of safety does not allow for deterioration and as a consequence higher levels of maintenance and refurbishment will be required to preserve the integrity of the structure.

Emergency preparedness

The design of the structure must include due consideration to the various types of emergencies that may occur and include within them technical solutions for the prevention of the event, as far as it possible to do so, supported by arrangements for ensuring the safety of all persons likely to be affected should one such event occur.

Consideration by the Designer should cover how extensive the impact of an emergency would be both in relation to the structure but also its impact on the surrounding environment, buildings, people, infrastructure, as well as the social and economic effects following on from the event. Whilst it is the duty of the client or end user to integrate his emergency procedures with those of the public services that are available, it is the task of the designer to ensure that the engineering controls in for example high hazard industrial processes, are designed to the highest standards available and, where necessary, developing new standards where existing standards may be insufficient to provide the necessary security.

Demolition and disposability

Hazards inherent in the structure that are not immediately apparent to those in the future who will be charged with taking down the structure must be noted in the health and safety file. These may include information on hidden stressors within the concrete walls and piles of the building, details of any sequencing necessary to ensure the stability of the structure as it is demolished, or hidden weak points that would interfere with controlled blasting.

The considerations at this point are in relation to factors that will not be immediately obvious to a competent demolition company and the designer is not required to go into all the details necessary for demolishing the structure.

Standards and codes

It is fundamental to the safety of the process that all designs are carried out by people who are competent to understand the requirements of the client, are knowledgeable of engineering principles, aware of the limitations of the materials that are to be used and comprehend the construction process and what is required to execute a safe structure.

Competency is a critical issue and it goes beyond mere technical capability to include the ability to comprehend and interpret the whole gamut of design codes and statutory rules that are applicable to design and construction. However, the range of applicable codes is not limited to technical matters but will include environmental statutes and the need to conduct environmental impact studies and the effects of pollution on the locality (short term and long term), traffic, vistas, noise, population, etc. There will be social policy regulations pertaining to equality issues and the requirement that new structures (notwithstanding limitations posed by their purpose) are accessible to all sectors of the community, and heritage laws that will limit what can be constructed, or how constructions should look within the landscape (see Chapter 1 *Legal principles* for more information).

Design teams should therefore ensure that they have the requisite specialist knowledge to ensure that due consideration is given both to the technical requirements of the structure and to the other legal requirements described. The legal and technical reasoning behind the submitted design(s) must be conveyed to the client and where they conflict with his desires they may be altered to the extent that they do not conflict with those legal and technical requirements.

Checking and supervision

Quality control is essential at all stages of the project, from design through to execution. The elements of quality control include:

- appropriate levels of supervision
- checking, including third-party checking of designs
- inspections, including third party inspections of the execution phases, and
- project review.

The level of supervision will depend upon the experience and qualifications of the persons carrying particular aspects of the design, the novelty of the design and the potential consequences should the structure fail.

Greater supervision by competent persons is to be given the less experience an individual has. This can decrease as his experience in particular aspects develop and he demonstrates a growth in competence; eventually, as he gains competence over a wide range of elements, supervision may reduce to a point when he gains the relevant professional recognition that will permit him to work unsupervised and in turn become a supervisor. However, this is not an absolute limit and very competent persons may return to a supervised status when working on innovatory projects, or when the finished project has a substantial impact on those who will use it and on the surrounding environment, business and social landscape.

Standing Committee on Structural Safety (SCOSS) has described high consequence structures as being of the nature of grandstands and public buildings whilst agricultural buildings may be regarded as low consequence structures and from this they describe the levels of supervision as in **Figure 4** (SCOSS, 2009).

The principles for inspection of the execution of the project follow similar lines with extended inspection at level 3 referring to third-party checking, and normal inspections at level 2 to be carried out in accordance with the quality control procedures of the organisation. Level 1 inspections are self-inspections.

Project reviews are required at various times to ensure that the project remains on course, that the strategic objectives are being met and that the designs are being effectively translated into quality structure. The points at which reviews take place will include the transition between phases of construction, changes in principal contractor, emergence of unforeseen factors, failures in the ability to translate the design into the structure, deviations from the original plan, alterations in the materials purchased, budgetary overspend requiring alteration to the project and any other time when significant changes are required.

Summary of main points

The core strategic objective in the construction of any structure is that it must be inherently safer in design such that it is non-injurious to those who will work on or use the structure. The role of the

Consequence Class & examples	Design Supervision levels	Characteristics	Min. recommended requirements for checking of calculations, drawings and specs.
CC3 High: Public Buildings, Grandstands	DSL3	Extended supervision	Third party checking
CC2 Medium: residential or office	DSL2	Normal supervision	Checking by different persons than those originally responsible and in accordance with the procedure of the organization
CC1 Low: agricultural buildings.	DSL1	Normal supervision	Self checking: performed by the person who created the design

Figure 4 Supervision Levels
Source: SCOSS Topic Paper SC/09/014

designer in accepting the Client's ideas and developing appropriate designs is to determine all the relevant factors that will negate this objective and to ensure that his designs incorporate technical solutions to these safety issues and to inform the contractor of any hazards that are not eliminated or controlled in the design.

Focusing on safety at this stage of the process will contribute substantially to the quality of the structure and the safety of both those who will build and who will use it, and building in costs for safety assessments during the design process will provide proportionately greater cost savings throughout the whole life of the project. The end result is a structure that stands acknowledged by all as a symbol of excellence in the built environment.

References

BS EN 1990: 2002 Basis of Structural Design, London: BSI.

Construction Skills. *The Construction (Design and Management) Regulations 2007, Industry Guidance for Designers*, 2007, King's Lynn: Construction Skills. Available online at: http://www.cskills.org/uploads/CDM_Designers4web_07_tcm17-4643.pdf

Huang, J. Safety Belts Save Two as Crane Tips at Taipei 101. *Taipei Times* 2004, 20 May, 4. Available online at: http://www.taipeitimes.com/News/taiwan/archives/2004/05/20/2003156219

Institution of Civil Engineers. *Design Risk Assessments and CDM 2007, ICE view*, 2007, Members Guidance Notes, London: ICE.

International Labour Organisation. *Seoul Declaration of Safety and Health at Work*, 2008, Seoul, Korea.

Malone, D. Can we learn to love uncertainty? *New Scientist* 2007, **195**(2615), 46–47.

Moore, D. A., *Incorporating Inherently Safer Design Practices into Process Hazard Analysis*, USA, 2000.

Standing Committee on Structural Safety. *The Assumptions behind the Eurocodes*, Topic Paper SC/09/014, January 2009.

Work and Pensions Committee. *The Role of the Health and Safety Commission and the Health and Safety Executive in Regulating Workplace Health and Safety*, Third Report of Session 2007–08, 2008, London: The Stationery Office.

Referenced legislation

Construction (Design and Management) Regulations 2007. Statutory instruments 320 2007, London: HMSO.

EEC. Council Directive 92/57/EEC on the Implementation of Minimum Safety and Health Requirements at Temporary or Mobile Construction Sites (eighth individual Directive within the meaning of Article 16 (1) of Directive 89/391/EEC), 24 June 1992.

Health and Safety at Work Act 1974. Elizabeth II. Chapter 37, London:HMSO.

Health and Safety at Work (Northern Ireland) Order 1978. Statutory instruments 1978 1039, London: HMSO.

Management of Health and safety at Work Regulations 1999. Statutory instruments1999 324. London HMSO.

Further reading

Carroll, B. and Turpin, T. *Environmental impact assessment handbook*, 2nd edition, 2009, London: Thomas Telford.

Construction Skills, industry-produced guidance documents for dutyholders. Available online at: http://www.cskills.org/supportbusiness/healthsafety/cdmregs/guidance/index.aspx

Construction Skills and Health and Safety Executive. *Construction (Design and Management) Regulations 2007 Industry Guidance for Small One-Off and Infrequent Clients*, 2007, King's Lynn: Construction Skills. Available online at: http://www.cskills.org/supportbusiness/healthsafety/cdmregs/guidance/Copy_3_of_index.aspx

Construction Skills and Health and Safety Executive. *Construction (Design and Management) Regulations 2007 Industry Guidance for CDM Co-ordinators*, 2007, King's Lynn: Construction Skills. Available online at: http://www.cskills.org/supportbusiness/healthsafety/cdmregs/guidance/Copy_2_of_index.aspx

Construction Skills and Health and Safety Executive. *Construction (Design and Management) Regulations 2007 Industry Guidance for Principal Contractors*, 2007, King's Lynn: Construction Skills. Available online at: http://www.cskills.org/supportbusiness/healthsafety/cdmregs/guidance/principal.aspx

Construction Skills and Health and Safety Executive. *Construction (Design and Management) Regulations 2007 Industry Guidance for Contractors*, 2007, King's Lynn: Construction Skills. Available online at: http://www.cskills.org/supportbusiness/healthsafety/cdmregs/guidance/Copy_4_of_index.aspx

Construction Skills and Health and Safety Executive. *Construction (Design and Management) Regulations 2007 Industry Guidance for Workers*, 2007, King's Lynn: Construction Skills. Available online at: http://www.cskills.org/supportbusiness/healthsafety/cdmregs/guidance/workers.aspx

Hendershot, D. C. and Post, R. L. *Inherent Safety and Reliability in Plant Design*, 2000, presented to Process Safety Center Annual Symposium, Texas.

International Labour Organisation. *Seoul Declaration of Safety and Health at Work*, 2008, Seoul, Korea.

Kletz, T. A. What You Don't Have, Can't Leak, *Chemistry and Industry* 1978, **6**, 287–292.

McAleenan, C. and McAleenan, P. *Safety in Design – a Risk Assessment Approach*, June 2004.

McAleenan, P. and McAleenan, C. *Design Sfety Analysis and Control – Explained*, November 2004. Available online at: http://websafety.com/Exchange/index.htm Paper 9.

Morris, M. and Simm, J. (Eds). *Construction Risk in River and Estuary Engineering: A Guidance Manual*, 2000, London: Thomas Telford.

Ove Arup and Partners. *CDM 2007 – Construction Work Sector Guidance for Designers* (C662), 2007, London: CIRIA.

Websites

British Standards Institution (BSI) http://www.bsigroup.com

Commission for Architecture and the Built Environment (CABE) http://www.cabe.org.uk

Health and Safety Executive (HSE) http://www.hse.gov.uk

Eurocodes Expert – making Eurocodes easier http://www.eurocodes.co.uk

Health and Safety Executive Northern Ireland (HSENI) http://www.hseni.gov.uk

Institution of Civil Engineers (ICE), Health and safety http://www.ice.org.uk/knowledge/specialist_health.asp

Chapter 10

Procurement

Ciaran McAleenan Expert Ease International, Lurgan, Northern Ireland, UK

doi: 10:10.1680/mohs.40564.0113

CONTENTS

The field of procurement is a complex yet interesting aspect of the construction process. If you get it right then the project is set on track for smooth delivery. Get it wrong and there is heartache and strife almost guaranteed before the contract ever commences. As a competent engineer you will play a vital role, alongside other construction professionals including quantity surveyors to advise and guide your Client towards the best style of contract and best procurement method to suit the construction project. In the years since Michael Latham and John Egan reported on the state of the construction industry, there has been a much greater emphasis on the need for a partnering approach to construction procurement. Whether this has yet been achieved is still a hotly debated topic and one that you can make your own mind up about as your own experiences expand. Needless to say all of these issues are explored within this chapter, with a particular reference to health and safety and how it integrates into the wider procurement process.

Box 1   Key learning points

In this chapter you will be introduced to:

■ Health and Safety Executive's take on the role of the Client.

■ Partnering and what it means for Clients.

■ Public sector approaches to procurement.

■ Health and Safety pre-qualification processes.

■ Your role as a professional advisor.

Introduction

On the face of it, construction procurement is a straightforward process. A Client wants a building or a structure, seeks bids from prospective contractors and selects the most economically viable bidder to carry out the work. But life in the construction industry is not quite so simple: there are statutory obligations on a Client to ensure that the contractors and for that matter all professionals they engage are competent, appropriately informed and adequately resourced, and it is that requirement that has taxed many of us in the industry over many years. Strangely enough, you might not be surprised to hear that after all these years of discussing and debating we still haven't reached a consensus. I fully expect that this will be the case for some time to come. Procurement as discussed here is from the point where the designs are completed through to the commissioning and hand-over of the structure, taking account of the financial and contractual relationships.

Now in the course of this chapter I will refer to various restraints placed upon and opportunities provided by the public sector – since the public sector is the largest client of the United Kingdom (UK) construction industry. However, the purpose of the chapter is to discuss the principles behind the competency and resources requirements as they affect health and safety in construction procurement, across all sectors and to present a workable solution that meets the needs and aspirations of all the key Duty Holders. In short, how the procurement aspect of the Construction (Design and Management) Regulations 2007 (CDM 2007) requirements can be demonstrably met by a Client. CDM 2007 in Northern Ireland is known as the Construction (Design and Management) Regulations (Northern Ireland) 2007 and is broadly similar to the Great Britain version. Both jurisdictions use the same Approved Code of Practice (HSE, 2007a). The CDM 2007 requirements on a client are that they are not permitted to:

> … appoint or engage a CDM co-ordinator, designer, principal contractor or contractor unless he has taken reasonable steps to ensure that the person to be appointed or engaged is competent …

Elsewhere in this book you can read about the responsibilities of the Key Duty holders in construction design and management (see Chapter 3 *Responsibilities of key duty holders in construction and design management*).

Now you may know that CDM 2007 requirements have their background in European Directives, which in essence means that there will be similarities in what is discussed here with the requirements in many other European countries. In addition, public sector clients are governed by public contracts legislation, which requires them '… to treat economic operators [contractors] equally and without discrimination and to act in a transparent and proportionate manner …' Legislation or not, I would expect this of any client, and any professional engineer engaged by a client to manage the procurement process will similarly be duty bound by their professional code of ethics.

Health and Safety Executive (HSE) UK in their leaflet *Want Construction Work Done Safely?* (HSE, 2007b) advising Clients on the introduction of CDM 2007 stated that:

> As a client, you have a big influence over how the work is done … CDM 2007 is not about creating unnecessary and unhelpful processes and paperwork. It is about choosing a competent team and helping them to work safely and efficiently together. Give them enough time and resource and you will get the building you want, when you want it and on budget.

In that regard, Contractors must have been able to satisfy the Client that they have the resources and competence to manage health and safety in compliance with CDM 2007, before being invited to bid. That introduces the concept of pre-qualification, which is discussed below.

As the chapter progresses we will examine some of the reasons for inclusion/exclusion set out in the legislation and where we will major is on professional and technical competence in the field of health and safety.

Partnering

Latham (1994) advocated that the client be at the core of the construction process. The way to achieve client satisfaction, he suggested, was through team work and cooperation in particular; partnering, where the ranges of expert opinion and advice come together to ensure that project risks are identified as early as possible by all the key participants and addressed collectively. This arrangement would continue throughout the life of the construction process. This theme was taken up by Sir John Egan in 1998 in his report *Rethinking Construction* where he showed that partnering leads to effective and efficient projects. Even now there are varying views as to the value of partnering, despite the development of specific partnering contract documentation. The process has been endorsed by the public sector and is evident in many large scale public contracts, however it is not as widespread or respected as either Latham or Egan had envisaged.

In the Foreword to the 2001 National Audit Office report, *Modernising Construction*, Sir Michael Latham continues to promote partnering, stating:

> It looks for reasonable margins built up by the whole team on an open book basis. All are signed up to mutual objectives through a charter for the project. All agree on effective decision making procedures. Problems are to be resolved collaboratively by the entire team, not shoved off onto those least able to cope with them. Continuous improvement and benchmarking are crucial. Partnering can be for a specific project or on a longer term strategic basis. It can achieve real cost savings and client satisfaction.

Commitments and initiatives

Many initiatives and directives have emerged over the years, which, while not always primarily driven by health and safety needs, have had profound effects on how health and safety is managed within the construction industry. Latham (1994)

and Egan (1998) examined relationships between Client and contractors and advocated a less adversarial approach to project procurement: 'Partnering'. Indeed it is widely held that the New Engineering Contract (NEC) documents were born out of this desire.

Within public sector procurement, the Office for Government Commerce (OGC, 2007), reacting to Latham and Egan, developed and continues to promote its *Achieving Excellence in Construction Procurement Guide* documents as best practice in public procurement. Their Achieving Excellence Guide Number 10 deals directly with the integration of health and safety into the procurement process, and, although dealing directly with public sector requirements and incorporating restraints that are unique to public procurement policy, provides information that could be readily adapted to the private sector.[1]

While the HSE UK's consultation document *Revitalising Health and Safety in Construction* (2003) discusses integrated teams, referred to Achieving Excellence Guide Number 10 and highlighted the work of the Strategic Forum (2009), its recognition of the value of Client-led health and safety initiatives within the procurement process was more implicit than explicit. That said, HSE UK did bring forward (as promised) new CDM legislation in 2007, which, although not fundamentally different from the earlier 1994 version, does place a greater emphasis on the Client's statutory obligations to procure competent and adequately resourced designers, contractors and coordinators.

The OGC (2005) produced common minimum standards for the procurement of built environments in the public sector and one area of note is given in **Table 1**.

In November 2004, leading bodies from across Europe signed the Bilbao Declaration at the European Construction Safety Summit, committing them to specific measures to improve the sector's safety and health standards, including; 'Integrating health and safety standards into procurement policies …'

The declaration expressly states that:

> … safety and health problems encountered during construction and operation could be avoided by ensuring that due consideration is given to these issues during the design and procurement process. (European Agency for Safety and Health at Work, 2004)

A major section of the Bilbao Declaration is devoted to discussing the links between health and safety and procurement, and in particular it gives praise to the UK for its public sector procurement guidelines.

The Strategic Forum for Construction [Great Britain] in its Targets 2012 (2009) document set down headline targets in, as they see it, the six key areas vital to delivering construction projects on time, safely and to budget. They represent the principles which it is intended will underpin all construction projects in order to achieve a better industry and exceed current

---

[1] It is not my intention to offer a detailed critique of existing documents. As a professional engineer you will be able to assimilate the discussions within this chapter and form your own views on the usefulness of existing publications.

Standard	Background	Further information
Pre-qualification and tendering processes should be appropriate for the project, meeting legal obligations and avoiding unnecessary bureaucracy and costs for suppliers	Construction firms incur significant nugatory costs from pre-qualifying to bespoke buyer formats, which can themselves go beyond legal requirements for the pre-qualification phase of procurement. Consistent use by the public sector of a single national pre-qualification database was a recommendation of Sir Michael Latham's 1994 report *Constructing the Team*. It remains a strong construction industry desire	Clients can move to meet this aspiration by making use of 'Constructionline', a government-owned, pre-qualification tool, a central repository of current, accurate, core data that mitigates for suppliers and buyers the expense of providing or soliciting commonly required information. Constructionline is being developed to support emerging best practice in construction procurement, within the legislative and regulatory procurement framework. It displays data on firms' health and safety performance and processes, and on workforce skills. It also helps suppliers seeking to form integrated teams to identify suitable partners: http://www.constructionline.co.uk

Table 1 Extract from Office of Government Commerce – Common Minimum Standards (OGC, 2006)
© Crown Copyright reproduced with permission from Her Majesty's Stationery Office.

best practice. In the introduction to the first of the six targets it states that:

A successful procurement policy requires ethical sourcing, enables best value to be achieved and encourages the early involvement of the supply chain. An integrated project team works together to achieve the best possible solution in terms of design, buildability, environmental performance and sustainable development.

One of its other headline targets states that health and safety is integral to the success of any project, from design and construction to subsequent operation and maintenance and sets a 2010 target to:

Reduce the incidence rate of fatal and major injury accidents by 10% year on year from 2000 levels.

And a 2012 target for a:

10% reduction year on year in the incidence rate of fatal and major injuries from 2010 levels.

Setting aside the fact that the targets are nowhere near to being achieved, it is important to note that work of bodies such as the Strategic Forum needs to continue and perhaps heighten its push towards better health and safety standards across the entire construction industry. Modern construction can't just be about profiting from partnering. It must be about delivering projects, within time and budget that do not adversely affect the health, safety and welfare of all those involved in its successful delivery. It is also worth remembering that the Institution of Civil Engineers' (ICE) code of conduct (2008) requires that:

All members shall have full regard for the public interest, particularly in relation to matters of health and safety, and in relation to the well-being of future generations.

Advice is given to supplement this conduct rule, which includes the following:

Producing competitive bids should not result in the inappropriate exposure to hazard of any person at any time …

So regardless of what initiatives come and go there is a long standing commitment from our profession to ensure that health and safety is central to our competence as engineers. So remember this the next occasion you are called upon to devise a competence assessment process for a procurement competition or to assist Clients in determining the competence of their appointees.

First principles

The question is how can Clients and contractors work together to meet the objectives set down by Latham and Egan to procure construction projects that are appropriately resourced and safely executed and in so doing meet the requirements of European driven health and safety and procurement legislation. The chapter will consider:

1 What role do professional engineers play in ensuring the consistent delivery of this objective?
2 Is partnering working and what future has it?

Let us consider first principles for a moment. As a minimum in all construction related procurement competitions, it is a necessity for construction Clients to test all applicants' ability to fulfil their health and safety responsibilities. What is going on with construction procurement is a transaction between two parties; a contract. And in that transaction the first party (the Client) agrees to pay an agreed amount for an item constructed by the second party (the Contractor). What the Client has to establish is whether the Contractor is fit to deliver on the end of the contract, prior to engaging them. In other words do they have the capability and the capacity?

Fundamentally there are two things a prospective Client needs to check:

1 Does the contractor have the necessary competence?
2 Can the contractor provide satisfactory and relevant references from previous clients?

This is the essence of any transaction and if you were to analyse the health and safety legislation and associated codes of practice and industry guidance in this regard that is what it all boils down to. I would go further and suggest that public contract legislation has the same core focus.

In a court room the criminal trial sets out to prove beyond all reasonable doubt while the civil trial works on the balance of probabilities. In establishing whether prospective bidders have proven that they have the competence and resources, you should be working on the balance of probabilities. To strive to prove competence beyond all reasonable doubt sets you on the path of developing extremely restrictive and thoroughly bureaucratic processes, which inevitably will seriously impact on your or your Client's ability to award contracts.

Meeting the procurement challenge

HSE UK in promoting CDM 2007 to construction Clients stated, with regard to the appointment of the right people:

> You are more likely to get what you need if you make sure those who design and build are competent have sufficient resources and are appointed early enough, so the work can be carried out safely. (HSE, 2007b)

Your challenge as a professional engineer appointed to assist the Client is to ensure that the Client's statutory obligations are fully met, while engaging the contractor or other professional, such as the CDM coordinator, in meaningful health and safety scrutiny. Health and safety competence submissions need to describe how the individuals or organisations have developed their health and safety management arrangements and to provide the Client with evidence that its health and safety management arrangements were complied with in previous relevant projects.

The Approved Code of Practice[2] (ACoP) (HSE, 2007a) associated with CDM 2007 states that organisations that are bidding for work should put together a package of information that shows how their own policy, organisation and arrangements meet the standards. And the standards are defined within the ACoP, in what is referred to as a set of 'core criteria'. These criteria, agreed by industry and the HSE, have been put forward to ensure that there is consistency in the manner in which companies are assessed. Now you may be aware that it has been discussed in various forums that there are concerns and difficulties experienced, particularly among public sector clients as to how consistency of approach can marry up with the European public procurement legislative requirements for equal treatment of prospective bidders, particularly those bidding from other European countries. The Public Contract Regulations 2006 state that:

A contracting authority shall

(a) treat economic operators equally and in a non-discriminatory way; and

(b) act in a transparent way.

Here we have a dilemma that is more imagined than real. Let me explain. Equal treatment does not mean treat everyone exactly the same. Equal treatment means that we do not disadvantage any prospective bidders because of their circumstances; for example as a Client it would be wrong to insist that all bidders held a particular qualification, available only in Great Britain, when the real issue is whether the bidder is competent in a certain field. Equal treatment would be to define the standards of competence required and ask each bidder to demonstrate how it meets that standard. Now for you as the assessor this might mean a bit more work as you may have to analyse a diverse range of evidence to reach a meaningful and correct determination as to the adequacy of any prospective bidders' competency submission. That is your responsibility and that of your Client. It is worth remembering that in assessing adequacy there is always a degree of subjectivity; that is, expert opinion based on your own experiences and competency, rather than a biased view. Therefore, if you are unaware of some aspects of health and safety competence measures then the bidder cannot be penalised. It is for you to ensure that you are able to deliver a correct and competent verdict on this through appropriate research into the evidence presented, or if this is outside your gift then you must hand the task to someone who can. That is treating all prospective bidders in an equal and non-discriminatory way. Plus do not forget that CDM 2007 also requires of you as an individual to be sure that you are competent to deliver any commission you accept. Ethically you can do no less.

Referring back to the Bilbao Proclamation, it is worth your while being aware of the following, particularly if you are involved with public sector procurement. The EU Directive 2004/18/EC ... Article 27 requires:

> A contracting authority ... shall request ... candidates in the contract award procedure to indicate that they have taken account ... of the obligations relating to employment protection provisions and the working conditions which are in force in the place where the works are to be carried out ...

The Public Contracts Regulations 2006 implement, for England, Wales and Northern Ireland, Directive 2004/18/EC of the European Parliament and Council. Now since the employment protection provisions in the UK construction sector, with particular regard to health, safety and welfare, are laid down in CDM 2007, there can be no doubt that the two pieces of legislation don't clash. Rather they complement each other and in that regard should make for a smoother procurement process.

On the question of transparency, this simply requires that you state, up front, what it is you expect prospective bidders to provide as demonstrable evidence of their health and safety competence and capability and equally make it clear what criteria you will use to judge the adequacy of that evidence.

---

[2] You may well be reading this in a jurisdiction other than the UK, where the CDM 2007 or its associated ACoP does not apply. In such circumstances look for the equivalent documentation from your local jurisdiction and if none exists then it is worth considering the merits of the UK CDM 2007 as current best practice.

Prove it ...

Pre-qualification in procurement is an area where many minds have battled and many disputes have arisen over the years. As I progress in this field my own view has changed and will probably continue to, as will yours as time goes by. After all, what is continuous improvement without re-examination of what you are doing to achieve that ultimate health and safety aim where no one is killed or seriously injured in the construction of or use of our structures? In that regard pre-qualification questions asked of contractors over the years have ranged from a small number (say eight to ten) to 40-plus questions, such as those published on the OGC's website.

It is worth bearing in mind that clients are in a strong position to use their procurement processes to instil a positive health and safety attitude across the industry. Once again I will refer you to the public contract statutory obligations, simply because they are a good way for any Client to behave, public or private. The Public Contract Regulations 2006 Clause 30 (2) state that:

A contracting authority shall use criteria linked to the subject matter of the contract to determine that an offer is the most economically advantageous including quality, price, technical merit, aesthetic and functional characteristics, environmental characteristics, running costs, cost effectiveness, after sales service, technical assistance, delivery date and delivery period and period of completion.

Remember that there is the dual requirement of transparency and equal treatment, so, in addressing the requirement to determine the most equally advantageous bid, you must specify what it is that you want to see from the Contractor, with regard to health and safety. Now if you re-read Clause 30 (2) from the Public Contract Regulations again you will not see a mention of health and safety in the list for determining economically advantageous bids, but that does not mean it is not to be considered, as indeed has been argued in the past. Rather you have to consider it as an aspect of technical merit, since bidders' professional and technical ability could not be judged as anything other than inadequate if health and safety wasn't integral to how they carried out their business. Remember also that EU Directive 2004/18/EC Article 27 refers to the need to determine how employment protection provisions and the working conditions are addressed by prospective bidders.

So back to what to ask in pre-qualification and how to assess the adequacy of the evidence presented. Now I have indicated that pre-qualification question sets have existed for some time and have consisted of many questions. I now am advocating that one question is sufficient:

Prove how your organisation has met all of its obligations under the Construction (Design and Management) Regulations on recent similar type construction projects.

(The judging criteria: The minimum standard for health and safety, which a prospective bidder must produce is that which is set out in Column 2, Appendix 4 of the CDM 2007 ACoP.)

This puts an onus on the Client to determine what evidence it will seek to establish whether the Contractor can demonstrate competence, based on the guidance in CDM ACoP Appendix 4 and that is a key component of your job. I would like to emphasise to you at this point that prevention is the priority and your responsibility is to develop a pre-qualification approach that embodies that principle, while at the same time allows you to judge the complete package submitted. In other words health and safety should be an integral aspect of the overall judgement of prospective bidders' competence (i.e. professional and technical competence alongside financial capability). A successful submission must demonstrate that the Contractor can construct a safe, healthy and effective structure.

Now when you ask the prospective bidders to 'prove it' you then have to have the competence to assess what is presented to you in order to reach a point where you are satisfied that the bidder has in fact proved, within the balance of probabilities, that they are competent to execute the construction project. That is to say, enough evidence has been presented. And what I am getting at here is the sufficiency of the evidence, not the existence of every single piece of evidence. Let me illustrate. If a burglar is caught on camera breaking into and robbing your house would you need to see camera footage of several other robberies he had carried out before you could deem him a thief? Probably not, but would that be enough to say he had been responsible for every burglary in the district? Not without some additional proof. Sufficiency of evidence! Now I did say balance of probabilities was the key, so think carefully how much evidence would help you determine that the various core criteria were adequately covered. And for that matter could a single piece of evidence cross a range or criterion?

When we discussed the equal treatment and determined that that does not mean treating everyone the same, rather that you do not discriminate against bidders, whose competency evidence base is different. Here then lies a challenge for you. To understand what is required of you under CDM 2007 and the associated ACoP, I say this here and stress the point because on many occasions I have come across a lot of people who look to the evidence examples given in Column 3 of Appendix 4 of the ACoP and see them as absolutes. If you are not careful and take this approach, you could end up with the allegations of restrictive practice and/or barriers to trade, since the examples given are often Great Britain attainable and, particularly in the European context, this would not be an acceptable practice. In addition this has led to specific trade cards and qualifications (named only as examples) being hyped as the legal requirement, when in many instances they may be wholly inappropriate for the type of construction project you are procuring. So, for your own sake and for the sake of accuracy, focus on Columns 1 and 2 of Appendix 4 and consider making your own Column 3, based on the type of construction project you are embarking upon. This could be a one-off or a series of projects, but in either case the work done to develop your own project specific examples will make the assessment process much smoother and is more likely to get you a range of bidders, any one of which you would be happy to employ to construct your project.

Getting back to considering the sufficiency of and the appropriate types of evidence, it is important in this regard to have a degree of flexibility. The thing to remember is that the requirement is to identify the evidence that is required and not to be too prescriptive in the manner in which it is to be presented. Bear in mind also that some evidence has a 'shelf life' therefore it is reasonable to look for demonstrable evidence from recent similar projects. Equally it is not unreasonable to use prior knowledge. Let's assume that a prospective bidder has been working for you on another recent similar construction project and as you monitored performance on that contract there was no appreciable health and safety non-conformance. Surely then it is reasonable for that evidence to be incorporated into the adequacy assessment of the procurement exercise you are currently engaged in.

Also let us assume that a Contractor has recently had his health and safety management system independently audited. Is it reasonable to accept that this will tick the evidence boxes for quite a few of the core criteria set out in the CDM 2007 ACoP? In that regard I would be inclined to say yes with certain provisos, such as:

1 Was the audit company competent?
2 Was the audit construction focused and did it include on-site compliance?
3 Was the audit to a recognised management specification; for example (but not exclusively) ILO OSH 2001 or BS OHSAS 18001?

If you are willing to accept an independent audit then you must be prepared to and be capable of scrutinising the auditor's report and the associated management action plan to ascertain the adequacy of response, with regard to whether you would be prepared to deem the evidence as acceptable. You will also have to determine and make clear in the pre-qualification prospectus how critical this audit is and how far it could go to satisfy the minimum standards.

Finally to discuss flexibility for you as the assessor and for the prospective bidder, having to address competence across a range of professional and technical disciplines, such as health and safety, quality and environmental, there is an opportunity to allow a portfolio of competency evidence to be produced together with a cross-reference matrix, indicating how each piece of evidence enclosed demonstrates competence in the particular disciplines. In that way you are acknowledging that health and safety is integral to the overall management and execution of the project. There is no reason why you couldn't weight each of the aspects according to how important you judge them to be, remembering of course to make this known to the prospective bidders.

Public-private partnerships/ private finance initiatives

It would be remiss of me not to briefly discuss public-private partnerships, or private finance initiatives as they are also known, since they have been around construction projects for a long time and, as their name suggests, they are a means of procurement practised within the public sector ostensibly where more traditional means of project funding are not available. In other words it is like purchasing a major structure or building, such as a road, school or hospital on credit, to be paid for over a substantial period (e.g. 25 years), where the private sector partner gets paid to maintain the structure during that period. Other terms used for the same process are: design, build, finance and operate (DBFO) and build own operate and transfer (BOOT).

From a health and safety perspective the Client will still have all the same obligations during project procurement to ensure that competent Designers and Contractors are engaged; however, where it differs is that once the contract is in place the CDM Client responsibilities are transferred to the DBFO company. HSE, in the Approved Code of Practice for the CDM Regulations, refers to this arrangement as a 'Special Purpose Vehicle' and they caution participants that if they do not set it up right the public sector could remain holding some of the duties that should otherwise have been transferred. However, that is not to say that the public sector Client is to be absolved of all duties and obligations, since as it pays for the project it has a role in its successful outcome. Additionally, the public sector Client will be using the structure or building in the intervening period and may well engage in some alterations, outwith the PPP contract. Therefore it is essential in the establishing of PPP schemes that a Relationships Document is prepared and signed up to by all parties, setting out all the roles and responsibilities on the project and the means for resolving any disputes. This is particularly important with regard to health and safety but holds just as true for every other management facet.

It is an extremely complex process, requiring vast amounts of expertise and experience, so my best advice is to engage the best advisors to assist in the process.

Summary of main points

The CDM 2007 places responsibilities on both Clients and those professionals they employ to deliver their construction projects. Clients are required to ensure that the Contractors they employ have the resources and competence to manage health and safety in compliance with CDM 2007. As a minimum, in all construction related procurement competitions, it is therefore necessary to test all prospective bidders' ability to fulfil their responsibilities under CDM 2007. This is not a new requirement or new feature in procurement, introduced with the recent legislative changes, but what is new is the emphasis the HSE has put on Clients to make improvements in construction health and safety. This requires that you, as a professional advisor to the Client, have to be capable of making a judgement as to whether the evidence provided by prospective bidders demonstrates that they meet the standard to be achieved.

References

Egan, Sir John. *Rethinking Construction. The Report of the Construction Task Force to the Deputy Prime Minister, on the Scope for Improving the Quality and Efficiency of UK Construction*, 1998, London: Department of Trade and Industry.

European Agency for Safety and Health at Work. *Bilbao Declaration – Building in Safety*, European Construction Safety Summit, Bilbao, November 2004. Available online at: http://osha.europa.eu/en/publications/other/20041122

Health and Safety Executive (HSE). *Revitalising Health and Safety in Construction. Discussion Document*, 2003, London: HSE. Available online at: http://www.hse.gov.uk/consult/disdocs/dde20.pdf

Health and Safety Executive (HSE). *Managing Health and Safety in Construction. Construction (Design and Management) Regulations 2007 Approved Code of Practice*, 2007a, London: HSE Books. Available online at: http://books.hse.gov.uk/hse/public/saleproduct.jsf?catalogueCode=9780717662234

Health and Safety Executive (HSE). *Want Construction Work Done Safely? A Quick Guide for Clients on the Construction (Design and Management) Regulations 2007*, 2007b, London: HSE.

Institution of Civil Engineers (ICE). *ICE Code of Professional Conduct, 2008*, London: ICE. Available online at: http://www.ice.org.uk/myice/myice_council_rules.asp

International Labour Office (ILO). *Guidelines on Occupational Safety and Health Management Systems*, ILO OSH 2001, Geneva: ILO. Available online at: http://www.ilo.org/public/english/protection/safework/cops/english/download/e000013.pdf

Latham, Sir Michael. *Constructing the Team. Joint Review of Procurement and Contractual Arrangements in the United Kingdom Final Report*, 1994, London: HMSO.

National Audit Office. *Modernising Construction. Report by the Comptroller and Auditor General HC 87 Session 2000–2001: 11 January 2001*, 2001, London: The Stationery Office. Available online at: http://www.nao.org.uk/publications/0001/modernising_construction.aspx

Office of Government Commerce. *Common Minimum Standards for the Procurement of Built Environments in the Public Sector*, 2005, London: OGC. Available online at: http://www.ogc.gov.uk/construction_procurement_common_minimum_standards_for_the_built_environment.asp

Office of Government Commerce. *Health and Safety. Achieving Excellence in Construction Procurement Guide 10*, 2007, London: OGC. Available online at: http://www.ogc.gov.uk/documents/CP0070AEGuide10.pdf

Strategic Forum for Construction. *Strategic Forum Targets 2012*, 2009. Available online at: http://www.strategicforum.org.uk/targets.shtml

Referenced legislation and standards

BS OHSAS 18001: 2007 Occupational Health and Safety Management Systems, London: British Standards Institution.

Construction (Design and Management) Regulations 2007 Reprinted March 2007. Statutory Instruments 320 2007, London: The Stationery Office.

Construction (Design and Management) Regulations (Northern Ireland) 2007. Statutory rules of Northern Ireland 291 2007, London: The Stationery Office.

European Parliament and the Council of the European Union. Directive 2004/18/EC of the European Parliament and of the Council of 31 March 2004 on the Coordination of Procedures for the Award of Public Works Contracts, Public Supply Contracts and Public Service Contracts. *Official Journal of the European Union* 2005, L **134**(30.4.2004), 114–240. Full details and amendments available online at: http://eur-lex.europa.eu/LexUriServ/LexUriServ.do?uri=CELEX:32004L0018:EN:NOT

Public Contracts Regulations 2006. Statutory Instruments 5 2006, London: The Stationery Office (for England, Wales and Northern Ireland).

Public Contracts (Scotland) Regulations 2006. Scottish statutory instruments 1 2006, London: The Stationery Office.

Further reading

Office for Government Commerce. Achieving Excellence in Construction Procurement Guide documents. Available online at: http://www.ogc.gov.uk/ppm_documents_construction.asp

Websites

Health and Safety Executive (HSE) http://www.hse.gov.uk

Institution of Civil Engineers (ICE) Health and safety http://www.ice.org.uk/knowledge/specialist_health.asp

NEC contracts http://www.neccontract.com/

Office for Government Commerce (OGC) http://www.ogc.gov.uk

Strategic Forum for Construction http://www.strategicforum.org.uk

Section 4: Health hazards

ice | manuals

Chapter 11

Controlling exposure to chemical hazards

Philip McAleenan Expert Ease International, Downpatrick, Northern Ireland, UK

doi: 10:10.1680/mohs.40564.0123

CONTENTS

Chemicals and hazardous substances are found throughout construction and demolition sites. Some are provided to allow the work activity to take place, while others are generated in the process of conducting the activity. Other substances are contained within the building materials and are released when they are worked on. The control of chemical and hazardous substances begins at the earliest stages of the design process and continues during the pre-construction phase. On site, project managers ensure that the health effects of exposure to chemicals and the necessary controls are communicated and appropriate, and that well maintained equipment is provided to workers.

This chapter covers the health effects of exposure to a number of different types of hazardous substances commonly found on construction sites, how they enter the body and which organs are likely to be affected. It covers a range of controls from engineering, to safe working methods and the use of personal protective equipment (PPE) and towards the end summarises some of the key legislation and forthcoming changes that apply to chemical safety. It cannot cover the full range of chemicals and hazardous substances that will be found on construction sites – these number in their thousands – but it does provide guidance on identifying and seeking relevant information on them.

Box 1 Key learning points

Readers will understand the hazards and risks posed by chemicals and other substances found on construction and demolition sites, know the function of safety labels and materials safety data sheets, and be able to assess and develop controls for the safe use and storage of substances.

Introduction

Working with chemicals and other hazardous substances, no matter how innocuous they may appear to be, can and does pose serious risks to health with consequences that are often fatal, in some cases many years after the initial exposure, e.g. asbestos which is responsible for 100 000 deaths worldwide annually (Jukka Takala, International Labour Organisation (ILO), quoted in *Hazards Magazine* 2010) of which some 4000 (cancer and mesothelioma) are in the United Kingdom (UK) (HSE, 2010).

Other chemicals can create health conditions that may make it necessary to change the type of work that a person does, for example exposure to a sensitising agent that causes dermatitis means that any task which involves exposure to that agent, even in tiny quantities, will bring about the allergic reaction thus rendering the job unsuitable for the affected worker. Sometimes the nature of the hazardous substance affects the lungs, leading to chronic obstructive pulmonary disease (COPD) a condition that in its worst form makes any attempt to move extremely laborious and the effected person may spend the rest of their days housebound.

Chemicals also impact on the environment, adversely affecting soil, watercourses, plant and animal life. Spillages, waste disposal and uncontrolled dispersal of hazardous dusts and fibrous materials can have immediate and long term consequences, with some contaminants remaining active for decades, even centuries, whilst heavy metal contamination continues to be a source of harm indefinitely; in 2005 excavations for Channel 4's 'Big Roman Dig' at Charterhouse near Bristol were halted due to soil contamination from lead processing 1900 years previously, (Ashtead Technology, undated), (see Chapter 5 *The different phases in construction – design in health and safety to the project life cycle* for more information on environmental considerations).

Chemicals and other hazardous substances are present throughout all construction sites. They are introduced by the process of construction and many different types of work activities involve the use of or contact with them. Additionally, some projects come into contact with pre-existing hazardous substances such as asbestos in old buildings, bio-hazardous substances such as bird droppings, fungi, rat urine and old contaminants in the soil that are exposed when foundations and drains are opened. This chapter will examine some of the most common types of hazardous materials that construction workers are exposed to, how they enter and affect the body, their impact on the environment, the legal requirements to control exposure and some details on those controls.

Where hazardous substances are found

Chemical and hazardous substances are found throughout construction and demolition sites. They exist as discrete substances such as solvents, asphalt and cement, or are contained within products and released on use, such as fumes from soldering and burning processes and silica dusts from cutting concrete and stone. They may exist already within the site prior to work commencing as soil and ground contaminants, or may be a by-product of construction processes.

They will be found in work areas, storage facilities and in the site offices. They include gases, vapours, fumes, liquids, dusts and solids, and will therefore be found in hollows and dips on the ground and floors, in wells and traps, in ceiling voids and in the atmosphere.

They are used for a variety of purposes, from concreting and bricklaying, surfacing, cleaning and degreasing, welding and soldering, painting and removal of paint, insulation, sheeting and panelling, sandblasting, maintenance and repair work, and sampling and testing to name but a few processes where they will be used or where they will be a by-product of the process. **Table 1** lists some of the most common chemicals and hazardous substances found on construction sites, each of which will be described further in the section on main types of hazardous substance. But remember this is a non-exclusive list and your own investigations or that of the Designers should show specific information for your site. Check the pre-construction information.

Table 1 is a synopsis of commonly found substances on construction sites and is neither definitive in listing hazardous substances nor (in conjunction with **Table 2**) in describing the detail of effects and controls. Sources of expert advice are detailed in the 'References'.

How chemicals and hazardous substances enter the body

There are a number of routes of entry into the body and, for the unprotected worker, chemicals will enter in one or more of the following ways:

- inhalation
- ingestion
- absorption through the skin
- by way of hair follicles, sweat and sebaceous glands, and
- by way of the placenta to unborn children.

Some substances can penetrate the membrane of the cell wall and will affect unborn generations through mutagenic (i.e. causing genetic changes) and teratogenic (i.e. causing birth defects from prenatal exposure) processes. The former may also be carcinogenic to the exposed worker and the latter may have no adverse health effects on the exposed worker and may therefore not be picked up until the birth of children.

Mutagens include benzene, found in solvents and synthetic materials, and ionising radiations used for testing metals. Examples of teratogenic chemicals are dioxins, chemical by-products from burning organic materials in the presence of chlorine, e.g. in waste incinerators, or they are naturally occurring in the ground.

Inhalation

Inhalation occurs when substances are breathed in through either the nose or the mouth. Unprotected inhalation of dusts will bring potentially dangerous particles into the respiratory system and, depending on the size of the particles, the inhaled substances may:

- be filtered off by the nasal hairs or deposited in the upper respiratory tract to be spat, coughed or sneezed out of the body (greater than 10 μm in diameter);
- settle in the mucus covering the bronchi and bronchioles and are then wafted up by tiny hairs towards the throat, (5–10 μm in diameter); or
- reach and settle in the lung tissue (>5 μm in diameter).

By way of illustration, a strand of human hair is about 100 μm wide, while a cell is 8 μm in diameter.

Fibres such as asbestos, which predispose to disease, have a length to diameter ratio of 3:1 with a diameter of 3 μm or less; the longer the fibre the more damaging to the lungs it will be.

Particles that can be coughed up again are not necessarily expelled from the body but may re-enter by way of ingestion.

The respiratory system's function is to allow gas exchange to all parts of the body. The inhalation of gases, vapours and fumes means that they will be able to pass through the walls of the lungs into the bloodstream and be carried throughout the body, affecting other organs. These gases may be toxic and immediately harmful, possibly fatal, although other toxic gases will have a longer term accumulative effect or, when in sufficient quantities, be asphyxiating. Some will react with the moisture in the lungs and cause ulceration, and possibly pulmonary oedema, i.e. a build-up of fluid in the lungs. Other gases will react with the kidneys, the nervous system, and the reproductive organs causing a variety of short and long term health problems.

- Benzene (found in solvents) is immediately fatal when exposed to high concentrations, whereas low concentrations will affect breathing, brain function, heart rate and cause unconsciousness. Chronic (long term) exposure will affect the bone marrow and decrease red blood cell production, depress the immune system and cause leukaemia.
- Welding produces carbon dioxide (CO_2), an asphyxiant, ozone (O_3) an irritant in the lungs, and manganese fume which at even

Substance (contains)	Source	Form	Notes
Asbestos	Old premises, furnaces, lagging on pipes	Solid, fibres and dust	Licensed contractors must be engaged when ACMs are to be worked on or removed Disposable PPE RPE Type H vacuum cleaner or wet rag removal of dust
Asphalt and bitumen (solvents)	Road surfaces, roofs	Solid emitting fume Liquid	Solvents used to 'cut' asphalt may contain highly toxic chemicals, e.g. benzene, dioxane, toluene, which are toxic, carcinogenic and allergenic
Cement and cement-containing products	Throughout construction sites	Powder Wet preparation	PPE Avoid skin contact Dust masks
Hexavalent chromium	Stainless steel, electroplated metals, wood preservation, cement	Released as vapour	LEV, PPE WEL 0.05 mg/m^{-3}
Lead	Paint Soldering	Solid when dry Liquid in paint Dust and vapour in removal processes	PPE RPE 50 µg/dl in blood is the action level (25 µg/dl for women and 40 µg/dl for young people)
Man-made mineral fibre, MMMF, e.g. fibreglass, rockwool	Lagging and thermal insulation	Solid, fibres and dust	PPE and RPE Fibre count max. MEL, 2 f/ml., 5 mg/m^{-3}
Silica dust	Found in quarry products	Solid, dust	Wet cutting and grinding PPE and RPE MEL 0.3 mg/m^3
Solvents	Cleaners, de-greasing processes Paints and paint removers	Liquid Vapour	Good ventilation (LEV) PPE RPE
Wood and wood dusts (preservatives)	Throughout the industry	Solid Explosive as a dust	Ventilation, LEV Wet sweeping WEL of 5mg/m^3

Note: ACMs – asbestos containing materials; COPD – chronic obstructive pulmonary disease; f/ml – fibres per millilitre; LEV – local exhaust ventilation; MEL – maximum exposure level; mg/m^{-3} – milligrams per cubic metre; PPE – personal protective equipment; RPE – respiratory protective equipment; WEL – workplace exposure level; µg/dl – micrograms per decilitre

Table 1 Common hazardous substances in construction

low levels (e.g. >0.2 mg/m^3) will cause neurological damage to the central nervous system (CNS), lungs, kidneys and liver.

Ingestion

Substances entering the body by way of the digestion system will, depending upon their nature, enter the bloodstream via the stomach (as with alcohol) or through the large intestine (as with water). Whether or not the substance passes through the body will depend upon its solubility, particle size and physical state.

Soluble substances will pass through to the bloodstream and then onto the liver, which attempts to render toxins less toxic before excretion. Large amounts of toxins, or toxins that react in the liver to form other toxins, will damage the liver itself and may eventually destroy it.

The CNS is the part of the nervous system that includes the brain and spinal cord, and which coordinates the activity of all

parts of the body. Damage to the CNS will cause loss of function in parts of the body at and below the point damaged, and in the most severe cases the damage may be fatal. Although the CNS is protected against most toxins, some such as heavy metals are capable of penetrating and causing harm.

Harm to the CNS generally has more serious consequences than harm to the peripheral nervous system (PNS), i.e. those nerves that radiate from the CNS throughout the body. The PNS is more susceptible to damage that can cause pain, loss of sensation and even loss of muscular control. Quite often, damaged nerves in the PNS can regenerate but this is a slow process and affected persons may experience incapacities for a prolonged period of time.

Chemicals in the bloodstream of pregnant women will transfer through the placenta to the unborn child. Chemicals that are normally harmless to adults because they are in low and within permitted exposure concentrations or are deemed non-harmful, may be extremely harmful to the unborn child and for this reason employers need to be aware of and provide additional controls and protections for women employees of child-bearing age, and the corollary is that women who are pregnant or intending to become pregnant must inform their employers where exposure to chemicals is a real possibility.

As previously mentioned, the reaction of the body to the presence of foreign bodies in the respiratory system is to sneeze or cough to dislodge and expel the particles from the system. Coughing and then swallowing has the effect of transferring the offending particle to the digestion system and where dangerous particles such as asbestos are so ingested the fibre may lodge in the gut and stomach, creating the potential for stomach cancers in later years.

Many people inadvertently ingest hazardous substances through simple failure to properly wash after exposure, or when they go to eat, drink or have a smoke without removing contaminated clothing, or doing the same in a contaminated environment.

Removal of contaminated clothing after the work activity has ended, even temporarily, is essential if the contamination is not to be introduced to the wearer or to others who he comes into contact with. Carrying the contamination home exposes other family members and there are recorded cases of mesothelioma of partners of asbestos workers who came into contact with the fibres when handling and washing work clothing (Jeeves, 2006). Removal of work clothes must follow safe removal protocols in order to avoid transfer of contaminants to the body or other clothing.

Absorption

The skin is a protective organ that covers the whole body, but it is not an absolute barrier to the passage of harmful chemicals into the body. When damaged, for example when abraded or cut, or when chemical solvents have removed the protective oils in the skin, chemicals have easy and ready access to the bloodstream and from there are carried around the whole body, affecting the many organs described above. Entry for soluble chemicals and bacteria is possible through hair follicles, the sweat pores, and sebaceous glands and indeed the build-up of contaminated grime blocks the pores where the hazardous material is held in place for prolonged periods of time thus increasing the exposure and bringing it above the permitted workplace exposure level. This is particularly true of oily hydrocarbon substances which are carcinogenic. This problem may also be caused by the continual wearing of soiled overalls which again hold the substance in contact with the skin thus causing health problems in different bodily locations.

While the skin is generally regarded as a single organ of the body it is not uniformly the same throughout. It differs in density, thickness, sensitivity and proclivity to harm. The soles of the feet and palms of the hands are thicker because of the pressure applied in daily walking and use of the hands. They may therefore take more abuse than other parts of the skin which are not subject to the same stresses, but likewise the hands, because they take so much abuse are also at greater risk of harm through actual exposure to hazardous materials and because they are more likely to suffer cuts and abrasions.

The soft skin around the eyes is particularly vulnerable and may easily be damaged by rubbing, especially when working is dusty environments. The skin of the genital area is vulnerable to effects of carcinogens in soiled work clothing resulting from rubbing hands on the thighs of trousers or putting oily rags in pockets.

Young people are at greater risk from harm to the skin as, up to at least the age of 18, their skin has not fully developed to the adult stage.

Those who sweat a lot or have fair complexions are also more vulnerable to damage to the skin as are those with poor hygiene practices.

Effects of hazardous chemicals and substances on health

Table 2 provides an indication of some of the effects that exposure to hazardous substances can have on health. Not everyone will respond to exposure in the same way; some are naturally more susceptible than others and will have a reaction at a much lower level of exposure sometimes even within the permitted workplace exposure levels. Workers who have been sensitised to an allergen will often experience a reaction to minute quantities in the environment, and with repeated exposure the reaction becomes more rapid and severe. The likely types of reaction from exposure to hazardous substances include:

- Irritation of the nose and throat causing sneezing and coughing and possibly breathing difficulties. Long term effects include asthma and other allergies.

- Irritation of the skin and eyes causing rashes, dermatitis, burning and sore eyes.

- Loss of consciousness (and possibly death) as a result of being overcome by toxic fumes or ingestion of toxic substances.

- Infections from bacteria and other organisms.

- Long term effects, particularly after prolonged exposure, causing cancers and reproductive problems; however in some cases a single exposure may have the same result, as with asbestos, where the onset of health problems may not occur for between 15 and 60 years after exposure.

Respiratory diseases

Diseases of the respiratory tract and system will depend upon the nature and concentration of the substances inhaled, the duration and the rate of inhalation.

Irritation of the nose and throat may be due to the body's mechanism for removing particles from the tract, or due to a skin reaction to the substance inhaled (see below), or a combination of both.

Respirable dust is that dust in air which on inhalation can reach and may be retained in the lungs. This dust will cause a tissue reaction that will vary in nature and site depending upon the type of dust and its size. Some will have an immediate effect causing shortness of breath, coughing, possibly an injury to the lung and bleeding or build-up of other fluids (pulmonary oedema). The effects may wear off after removal from exposure, or after a few hours or days. Repeated exposure may increase the damage to the lungs, prolonging and worsening the effects and making recovery more and more difficult. Other substances have no discernible immediate effect but set in motion the development of severe, debilitating and fatal reactions weeks, months and even many years after exposure: mesothelioma may not manifest itself for 20–50 years after exposure to blue or brown asbestos, and silicosis brings on COPD, a narrowing of the airways with a reduced ability to breath comfortably and effectively.

Skin diseases

Dermatitis means an inflammation of the skin. The outer layer, the epidermis, has a protective function, thickening in parts of the body which are under constant stress and thus prone to injury (soles of the feet and palms of the hands) in order to afford extra protection. A moist layer secreted from the sweat and sebaceous glands covers the skin, helping to protect it from acids, alkalis, excessive water and to some degree from heat and friction by preventing the skin drying out. This natural grease can be removed by contact with solvents and repeated exposures will destroy the skin's ability to produce this grease.

Any part of the body may be affected by dermatitis but mostly it is on the hands, wrists and forearms where exposure is generally greatest. Damage follows from exposure to chemical and biological agents as well as physical agents. The most common form, irritant or contact dermatitis, results from contact with acids, cement, solvents, some metals and salts and most people will have an immediate reaction on contact. The severity and duration of the reaction will depend on the concentration and duration of exposure and repeated contact will worsen the reaction.

Sensitising dermatitis does not occur until the individual has been sensitised to a particular substance, i.e. the immune system becomes hypersensitive to an agent and the body reacts to the presence of that agent. This process involves an initial reaction in the blood that leads to dermatitis on subsequent exposure which may only require a very small amount of the substance. Sensitisers include chrome salts (paints and cement), nickel, cobalt, epoxy-plastics (paints and coatings), urea or phenolic resins, rubber additives, some woods and plants. Some substances will act as both irritant and sensitiser.

Sometimes medicines, contact with some chemicals such as dyes, wood preservatives, coal-tar and pitch, and contact with some plants can make the skin more sensitive to sunlight.

Skin cancers can occur when the skin is exposed to carcinogenic substances including sunlight, ionising radiations,

hydrocarbons and arsenic compounds. With regard to carcinogenic compounds the skin can become ingrained with the offending substance and in the absence of proper hygiene and regular washing cancers can develop, particularly where the skin is more sensitive. Substances on the hands can transfer to other parts of the body by contact and it is important to wash before as well as after visits to the toilet.

Within deeper layers of the skin the epidermis has pigment cells that produce the tan after exposure to sunlight. This protects the body from ultraviolet light which can cause acute and chronic health problems including burns, cancers such a melanomas and may compromise the immune system. Damage to these cells therefore leaves the skin unprotected from sunlight and ultraviolet radiation.

Toxins

Toxins are substances that cause harm by reactions with body tissues and their toxicity may be measured by their lethal dose (LD50) and lethal concentration (LC50). LD50 refers to the milligrams of toxin per kilogram of body weight (mg/kg) in a single dose that will be fatal for 50% of those exposed, whereas LC50 refers to the concentration of the inhaled substances that will have a 50% fatality rate within a stated period of time.

The effects of the toxins may be acute, i.e. the effects have a rapid onset and short duration, or may be chronic, i.e. gradual onset and prolonged effect. They may also be local, affecting only the point of contact or more generalised following absorption and distribution throughout the body by way of the bloodstream.

Toxins may act to compromise normal cell function, damage cell membranes, interfere with enzyme and immune systems, as well as RNA (ribonucleic acid) and DNA (deoxyribonucleic acid) activity. The pathological response may be irritant, corrosive, toxic, fibrotic (excessive collagen in the organ such as cirrhosis of the liver), allergic, asphyxiant, narcotic, anaesthetic or neoplastic (abnormal proliferation of cells causing tumour).

Organic materials

Decaying organic matter, plants and animals, produce by-products of decomposition that are toxic by inhalation and ingestion. They will be found in sewage systems, wells and sumps and in the soil, where the gases may be released when the ground is excavated.

Table 2 synopsises the effects of and routes of entry into the body of the common types of chemical and hazardous substances found on construction sites.

Main types of chemical hazard in construction

Having looked at the types of effects chemical hazards have on the human organism and how they may enter the body, we turn now to look in more detail at the more common types of chemical and hazardous substances illustrated in **Tables 1** and **2**.

Substance (contains)	Classification	Route to body	Organs affected	Effects
Asbestos	Carcinogen	Inhalation Ingestion	Lungs Digestion system	Mesothelioma (always fatal) Lung cancer (almost always fatal) Asbestosis (debilitating, possibly fatal) Diffuse pleural thickening (not fatal) No cure, 15–60 year lag between exposure and onset
Asphalt and bitumen (solvents)	Flammable Explosive Irritant Toxic (if containing hydrogen sulphide vapour)	Inhalation Contact	Skin, eyes and hair follicles Respiratory tract	Burns Rashes Skin cancers
Cement and cement-containing products	Sensitising Irritant Allergen (see Haxavalent chromium below)	Inhalation Absorption	Skin Respiratory system	(Alkaline) burns, possibly leading to amputation Dermatitis COPD
Hexavalent chromium	Carcinogen Sensitising irritant Allergen	Inhalation Respirable Absorption	Lungs Kidneys Intestines	Cancer Contact dermatitis, skin and nasal ulceration COPD
Lead	Toxic Women and youths under 18 are particularly at risk	Ingestion Inhalation	Stored in bone Kidneys Nerves and brain Reproductive system	Developmental delay in youth, children and foetus Kidney, nerve, brain damage, infertility
Man-made mineral fibre, MMMF, e.g. fibreglass, rockwool	Carcinogenic Irritant	Inhalation Irritating to skin	Skin, eyes Respiratory system	Cancer Irritant
Silica dust		Inhalation Respirable	Lungs	Silicosis (irreversible and continues after exposure has stopped) often fatal COPD
Solvents	Irritant Flammable, explosive	Inhalation Absorption through skin Ingestion	Respiratory system Eyes Skin Nervous system	Irritation, nausea, light-headedness Dermatitis Fatal at exposure to very high concentrations of vapour
Wood and wood dusts (preservatives)	Asthmagen Allergen Sensitiser	Inhalation Ingestion Absorption	Lungs Skin and eyes	Dermatitis Asthma

Note: COPD – chronic obstructive pulmonary disease

Table 2 Synopsis of effects of common hazardous substances on the body

The UK chemical industry produces some 95 000 different chemicals and chemical products annually (guidance on over 2000 chemicals that are of commercial interest can be found in the International Labour Organisation's (ILO) *Encyclopaedia of Occupational Health and Safety*, ILO, 2010). Many of these chemicals are to be found on construction sites in almost all trades activities, some as commercial products and others as by-products of work activities or processing of natural and man-made materials and include:

■ oils and greases used to maintain and operate machinery;

■ sealants and coatings on surfaces, vessels and chambers;

■ asbestos and man-made mineral fibres for insulation;

■ adhesives, solvents and paints;

■ fumes from welding, soldering and grinding activities;

■ dusts in cements, mortar, grinding and cutting stone; and

■ fumes from asphalt, bitumen and other quarry products.

It is outside the scope of this manual to deal with all the chemicals and hazardous substances that are present on construction sites. Detailed information can be obtained from a number of sources: the Material Safety Data Sheet (MSDS) provided with the chemical, the ILO's International Chemical

Safety Cards (ICSCs) available online, and *EH40, Workplace Exposure Limits*, published annually by the Health and Safety Executive (HSE, 2007c). This section provides an overview of the most commonly found chemicals or types of chemical and hazardous substance found on construction sites, and guidance on the control measures that should be considered. They follow the lists on **Tables 1** and **2** which are listed alphabeticaly rather than any priority-based order for ease of reading.

Asbestos

Asbestos is a naturally occurring fibrous silicate mineral that is mined in several countries. It has two types of minerals: amphibole minerals which are hard with straight and stiff needle-like non-soluble fibres and serpentine minerals which are softer with flexible curly white fibres. There are several forms, of which four – crocidolite, amosite, chrysotile and anthophylite – are mined for commercial use.

All forms are made up of long chain molecules of silicon and oxygen which are responsible for the fibrous nature of the material. Each has different physical and chemical properties depending upon other elements such as the calcium, magnesium or iron incorporated into the chemical structure.

These physical and chemical properties include:

- thermal stability and resistance
- chemical resistance
- high tensile strength
- abrasion resistance
- low electrical and thermal conductivity
- low biodegradability, and
- good sound absorption,

and have made asbestos an important constituent in over 3000 commercially available products including fireproofing and insulation, vinyl flooring, PVC (polyvinyl chloride) panels and cladding, and brake linings of vehicles and lifts. Asbestos can also be found in paints, coatings, sealants and adhesives.

Asbestos was in common use from the 1950s to the mid-1980s but it was not until 1999 that chrysotile, used in asbestos cement, was banned. Buildings constructed or refurbished during this period will likely contain asbestos materials, thus an estimated two-thirds of commercial buildings in the UK contain asbestos material, i.e. over 500 000 non-domestic buildings (HSE, 2003) (see HSEUK, undated, for a link to an HSE interactive diagram of asbestos-containing materials (ACMs) in buildings).

It is the very properties that make asbestos so useful that also make it extremely hazardous. All forms of asbestos are dangerous to health following inhalation of fibres. When fibres become lodged in the lung tissue, the body's defence mechanism tries to dissolve them by producing a type of acid. However, the chemical resistance properties of asbestos are such that these fibres are not dissolved but instead scarring occurs on the lung tissue that eventually may become so severe that the lungs cannot function.

The stiff fibrous nature of asbestos assists lodgement in tissue and resists the natural coughing mechanism for dislodging particles. Where coughing succeeds in dislodging fibres, swallowing may transfer the fibres to the digestion system where they may also become firmly lodged.

There is a long latent period between exposure and onset of ill health that varies between 10 and 40 years. Because there are no immediate ill effects, workers in many industries continued to work in highly contaminated environments before they became ill with a range of diseases, many of which are fatal. Currently [2010] some 4000 people in the UK and approximately 100 in Northern Ireland (NI) die annually from asbestos related diseases, and this number is expected to continue to rise for some time to come.

The main diseases associated with asbestos are:

- Asbestosis, a fibrosis (scarring) of the lung tissue that progressively develops COPD, with death often occurring from heart and lung failure (cardiopulmonary failure). It also increases the risk of lung cancer or malignant mesothelioma.

- Diffuse pleural thickening, an increase in the pleural fluid in the lungs which on re-absorption thickens the pleural membrane leading to breathlessness. It increases the risk of lung cancer and mesothelioma.

- Pleural plaques are localised areas of pleural thickening and are usually asymptomatic.

- Lung cancer.

- Malignant mesothelioma is a tumour that occurs on the pleural membrane, the peritoneum (in the abdominal cavity) or the pericardium (lining the heart). It is invariably fatal.

There is no cure for asbestos related diseases, and the risk of lung cancer is higher among those who smoke.

Asbestos and ACMs in good condition are safe as long as they remain undisturbed. It is not necessary to have a specialist or licensed contractor engaged when asbestos is to be worked on, unless it is for high risk material such as insulation on pipes or in asbestos insulating panels. Where asbestos is to be worked on, a safe system of work is required. Some features of safe working include keeping asbestos damp, using hand tools (power tools create more dust), using appropriate disposable personal protective equipment (PPE) including respirator (PPE must not be taken home for washing), cleaning up using a Type H (used for removal of hazardous products) vacuum cleaner or damp cloth, washing hands and face at every break from the task and at the end of the day, and putting asbestos waste in suitably sealed containers before taking it to a licensed tip (HSE, 2007a).

Chapter 18 *Demolition, partial demolition, structural refurbishment and decommissioning* provides further information on the removal and disposal of asbestos.

Asphalt and bitumen

Asphalt is derived from oil and comes in a variety of forms, from liquid to semi-solid and solid. It has a number of uses in construction including road surfacing, roof and pipe covering (poured liquid or as roofing felt) and other similar uses. The material is often mixed with solvents to soften or bring it to a liquid stage before application. The solvents, on evaporation, give rise to fire and explosion hazards, particularly when being used in poorly ventilated enclosed spaces.

Health hazards associated with asphalt result from contact with the material and may have acute (short term) effects such as skin irritation, rashes and acne-like conditions at hair follicles and skin pores. Heated asphalt creates fumes containing minute particles that will irritate the respiratory tract.

Additionally, hydrogen sulphide (H_2S) vapour, a very toxic gas, may be released if the asphalt contains solvents. H_2S has a long term (8 hours) workplace exposure limit of 5 ppm with a short term exposure limit (STEL) (10 minutes) of 10 ppm. It affects a range of body systems, particularly the nervous system and at low concentrations is an irritant, will cause coughing, shortness of breath and fluid in the lungs. In higher doses it prevents oxygen from binding and thus stops respiration. It is known for its 'rotten eggs' smell, although since olfactory fatigue occurs at high concentrations and at continuous low concentrations the sense of smell is not a reliable indicator of its presence or its dispersal.

The chronic effects of contact with asphalt or exposure to the fumes are increased rashes, which may take some time to clear up after exposure has ended, and various cancers. Solvents used in asphalt include:

- benzene (WEL 1 ppm) which can cause leukaemia and possible skin cancers;
- dioxane (WEL 25 ppm, STEL 100) which is very toxic to the liver and kidneys and may also cause cancer; and
- toluene (WEL 50 ppm, STEL 100 ppm), causes kidney and liver damage as well as dermatitis.

  (WEL – workplace exposure level; STEL – short term exposure limit)

Controls include substitution for a safer material, good ventilation and local exhaust ventilation (LEV) when in enclosed spaces, appropriate respiratory protective equipment (RPE) and PPE, and face and eye protection.

Cement and cement-containing products

Cement and cement-containing products are common throughout the construction industry although it is primarily used as a mortar and in concrete, and the harm that it poses is dependent upon the form and stage of production and use that it is in.

Basic cement is a compound containing calcium oxide (CaO), clay and gypsum (calcium sulphate dihydrate) (WEL for inhalable particles 10 mg/m^{-3}, for respirable particles 4 mg/m^{-3}). Further elements may be added to create a range of blended cements; these elements include furnace slag, fly ash (which contains substantial amounts of silicon dioxide, as well as calcium oxide), pozzolans (usually siliceous materials including silica fume) and waterproofing substances and colourings.

The chemical reactions between the different components and water make the strength and durability of the cement and concrete structures. These reactions also give rise to the heath issues that range from dermatitis to severe alkali burns that may result in amputation of one or more limbs.

Calcium oxide reacts vigorously with water and will therefore act as an irritant when in contact with moist skin, the eyes and the moisture within the respiratory and digestion tracts, causing coughing, nausea and breathing difficulties. Sufficient contact may turn into burns, especially in the nasal passages.

Wet cement causes alkali burns when it is held in continual contact with the skin. This may lead to ulceration of the skin that could take months to heal and in more severe cases may lead to amputation.

Silicon dioxide is a crystalline silica dust that when inhaled can lead to silicosis, bronchitis and (rarely) cancer. Those who are asthmatic or have other restrictions on their lung capacity are more susceptible on shorter exposures. Silicosis, like asbestos related diseases, may have a long onset time after exposure, e.g. 10–20 years. The WEL for silica is 0.1 mg/g^{-3}, a limit which is kept under review by the HSE.

With dusts it must be remembered that it is not only the person working on the task that is affected but also others in the vicinity, including members of the public on sites that are adjacent to pedestrian walkways or when construction work is conducted in occupied domestic premises.

Primary controls include avoiding direct skin contact with wet cement, appropriate PPE including gloves, eye protection and use of appropriate masks when dust is present, ensuring the cutting and grinding tools are fitted with water spray or dust extraction devices, immediate washing when contact is made with wet cement and seeking medical assistance following splashes to the eyes or prolonged skin contact that shows signs of burns.

Hexavalent chromium

Although listed separately as it is a vapour that may be released from other processes such as welding on electroplated metals or from the decomposition of wood preservatives, Hexavalent chromium is also generated in the cement manufacturing process through oxidation of trivalent chromium and thus workers inhaling cement dust or handling wet cement are at risk.

Hexavalent chromium is both a sensitiser causing dermatitis and a carcinogen.

Lead

Old lead based paints used on buildings and bridges until about 1970, lead piping in old structures, soldering and working directly with lead and lead alloys are all sources of lead based health hazards. Burning off the paints creates fumes, as does hot cutting and soldering, while dust is created from grinding and scraping actions to remove and/or smooth lead-containing surfaces (and therefore potentially hazardous to families when dust coated clothes are taken home).

ICE manual of health and safety in construction © 2010 Institution of Civil Engineers

Lead acts on, and is harmful to, all organs in the body, particularly the brain, kidneys and reproductive system. It can be inhaled or ingested and is quickly distributed by way of the bloodstream throughout the body with the greater percentage being deposited and stored in the bones and teeth, where it may later be released back into the blood. Unborn children are at particular risk.

Lead cannot be excreted and therefore further exposures have an accumulative effect, increasing the amount of lead in the body until it reaches levels in the blood that cause symptoms to appear (on average 60 µg/dl in adults, 45 µg/dl in children).

The symptoms of lead poisoning include lethargy and tiredness, abdominal pain and aggressiveness. Severe lead poisoning can lead to seizures and death. In children, lead poisoning will cause developmental delay, learning difficulties and behavioural problems.

Controls include working in designated areas that have LEV, dust control techniques including industrial vacuum cleaners with appropriate high efficiency particulate air (HEPA) filters, appropriate PPE especially RPE, washing and changing facilities and welfare facilities that are uncontaminated. Employers will also have in place health monitoring.

Man-made mineral fibre

Man-made mineral fibre (MMMF) products such as fibreglass, rockwool and ceramics have excellent thermal and sound insulation properties and are therefore found throughout construction sites, being handled regularly by laggers, plumbers and carpenters.

Although fibrous, MMMFs are unlike asbestos fibres in chemical make-up, physical characteristics and effects on the body. With the exception of aluminosilicate ceramic fibre (used in high temperature furnace linings), MMMFs are less harmful with larger fibres acting as an irritant on skin, eyes and the respiratory and digestive tracts, whereas dust of fibreglass poses a possible risk for lung cancer. Aluminosilicate ceramic fibre devitrifies at high temperatures to form hazardous crystalline silica which, when broken up, may create a dust and this has the potential to cause silicosis. High specification RPE is an essential when demolition works involve breaking up such furnaces. For loft insulation work and the like, the use of appropriate PPE including gloves to prevent skin contact and respirator masks of appropriate standard to prevent inhalation/ingestion are required.

Solvents

Solvents are liquids capable of dissolving other substances. In construction, hydrocarbons are often used as degreasing agents. They have a wide range of toxicological properties and can:

- cause dermatitis by removing natural protective grease from the skin;
- cause narcosis (unconsciousness) by acting on the CNS;
- damage the peripheral nerves, liver and kidneys;
- interfere with blood formation and cardiac rhythm; and

- harm the lungs (this is especially true of chlorinated solvents when they decompose in heat to form hydrochloric acid and phosgene, (infamous as a chemical weapon in World War I).

Hydrocarbons are also carcinogenic.

In their liquid state they are less dense than water and will float, but in their vapour state they are heavier than air and will tend to flow and settle at the bottom of dips, hollows, wells, sumps, etc. With the exception of chlorinated hydrocarbons they tend to be flammable and explosive, and will float away when water is used to extinguish them, thus spreading the fire.

Skin penetration and absorption will vary depending on a number of factors and the amount taken up by the body will often be more important in estimating the potential harm rather than the concentration to which the body is exposed.

Work with solvents should be conducted in a well ventilated area with appropriate RPE used when ventilation is inadequate. Likewise, appropriate PPE is to be used to avoid skin and eye contact, remembering that solvents may react with and reduce the effectiveness of the materials (especially in gloves) to protect. Some solvents, such as acetone, will rapidly degrade some materials such as latex, rendering such gloves totally unsuitable.

Wood and wood dusts

Wood and wood based products are common throughout the industry. Custom boards, such as particle and fibre boards, are manufactured with resins that contain formaldehyde. The resins do not normally produce vapours at normal temperature, but when they are heated formaldehyde is produced. Formaldehyde is an irritant on the eyes, skin and respiratory tract and is a sensitiser on prolonged contact, leading to dermatitis. It is not a proven human carcinogen (HSE, 2007b), it is however extremely flammable and explosive.

Wood dust is an allergen causing asthma and has a WEL of 5 mg/m^3 which must not be exceeded. Some hardwood dusts can cause cancer, especially of the nose. Controls include LEV, appropriate respirator masks and PPE.

Controls, storage and disposal

Safer alternatives

The common approach to safe working with chemicals is to consider whether the chemicals are required in the first place and, if not, eliminate them from the company's inventory.

However, where it is essential to have chemicals as part of the process, consideration should be given to using the safest alternative that will do the same job. Many companies are using chemicals that were the only ones available at the time they started using them without having periodically checked whether safer alternatives have come on the market since. Control of substances hazardous to health (CoSHH) assessments are a legal requirement when hazardous substances including chemicals are present in the workplace. Not only must the assessments include the consideration of safer alternatives but they must also be reviewed periodically with shorter review intervals for higher hazard substances. This means that newer alternatives that come on the market should be included in assessments.

Safer alternatives may also be considered at the design stage with specifications, e.g. bitumen emulsion instead of cut-back bitumen in road construction, or alternative designs and materials that eliminate the need for cutting when laying ornamental brickwork (see Section 3, Chapter 9 'Assessing safety issues in construction').

Engineering controls

Where fume is generated, the aim should be to have it carried out in an area where the ventilation will safely dilute and disperse it. For many gases and vapours a good flow of fresh air through the work area will suffice. Activities generating more hazardous fumes and vapour that cannot safely be carried out in the open must be conducted in an appropriate area equipped with local exhaust ventilation. Whether inside or in an open area, consideration must be given to where the dispersed fume is going and who may be affected by it, immediately or at a later time when they may enter a vapour trap that has not been ventilated.

Dust control in the open air involves the use of hand or low-power tools that generate little dust, and power tools equipped with water spray or dust collection devices. Where practicable, dust-generating activities may be conducted in a controlled area equipped with LEV and dust extractors.

Scouring, scraping or smoothing surfaces should be carried out in wet conditions to reduce dust generation.

Removal of old lead-based paints by heat should be carried out using low temperatures that are sufficient to loosen the paint without breaking it down to cause fume. Even at low temperatures avoid keeping the heat on the one spot for too long.

Application of chemicals to surfaces is best done by hand using a brush as this reduces the volume of particles in the air that would be generated if spraying was utilised.

PPE and RPE

All PPE and RPE must be appropriate for the hazard(s) being controlled. This includes providing PPE constructed of a material that will not react with the chemicals and degrade. The MSDS or reference to the ICSCs will give some information on suitable materials as will the supplier of PPE. Sometimes the material of the PPE will degrade anyway, and it is important that workers are provided with sufficient sets of PPE for the duration of the activity and instruction on the frequency for changing. Some PPE may need to be disposable after use, for example when working with asbestos, and they may require strict donning and removal protocols to minimise the risk of contact with contaminated outer surfaces.

Disposable boot covers are not recommended as they present a slip hazard. Where dangerous contaminants such as asbestos and silica dusts are present, laced boots are unsuitable as they are difficult to clean effectively. Cleaning of boots (and other

contaminated equipment) must be carried out on-site in an appropriately designated facility where the contaminants can be safely collected and disposed of. Under no circumstances should contaminated PPE be taken home for washing.

Respirator masks must be appropriate in respect of the particle size or gas to be filtered, and changed as frequently as necessary to ensure that they remain effective.

In some instances, where gases and fumes are generated, respirator masks are unsuitable for protection and RPE (breathing apparatus) must be provided. Where this is the case the area in which the work is taking place must be designated a confined space and the controls detailed in the Section 5, Chapter 16 'Confined spaces' come into effect.

Safe system of work

When working with any dangerous substance or on activities that will generate them, it is essential that a safe system of work has been established and communicated to all workers. This will include details on how the activity is to be carried out, the equipment, PPE and RPE that are to be used, the cleaning and disposal processes and any PPE donning protocols. The safe system will also include any emergency first aid that is to be followed in the event of an uncontrolled exposure.

All the above controls are used in conjunction with rather than as alternatives to each other. Priority should be given to safer alternatives and engineering controls.

Storage

Chemicals and composite substances react in a variety of ways when in contact with other chemicals, water, light and heat. It is important to be aware of the reactivity of each when considering where and how they should be stored and in what quantities.

It is preferable to have reduced quantities delivered to site, especially of higher hazard, e.g. flammable and explosive chemicals, to eliminate the need for bulk storage and thereby reduce the severity should an incident occur. Where large quantities are necessarily required, more frequent deliveries, e.g. daily deliveries, obviate the need to build large fit-for-purpose storage. This would be an appropriate consideration at the early design and planning stages.

Chemicals that react with other chemicals should be stored separately or, if in the same store, far enough apart and with a physical barrier to prevent them coming into contact should a spillage occur. Those that react with water should be stored in a dry insulated facility away from any water pipes and faucets.

Chemicals that react to heat and light either to degrade or break down into other hazardous substances should be stored in light- and heat-controlled facilities.

Substances that release vapours when heated must be stored in a facility that has been designed with suitable ventilation, extraction and containment, and external warning signals. Oils, including waste oils, and solvents must be contained within suitable explosion-proof vessels and stored in cool facilities.

All storage facilities must be secure against unauthorised access. Check the MSDS for storage advice or if you are not sure how chemicals are likely to react with each other.

Spill clearance and waste disposal

Incidents do occur from time to time and to minimise environmental damage as well as harm to other people the following safety controls should be considered where appropriate.

Tanks of oils and fuels should be contained with suitable bunding (i.e. containment) capable of holding a volume greater that the volume of the tank. Portable bunding should be used when transporting fuels, etc. around the site.

Spill kits and appropriate absorbent materials should be available close to any area where spillages may occur and the workers instructed to safely respond to the spillage as quickly as possible to prevent deep seepage into the soil or into watercourses. Spillages of highly evaporative or extremely hazardous substances may require a specialist clean-up team. This should be considered at as early a stage as possible and a clean-up team put on standby. In these circumstances increased controls to prevent spillage will take precedence.

Wet working to prevent dusts is rendered ineffective if the runoff is allowed to pool and dry out at which time the dusts will regenerate when disturbed by wind or people and vehicles crossing over it. Ensure that runoff is appropriately drained or collected whilst still damp and disposed off in an effective manner.

The collection of highly hazardous dusts, such as asbestos, must not be allowed to escape into the environment. As described above, large scale works involving the removal and disposal of panels must be carried out by specialist contractors. However, small amounts of dust generated by minor works must be cleaned up using a Type H vacuum cleaner, or dampened cloth and the waste, including cloth, disposable PPE, etc. placed in sealed containers/bags and taken to a licensed tip.

The MSDS and ICSCs will provide details on the most appropriate method for cleaning, collection and disposal of waste.

The control of major accident hazards (COMAH)

COMAH is a substantial subject in its own right with the subject matter primarily aimed at operators of major hazard sites and covering management, prevention and responses to major accidents; it is thus beyond the scope of this chapter. However it is important to note that construction that takes place on these sites will have an impact on the management and prevention procedures and, depending on the scale of the construction project, amendments or additions will need to be applied to the procedures with external bodies needing to be informed in the stages prior to commencement.

Major hazard sites are those that manufacture, process and store dangerous chemicals and substances in quantities that would pose a substantial harm to workers, people in the vicinity and the environment should a major accident occur. They are generally chemical manufacturing plants and storage warehouses, but will include nuclear and explosives sites, and other sites where defined threshold quantities of dangerous substances are kept or used (HSE, 2008).

The COMAH Regulations apply to such sites and place a duty on the site owners to demonstrate to the competent authority that their operation is safe. Lower tier sites must prepare a Major Accident Prevention Policy (MAPP) whilst higher tier sites must also produce a more precise safety report and both must produce on-site emergency plans and notify the local authority which in turn prepares the off-site emergency plan.

Construction and demolition works taking place on major hazard sites impact upon the MAPP, safety report and the emergency plans that exist for routine site operation; these must therefore be reviewed taking into consideration the proposed works. As the local authority and the competent authority have statutory duties in relation to major hazard sites, they may need to input to the review process; the competent authority may wish to carry out an inspection before agreeing the new plans, particularly if the proposed works are substantial in scale and duration. The principal requirement is to integrate the construction safety measures into the safety requirements of the Client's site and to design and conduct the work in accordance with the principle of preventing a major accident.

The competent authority is:

- in England and Wales, the HSE and the Environment Agency
- in Scotland, Scottish Environment Protection Agency
- in Northern Ireland, Health and Safety Executive Northern Ireland (HSENI) and Department of the Environment.

Statutory issues

Control of Substances Hazardous to Health Regulations (COSHH)

These regulations require employers to control the chemicals and hazardous substances that are used or are present in the workplace. Great Britain and Northern Ireland each has separate versions of COSHH regulations. The process of controlling exposure has seven key steps:

1. Assess how the substance is being used.
2. Decide on the appropriate precautions that are needed.
3. Prevent or adequately control exposure to the harmful effects of the chemical.
4. Ensure that the agreed control measures are being used and maintained.
5. Monitor employees' exposure.
6. Carry out appropriate health surveillance.
7. Ensure the employees are properly informed, trained and supervised.

The COSHH assessment must be carried out for each different use and method of application of the substance, as well as in respect of its storage.

Information useful to aid the assessment process can be obtained from a range of sources including:

- The Material Safety Data Sheets (MSDS) that come supplied with the substances. Many people mistakenly view this information sheet as sufficient; it contains basic chemical details, the risk and safety phrases and some suggestions on PPE, first aid and spill control which makes it a starting point rather than the end point of the assessment.

- Further details from your supplier if the MSDS is unclear or incomplete.

- Information directly from the manufacturer if it is not the direct supplier.

- International Chemical Safety Cards.

- Occupational health professional.

- HSE and HSENI.

- Trade bodies.

- Safety representatives.

- First aiders.

Assessments should be reviewed periodically to ensure that the controls remain effective, with more frequent assessments the higher the hazard presented by the substance.

A sample COSHH assessment for a paving company is available on the HSE website at http://www.hse.gov.uk/COSHH/riskassess/paving.htm

The HSE also provides an interactive website, COSHH Essentials, to assist with conducting assessments.

Globally harmonised system (GHS) of classification and labelling of chemicals

Under the auspices of the United Nations (UN), different countries agreed to create local legislation that will harmonise the ways in which chemicals are classified and labelled throughout the world. In January 2009 the European Union (EU) brought into force the Regulation on Classification, Labelling and Packaging of Substances and Mixtures (CLP Regulation) which adopted the GHS and set a transitional period to June 2015 to allow the existing systems to make the necessary changes. The Chemicals (Hazard Information and Packaging for Supply) Regulations 2002 (CHIP) and CHIP NI 2002 Regulations will be amended to take on board the new requirements.

The main changes will be:

- new scientific criteria to assess the hazardous properties of a chemical;

- two new pictograms for the hazard warning labels and a redesign of the existing labels, (details in Annex 1 of the UN GHS);

- the current hazard phrases and safety phrases to be become hazard statements and precautionary statements.

Registration, evaluation, authorisation and restriction of chemicals (REACH)

This new European regulation came into force in June 2007 with the aim of providing a high level of protection for people and the environment from the harmful use of chemicals and to ensure that manufacturers and importers of chemicals understand and manage the risk associated with their use. One of the requirements is that manufacturers and importers register their substances with a central European Chemicals Agency (ECHA) providing a standard set of data on each substance. The effect of failing to register and provide the data package on their chemicals is that they will no longer be able to legally manufacture or import them.

With an estimated 30 000 chemicals and preparations on the market, it will take time for them to be registered and the process is to be staged over 11 years.

An important aspect of REACH is the classification and labelling of the chemicals, and in that respect REACH will dovetail with CLP and CHIP.

Summary of main points

Working with chemicals and other substances, no matter how innocuous they may appear to be or how familiar an employee is with their use, can and often does pose risks to health. It is essential when using any substance in the workplace that attention is paid to the hazard information provided on the labels, the MSDSs and the COSHH assessment.

It is important that all on the worksite are familiar with and understand the hazards posed by substances they are exposed to, whether provided for use or generated by the work activity. Everyone at work is entitled to information, instruction and training on the hazards and safe working methods, to be provided with appropriate equipment including PPE that will safeguard against exposure.

Employees must follow the agreed safety procedures and use or apply all control measures appropriately and safety. Where equipment is defective or insufficient in any way it must be reported immediately to the relevant manager.

In the event of an accident involving hazardous substances, the emergency procedures must be followed immediately and the accident reported as soon as practicable.

Anyone who experiences health problems, e.g. respiratory difficulties, skin conditions or allergies, whether or not associated with the use of chemicals/substances, must report them immediately and if necessary seek medical advice before using substances at work.

References

Ashtead Technology. *Lead, Cadmium and Arsenic Halt Channel 4s Time Team in 'Big Roman Dig'*, undated. Available online at: http://www.edie.net/products/view_entry.asp?ID=1698andchannel=0

Hazards Magazine. Available online at: http://www.hazards.org/asbestos/ilo.htm

Health and Safety Executive (HSE). *Managing Asbestos, Your New Legal Duties*, 2003, London: HSE.

Health and Safety Executive (HSE). *EH40/Workplace Exposure Limits 2005*: Table 1 List of approved workplace exposure limits (as consolidated with amendments October 2007), London: HSE Books. Available online at: http://www.hse.gov.uk/COSHH/table1. pdf

Health and Safety Executive (HSE). em6 *Asbestos Essentials, Personal Protective Equipment (PPE)*, 2007a, London: HSE. Available online at: http://www.hse.gov.uk/pubns/guidance/em6.pdf

Health and Safety Executive (HSE). *Surface Engineering, COMAH Application and Main Duties*, 2007b, London: HSE. Available online at: http://www.hse.gov.uk/surfaceengineering/mainduties. pdf

Health and Safety Executive (HSE). *EH40/2005 Workplace Exposure Limits*, 2007c, London, HSE Books.

Health and Safety Executive (HSE). *Major Hazard Sites and Safety Reports, What You Need to Know*, 2008, London: HSE. Available online at: http://www.hse.gov.uk/comah/background/essentialinfo. pdf

Health and Safety Executive (UK). *Where Can You Find Asbestos?*, undated. Available online at: http://www.hse.gov.uk/asbestos/ essentials/building.htm

Health and Safety Executive (UK). *Health and Safety Statistics, 2008/2009*, 2010, London: HSE. Available online at: http://www. hse.gov.uk/statistics/overall/hssh0809.pdf

International Labour Organisation. Encyclopaedia of occupational health and safety (accessed online, 6 January 2010) http://www.ilo. org/safework_bookshelf/english?dandnd=170000102andnh=0

Jeeves, P. Wife was Exposed to Asbestos in Laundry, 2006, *Yorkshire Post* 18 July. Available online at: http://www.mesothelium.com/ wife-exposed.jsp

Takala, J. *International Labour Organisation* (ILO). *Hazards Magazine* 2010. Available online at: http://www.hazards.org/ asbestos/ilo.htm

Referenced legislation

Control of Major Accident Hazards Regulations 1999. Statutory Instruments 743 1999, London: The Stationery Office.

Control of Substances Hazardous to Health Regulations 1994. Statutory Instruments 1994 3246, London: HMSO.

CHIP regulations

Chemicals (Hazard Information and Packaging for Supply) Regulations 2009. Statutory Instruments 716 2009, London: The Stationery Office.

Chemicals (Hazard Information and Packaging for Supply) Regulations (Northern Ireland) 2009. Statutory rules of Northern Ireland 238 2009, London: The Stationery Office.

Regulation (EC) No 1907/2006 of the European Parliament and of the Council of 18 December 2006 Concerning the Registration, Evaluation, Authorisation and Restriction of Chemicals (REACH). *Official Journal of the European Union* 2006, **L396**, 1–849.

CLP regulation

Regulation (EC) No. 1272/2008 of the European Parliament and of The Council of 16 December 2008 on Classification, Labelling and Packaging of Substances and Mixtures, Amending and Repealing Directives 67/548/EEC and 1999/45/EC, and Amending Regulation (EC) No 1907/2006 s. *Official Journal of the European Union* 2008, **L353**, 1–1354.

Further reading

ACGIH. *Health Effects of Occupational Exposure to Emissions from Asphalt/Bitumen Symposium*, Dresden, Germany, 2006. Available online at: http://www.acgih.org/events/course/Asphalt06_Pre limPrgm.htm

American Federation of State County, and Municipal Employees (AFSCME). *Issues/Asphalt*, 1989. Available online at: http://www. afscme.org/issues/1363.cfm

Expert Ease International. *Asbestos at Work*, 2004.

Expert Ease International. *COSHH Awareness and Assessments*, 2006, Northern Ireland: EEI

Gore, D. and Sleater, A. Research Paper 99/81, Asbestos, 1999, House of Commons Library. Available online at: http://www.parliament. uk/commons/lib/research/rp99/rp99-081.pdf

Health and Safety Executive (HSE). *Man-made Mineral Fibres (MMMF)*, OC267/2, 2004.

International Programme on Chemical Safety (ILO). *International Chemical Safety Cards, ICSCs*, 1998. Available online at: http:// www.ilo.org/safework_bookshelf/english?dandnd=170000102 andnh=0

United Nations (UN). *Globally Harmonized System of Classification and Labelling of Chemicals*, 2005, Geneva: UN. Available online at: http://www.unece.org/trans/danger/publi/ghs/ghs_rev01/ English/00e_intro.pdf

Websites

Chemical Industries Association http://www.cia.org.uk/newsRoom. php?id=97

European Chemicals Agency (ECHA) http://echa.europa.eu/

European Commission, Enterprise and Industry, Chemicals, REACH http://ec.europa.eu/enterprise/sectors/chemicals/reach/index_ en.htm

Health and Safety Executive (HSE). http://www.hse.gov.uk

Health and Safety Executive (HSE), CLP Regulation adopting in the EU the Globally Harmonised System (GHS) http://www.hse.gov. uk/ghs/eureg.htm

Health and Safety Executive (HSE), COSHH Essentials http://www. coshh-essentials.org.uk/

Health and Safety Executive (HSE), Example COSHH risk assessment – Paving company http://www.hse.gov.uk/COSHH/ riskassess/paving.htm

Health and Safety Executive Northern Ireland (HSENI) http://www. hseni.gov.uk

Health and Safety Executive (HSE), United Nations Globally Harmonised System of Classification and Labelling of Chemicals http://www.hse.gov.uk/ghs/index.htm

Institution of Civil Engineers (ICE), Health and safety http://www. ice.org.uk/knowledge/specialist_health.asp

International Labour Organisation http://www.ilo.org

UK REACH Competent Authority http://www.hse.gov.uk/reach/ compauth.htm

ICE manual of health and safety in construction © 2010 Institution of Civil Engineers   www.icemanuals.com   133

ice | manuals

Chapter 12

Controlling exposure to biological hazards

Akinwale Coker Department of Civil Engineering, Faculty of Technology, University of Ibadan, Nigeria
and **Mynepalli K. C. Sridhar** Division of Environmental Health, Department of Community Medicine,
College of Health Sciences, Niger Delta University, Wilberforce Island, Bayelsa State, Nigeria

doi: 10:10.1680/mohs.40564.0137

This chapter deals with controlling exposure to biological hazards in the construction industry. The biological hazards are defined and the natural and man-made sources are documented where a hazard is likely to occur. Construction industry predisposes the workers to respiratory and cardiovascular problems due to dust, injuries from falls, sharp instruments, and attacks by venomous stings and bites, vectorborne infections and bioterrorism. Further, recognisng a biohazard, routes of entry, biosafety levels, and biohazard preventive measures are discussed. Hazards control measures include elimination, engineering controls including retrofit technologies, administration controls, and provision of adequate and appropriate personal protective equipment. Rodent control is very important in the construction industry. Common disinfectants against biological agents are suggested with procedures for preparation and use for routine and emergency situations particularly to manage the spills. A part of the chapter also deals with biowaste management, monitoring and assessment of biohazards, risk assessment, and environmental impact assessment. Roles and responsibilities of stakeholders are highlighted. Some case studies are given on the biohazards experienced in Nigeria and the USA, particularly in the road and building construction, laboratories, and a university research facility.

CONTENTS

Box 1   Key learning points

- What are biological hazards?
- Common biological hazards reported in construction industry.
- Identification of hazards.
- Assessment of degree of hazards from biological sources.
- Risk assessment.
- Environmental Impact Assessment.
- Levels of biosafety.
- Hazard prevention and control.

Biological hazards

Biological hazards occur in many sectors of employment and occupations. The Control of Substances Hazardous to Health (COSHH) and Approved Code of Practice (ACoP) have the following general definition of a biohazard:

> A biological hazard (biohazard) is any microorganism, cell culture or human endoparasite, including any which have been genetically modified that can cause infection, allergy, toxicity or otherwise, create a hazard to human health. (Biohazard Cleaning, 2008)

The construction industry is potentially hazardous. Equally hazardous is the management of 'bio-facilities'. According to the USA Occupational Safety and Health Administration (OSHA), some 6.5 million people work in the construction industry. The rate of fatal injuries occurring in the construction activities is higher than the national average for all industries put together (Pattron, 2005). In Britain, about a 2.2 million people work in the construction industry, making it the country's biggest industry. It has been identified as one of the most dangerous, with over 2800 deaths recorded in the last 25 years. In the City of London, over 300 cases are reported annually on workplace accidents, diseases or dangerous episodes. Legionnaire's disease occurrence is about 250 live cases at any given time. Some 500 inquiries are received on health and safety matters (City of London, 2009). Limited time, cheap labour, rush for rapid infrastructural development and corrupt practices often lead to compromising health and safety at the expense of workers. As a result, there is a progressive increase in the incidence and prevalence of construction injuries, worldwide.

A few terminologies are in vogue for describing biohazards. A hazard is defined as any physical, chemical or biological entity that has the potential to affect human health and well-being and the environment (Goetsch, 1999). Environmental hazard is a generic term for any situation or state of events that poses a threat to the surrounding environment. It may be natural or man-made. A biological hazard or biohazard is an organism, or substance derived from an organism, that poses a threat to human or animal health. Biohazardous materials/organisms include all infectious agents (bacteria, chlamydia, fungi, parasites, prions, rickettsia, viruses, pollen, etc.) which can cause disease in humans, or cause significant environmental or agricultural impact. Biosafety implies principles, technologies and practices that are implemented to prevent the unintentional exposure to pathogens and toxins, or their accidental release into the work or living environment.

Sources of biological hazards

Various sources of hazards – physical, chemical or biological – in the construction industry are given in **Figure 1** and **Table 1**. The sources of microbial infections may be from:

■ Workers in medical, healthcare and biotechnological laboratory facilities, and visitors to them, become infected with pathogenic microorganisms or genetically modified microorganisms.

■ Exhaust air, waste water and waste from bio-facilities, infecting residents living around the facilities and contaminating the environment.

■ Those existing latently in the natural environment (or settling on animals) proliferating and dispersing due to the change of environmental conditions. HIV/AIDS can survive in blood outside of the body for three weeks, hepatitis B and C can survive in these conditions for three months (source: Biohazard UK Ltd, 2008).

■ Microorganisms or toxins existing latently in ingested foods and drinks can cause/lead to infection, poisoning, allergy, cancer, etc.

On occasions the administration of biologics – vaccines, antitoxic sera, blood component preparations, antibiotics, etc. – causes a fatal side reaction – headache, fever, anthema, convulsion, paralysis, etc.

Sharp injuries and the associated infections are common in buildings. Some of the common places where such sharp injuries can occur are: noticeboards, letter boxes, staircases/hand banisters, radiators, refuse bags, pipework, air vents, lift wells, toilets and cisterns, bedding, under floorboards, ceiling voids, under carpets, and door handles/door frames.

The sources of these biological hazards may also occur from diagnostic, patient care and research laboratories, through poorly constructed scaffolds, falls, improperly constructed trenches, injuries arising from cranes and forklifts, electrical shocks, exposures to dusts, bacteria, viruses and fungi, and failure to use personal protective equipment (PPE). Some of the factors responsible are rapid rate of development (particularly in developing countries), use of foreign under-trained and overworked labour force, use of sub-standard materials

in the construction, improper supervision of workers, taking shortcuts in execution of the work, absence of articulated health and safety programmes, absence of standard operating procedures, and lack of training (Pattron, 2008; Unionsafe, 2009).

The environment in most construction sites is dusty and has the potential to cause a series of respiratory and cardiovascular problems. Air quality has been linked to the development of cardiovascular diseases (Brunekreef and Holgate, 2002; Schwartz, 1999). Dust is an ideal medium for the transport of bacteria, viruses and fungi that can be easily inhaled. Dust itself is a respiratory irritant and can cause congestion, chronic and acute respiratory infections, and death. Workers also may be carriers of a wide range of potential biological hazards (Goetsch, 1999). Further information on this can be found in other chapters in Section 4 *Health hazards*.

In addition, outdoor workers may be exposed to many types of venomous wildlife and insects. Venomous snakes, spiders, scorpions and stinging insects can be found in various geographic regions. They are more dangerous to workers who have existing allergies to the antigens from certain animals. Anaphylactic shock is the body's severe allergic reaction to a bite or sting and requires immediate emergency care. Thousands of people are stung each year, and as many as 40–50 people in the USA die each year from severe allergic reactions. Venomous snakes include rattlesnakes, copperheads, cottonmouths/water moccasins and coral snakes. Stinging insects include bees, wasps, hornets and fire ants. Venomous spiders include black widows, brown recluse spiders and hobo spiders. Poisonous plants found in the USA include poison ivy, poison oak, and poison sumac. These plants can cause allergic reactions if the leaves or stalks are bruised and come in contact with workers' skin. These plants can also be dangerous if they are burned and their toxins are inhaled by workers (NIOSH, 2002, 2008). Although fatalities from adder (poisonous snake) bites are not common in the UK, envenomation has been known to cause significant morbidity (Harborne, 1993). In the UK, insect bites and stings are usually no more than a nuisance, causing a few red, itchy spots or lumps. In some people they can cause severe allergic reactions. Common biting insects include bedbugs, flies, fleas, midges, mites, mosquitoes and ticks. Common stinging insects include bumblebees, honeybees, hornets and wasps (Bupa, 2009).

In the UK, ragwort is a common noxious weed. It is a poisonous weed of extensively farmed grassland and unmanaged ground which regularly causes losses of stock, especially cattle and horses. Giant hogweed is another fairly common invasive weed in the UK. It is notorious for producing a virulent contact dermatitis on the human skin and it is one of the causes of phytophotodermatitis.

Occupational hazards

The construction industry is one of the major occupational groups and abides by the occupational health and safety regula-

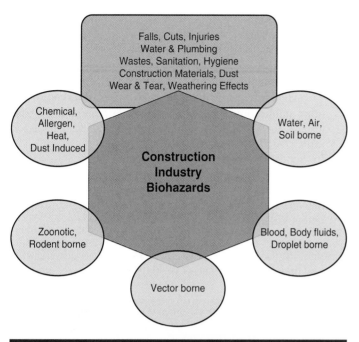

Figure 1 Biohazards in construction industry

Box 2 Common vectors and the diseases transmitted in the construction industry

Mosquitoes (species of *Anopheles, Culex, Mansonia Aedes*): malaria, filariasis, viral encephalitis, dengue, yellow fever, *Chikun Gunya, West Nile Fever and others

Housefly (*Musca domestica*): typhoid, diarrhoea, cholera, *polio, gastro-enteritis, *anthrax, etc.

Sand fly: leishmaniasis, oriental sore/ sand fly fever

Tsetse fly (*Glossina palpalis*.) – *Trypanosomiasis (sleeping sickness)

Louse – epidemic typhus, relapsing fever

Bed Bugs (*Phlbaetoma* sp.)

Rats (flea): *endemic typhus, *bubomic plague, bluetongue vvirus (*Culicoides obsoletus, C. Pulicaris*), fleas (species of Xenopsylla, Pulex, Ctenocephalides)

Black fly (*Simulium damnosum*): *onchocerciasis

Cyclops: *Dracunculus medinensis* (*guineaworm*)

Reduviid bug: *chagas disease

Hard ticks: *typhus, encephalitis

Soft ticks: Q fever, relapsing fever

Itch mites: *scabies

Cockroaches: enteric pathogens

Note: Infections marked *are not common or controlled in UK.

tions (see Chapter 1 *Legal principles* for further information). Occupational health is the promotion and maintenance of the highest degree of physical, mental and social well-being of workers. Occupational hygiene is concerned with the identification, measurement, appraisal of risk and control to acceptable standards of physical, chemical and biological factors arising in or from the workplace, which may affect the health or well-being of those at work. For any worker, a pre-employment medical examination ('fitness to work') is conducted to establish the individual's physical and emotional capacity to perform the job (also see Chapter 8 *Assessing health issues in construction* and Chapter 13 *Controlling exposure to physical hazards*). The control of hazards is an important factor in the working environment. The factors that should be considered in maintaining good occupational practices are: space allocation, temperature, ventilation, illumination, housekeeping and maintenance, washing facilities, sanitary facilities, first aid facilities and welfare.

Vectorborne hazards

Certain construction activities such as water resources development and road construction have resulted in vectorborne infections, some of which have transformed from zoophyllic (having an attraction to animals) to anthropophillic (human-seeking/preferring). Vectors are normally invertebrate animals, usually arthropods, but they may also include fomites, which are defined as 'any inanimate object that may be contaminated with disease-causing microorganisms and thus serves to transmit disease', or rodents, which may harbour vectors (Coker and Olutoge, 2006). Some of the insects are mechanical vectors which simply transmit the germs.

Societal changes such as human encroachment on natural disease foci, modern transportation and containerised shipping have added to the resurgence of vectorborne diseases which were considered to be under control in many countries but are now rampant in developing countries. Puddles of water at construction sites are prime breeding grounds for various types of dangerous insects and some of them act as vectors of disease. Certain simple environmental upkeep measures are to ensure such things as old tires, cans, pottery containers, etc. are kept free of water or turned upside down so that puddled water is removed and vector breeding reduced to some extent. Currently, climate change is enhancing the movement of these vectors to other areas hitherto unaffected. Some of the common vectors and the transmitted diseases are given in Box 2.

Hazards from bioterrorism

Bioterrorism is a growing threat. Bioterrorism is intentional release or dissemination of biological agents (bacteria, viruses or toxins); these may be in a naturally occurring or in a human modified form. The agents are categorised into A, B and C.

Category A contains: anthrax, smallpox, botulinum toxic, bubonic plague, viral hemorrhagic fever, and tularaemia.

Category B contains: brucellosis, *Clostridium perfringens*, *Salmonella* species, *E. Coli* O157 H7, shigella, staphylococcus, glanders (*Burkholderia mallei*), melioidosis (*Burkholderia pseudomallei*), psittacosis (*Chlamydia psittaci*), Q fever (*Coxiella burnetii*), ricin toxin from *Ricinus communis* (castor beans), staphylococcal enterotoxin B, typhus (*Rickettsia prowazekii*), viral encephalitis (alphaviruses, e.g.: Venezuelan equine encephalitis, eastern equine encephalitis, western equine encephalitis), water supply threats (e.g. *Vibrio cholerae*, *Cryptosporidium parvum*).

Category C agents are emerging pathogens that might be engineered for mass dissemination because of availability; they are easy to produce and disseminate, or may possess a high mortality or major health impact

Historically, bioterrorism has been documented from primitive throwing of human faeces against enemies to modern genetic engineered devices. Terrorists could also employ a biological agent that would affect agricultural commodities over a large area (e.g. wheat rust or a virus affecting livestock), thus destroying national economy. Victims of biological agent attack may serve as carriers of the disease with the capability of infecting others. The UK, to date, has not experienced any bioterrorism attacks of serious magnitude. Most notable incidents are:

(a) In June 1993 the religious group Aum Shinrikyo released anthrax in Tokyo, although it was unsuccessful.
(b) In 1984, followers of the Bhagwan Shree Rajneesh attempted to control a local election by incapacitating the local population; this was done by infecting salad bars in 11 restaurants, produce in grocery stores, doorknobs and other public domains with *Salmonella typhimurium* bacteria in the city of The Dalles, Oregon, USA. The attack infected 751 people with severe food poisoning.
(c) In September and October 2001, several cases of anthrax broke out in the USA. This was caused deliberately through letters laced with infectious anthrax which were delivered to news media offices and the US Congress; the letters killed five (Atsui, 2007; CDC, 2008).

Some of the indicators used in identifying such events include: unusual occurrence of dead or dying animals, unusual

	Source	Problem	Remarks
1.	Contaminated water supply	Waterborne infections, e.g. cholera, typhoid, dysentery	
2.	Dead legs in the plumbing system and aerosols from showers	Legionella	
3.	Recreational water (heavily contaminated with pathogens)	Increased risk of gastrointestinal and other infectious illness	Usually self-limiting
4.	Allergens (grass pollen grains, excreta from house dust mites).	Attacks of asthma or 'hay fever' (allergic rhinitis)	High exposure to these allergens early in life, increases the risk of suffering from asthma later in life
5.	Large wood dust particles (from oak, western red cedar, blackwood); timber contaminated with fungi (moulds)	Trapped in the nasal passage and may cause nasal cancer; chronic lung disease by reducing lung function; occupational asthma; allergy	■ Control of wood dust through dust extraction system ■ Personal protection (dust masks, eye protecion) when machining wood ■ Work area must be cleaned up every day and the wood dust safely removed ■ Work area to be kept well ventilated
6.	Plumbing works	Risk exposure to hepatitis A and infectious bacteria, viruses and protozoa such as Giardia Symptoms of hepatitis A are fever, headache, nausea, pain in the abdomen, dark urine and jaundice	Decontaminate the area Vaccinate the workers against hepatitis A
7.	*Cryptococcus neoformans*, species of fungi, often present in the excrement of pigeons	Remain alive for months in dried bird faeces; infection occurs in lungs with pneumonia-like symptoms and at times like meningitis	■ Adequate protection to workers ■ Appropriate respiratory equipment such as a P2 half facemask with a suitable filter ■ A long-sleeved shirt, trousers, gloves and fully enclosed shoes should also be worn
8.	Waste water treatment plants, (septic tanks, trickling filters, activated sludge, waste stabilisation ponds, sludge digesters)	Faeco-oral transmission and aerosolborne infections. The most problematic organisms (with latency periods in parenthesis) are: salmonella (6–2 h), tetanus (8 days), shigella (1–3 days), leptospirosis (4-10 days), *E. coli* (3 days), tulareisis (3–5 days), yerisinia (1–14 days), hepatitis A (30 days), B (60–90 days), and C (6–9 weeks), HIV (1–2 years), poliomyelitis (6–20 days), *Entamoeba hystolytica*, *Giardia lamblia* (2–4 weeks) and others	Newly employed and less than 2 years of work experience are affected most; hepatitis C virus is a growing concern among sewer workers but not related to occupational risk; infections with hepatitis E virus and *Helicobacter pylori* are yet to be confirmed among sewer workers

Table 1 Causative agents of biohazards in the construction environments

casualties, unusual illness for region/area, definite pattern inconsistent with natural disease, unusual liquid spray which can be liquid or vapour, and spraying and suspicious devices or packages. Early detection and rapid response to bioterrorism depend on close cooperation between public health authorities and law enforcement.

Recognition of a biological hazard

Hazard may be identified from:

- complaints from the workers either directly or through supervisors, or 'walk-through' site survey.

Biohazards may be identified by:

- identification of a credible threat;
- evidence of bioterrorism (devices, agent, clandestine lab);
- diagnosis through identification of the disease agent;
- detection through surveillance data.

Often, when people are exposed to a pathogen, they may not be aware, and those who are infected, may not feel sick for some time. This delay between exposure and onset of illness is called the incubation period which is characteristic of infectious diseases (**Table 1**). The incubation period may range from several hours to a few weeks, depending on the exposure and the pathogen dose. Some of the characteristics of biohazards follow.

- It takes a long time to recognise the pathogens or their products (toxins) directly causing biohazards as true causes.
- Infection is likely to spread through sensitive organisms or vectors.
- Even if pathogens infect people, infection may not be visible immediately and the infected person may be a carrier unknowingly.
- In some of the emerging diseases, the time elapsed between infection and diagnosis may cause the conditions of the disease to worsen because proper cause-removing measures and treatment procedures cannot be taken and the cause-and-effect relationship is unclear.

Routes of entry

Biohazards may be transmitted to a person through inhalation, injection, ingestion or physical contact (Figure 2). Out of the 75 trillion cells in the human body, 25 trillion make up the blood which readily picks up the biohazardous materials. The eye is a sensitive organ which can pick up an infective agent through foreign particles which serve as mechanical agents and, when the eye is rubbed with hands, the infective agents get settled. The lung is another organ (with 75 square yards (62.71 sq. m) of surface area) with rapid blood flow which can pick up the biological materials. Through coughing, sputum formation and macrophage cell production the lungs try to eliminate the germs and particles.

The skin is the largest organ of the body with about 19 square feet (1.765 sq. m) of surface area. One square inch (6.451 sq.cm) of skin contains 72 feet (21.95 m) of nerves and 15 feet (4.572 m) of blood vessels. Skin is an effective barrier for infections. The thickness of skin varies from 0.5 mm on eyelids to 3 or 4 mm on the palm and sole of the foot. Injuries can permit the skin to absorb the infective agent even though the skin may have 60 layers of cells beneath. The digestive system is another major route and many pathogenic organisms enter through faeco-oral transmission. Washing hands as frequently as possible, preferably with soap or disinfectants, can minimise the biohazard transmission.

The construction industry is currently exempt from the US Federal Bloodborne Pathogens Standard [OSHA(c)]. However, the General Duty Clause [Section 5(a) (1)] of the OSH Act requires that contractors provide a workplace free from known hazards such as those that are bloodborne.

Biosafety levels (BSL)

The biosafety levels (BSL) were classified in 1974 by the US Centers of Disease Control and Prevention based on a combination of procedures and facility design. In the European Union they fall under Council Directive 90/679/EEC of 26 November 1990 on the protection of workers from risks related to exposure to biological agents at work. There are four levels of biosafety where practices, safety equipment, and facility design and construction are deployed appropriately to minimise biohazards:

Level 1 (BSL-1) – for sites suspected to have defined and characterised strains of viable microorganisms not known to cause disease in healthy individuals; does not need any special primary or secondary barriers other than a sink for handwashing. Disinfectants must be properly used.

Level 2 (BSL-2) –moderate-risk agents that are present and associated with human disease of varying severity; infectious agents include, *E. coli,* many salmonella, some fungi such as ringworm, California encephalitis viruses, human herpes simplex viruses, many influenza viruses, transmissible gastroenteritis of swine, mouse hepatitis virus and a few parasites.

Preventive measures include: isolation of source areas from other activities, biohazard sign, room surfaces, impervious and readily cleanable equipment (should include an autoclave), certified high efficiency particulate air (HEPA*) filtered class I or II biological safety cabinet for organism manipulations, and PPE to include aprons to be worn in the infection areas, and gloves worn when handling infected materials. All contaminated material to be properly decontaminated. A safety plan must be prepared and registered with the appropriate governmental arm.

Level 3 (BSL-3) – infectious agents that usually cause serious human or animal disease, or which can result in serious economic consequences, but do not ordinarily spread by casual contact

---

* HEPA is a registered trade mark. HEPA filters remove at least 99.97% of airborne particles 0.3 micrometers (µm) in diameter.

from one individual to another (high individual risk, low community risk), or that can be treated by antimicrobial or antiparasitic drugs. Examples include bacteria such as anthrax, Q fever, tuberculosis, and viruses such as hanta viruses, human immuno-deficiency viruses (HIV), eastern and western equine encephalitis viruses.

Preventive measures include specialised design and construction of critical areas, with controlled access, double door entry and body shower; all wall penetrations must be sealed; ventilation system design must ensure that air pressure is negative to surrounding areas with no recirculation of air; air exhausted through a dedicated exhaust or HEPA filtration system; minimum furnishings that can be readily cleanable and sterilisable (fumigation); risk area windows sealed and unbreakable; and provision of backup power. Provision to be made of dedicated handwashing sink with foot, knee or automatic controls, adequate PPE (aprons, head covers and dedicated footwear, gloves and appropriate respiratory protection), showers before exit, animal wastes to be disposed of as infectious materials, activities involving infectious materials to be conducted in biological safety cabinets or other appropriate containment devices.

Laboratory staff must be fully trained in the handling of bio-hazardous materials, in the use of safety equipment, disposal techniques, handling of contaminated waste, and emergency response. Standard operating procedures must be provided and posted within the laboratory outlining operational protocols, waste disposal, disinfection procedures and emergency response. The facility must have a medical surveillance programme for all personnel working in the containment laboratory and an accident report system. There is also a need for primary and secondary barriers to protect personnel and the environment (infectious aerosols possible); you must also prepare a safety plan and register with governmental arms; monthly activity reports required.

Level 4 (BSL-4) – for dangerous and exotic agents that pose a high individual risk or life-threatening diseases; may be transmitted by way of the aerosol route; no available vaccine or therapy. Infectious agents are all viruses, such as ebola viruses, herpes B virus (monkey virus), foot and mouth disease. Facilities to be provided are highly specialised, secure with an air lock for entry and exit, positive pressure ventilated suits, and a separate ventilation system with full controls to contain contamination. Only fully trained and authorised personnel may enter the Level 4 containment laboratory. On exit from the area, personnel will shower and re-dress in street clothing; personnel to be provided with one-piece, positive-pressure-ventilated suits.

Biohazard preventive measures

Depending on the nature of construction work, a contractor with the help of a Consultant will be able to assess the nature of the hazard that is likely to be there and make provision for recommended preventive measures. These measures may be environmental, medical or others. Under the Health and Safety at Work Act 1974 and the Management of Health and Safety at Work Regulations 1999, employers have a legal duty to protect the health of employees and anyone else, e.g. members of the public who may be on their premises. Specific legislation on hazards that arise from working with

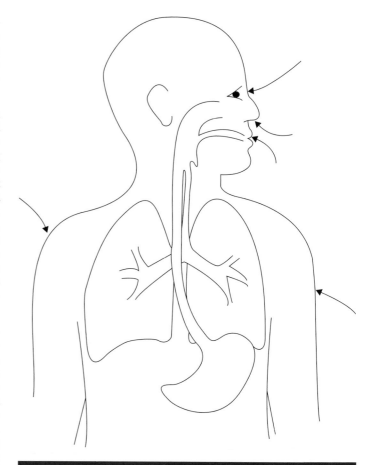

Figure 2 Routes of entry of biological agents

biological agents such as bloodborne viruses is contained in the COSHH Regulations 2002. Under COSHH, employers have a legal duty to assess the risk of infection for employees and others affected by work. Suitable precautions and adequate information, instruction and training must be provided.

Environmental measures – dusty environments must be wetted frequently to reduce dust to acceptable levels. Use of breathing apparatus and dust masks; regular testing of workers to ensure health and fitness to work; workers must be given a complete medical examination to determine suitability to work before or as needed to rule out the possibility of disease; workers who are found to be suffering from bacterial, viral and fungal infections should be isolated and treated before returning to work.

Medical measures – people working in wastewater treatment plants and infectious areas may observe certain additional precautions besides personal hygiene and environmental sanitation. These include:

■ medical history – limb mobility, skin disorders, asthma, disorders that could produce unconsciousness (e.g. diabetes, stroke);

■ physical examination;

- liver, kidney and hematologic function;

- immunisation review;

- strongly recommend: tetanus–diphtheria (booster every ten years);

- optional: polio and typhoid fever. These must be discussed on a case-by-case basis with an occupational physician before taking any action;

- not recommended: hepatitis B virus (HBV), cholera.

Hazard controls

Hazard control is the basis of biosafety. It is a rule of thumb to keep exposures or risk of hazards to the lowest possible. Hazard control can be temporary or permanent. Engineering controls often offer more permanent solutions which are sustainable. Controls are usually specified in the legislation and in most cases organisations will have hazard specific guidelines. The employer has a duty of due diligence and is responsible for taking all reasonable precautions, under the particular circumstances, to prevent injuries, accidents or infections in the workplace. When the guidelines are not clear, occupational hygienists or safety professionals may be consulted for 'best practice' or 'standard practice'.

Methods to adopt for hazard control which should be applied at the source of hazard, along the route of transmission and at the worker involved are:

- **Elimination** (or substitution): remove hazard from the workplace.

- **Engineering controls**: designs or modifications to plants, equipment, ventilation systems and processes that reduce the source of exposure. These are very reliable if they are well designed, used and maintained properly.

- **Administrative controls**: controls that alter the way the work is done, including timing of work, policies and other rules, and work practices such as standards and operating procedures (including training, housekeeping, equipment maintenance and personal hygiene practices).

- **PPE**: this primary barrier is the equipment worn by individuals to reduce exposure.

- **Process control**: involves changing the way a job activity or process is done to reduce the risk. Monitoring is required before and also after the changes are implemented to make sure the changes result in lower exposure. Examples include:

  - using wet methods rather than dry when drilling or grinding to prevent dust from being created;
  - using steam cleaning instead of solvent degreasing (but ensure low potential high temperature/heat stress);
  - use electric motors rather than diesel ones to eliminate diesel exhaust emissions;
  - float 'balls' on open-surface tanks that contain solvents used for degreasing operations;
  - instead of conventional spray painting, try to dip paint with a brush, or use 'airless' spray paint methods. These methods will reduce the amount of paint that is released into the air;

  - decrease the temperature of a process so less vapour is released;
  - use automation – exposure will be reduced;
  - use mechanical transportation as against manual.

- **Enclosure and/or isolation**: these methods aim to keep the hazard in and the worker out (or vice versa). An enclosure keeps a selected hazard physically away from the worker, e.g. abrasive sand blasting cabinets or remote control devices. Isolation places the hazardous process geographically away from the majority of workers or public.

- **Ventilation**: ventilation is a method of control that strategically 'adds' and 'removes' air in the work environment. Local exhaust ventilation is very adaptable to almost all chemicals and operations. It removes the contaminant at the source so that it cannot disperse into the workspace.

- **Retrofit option**: the options are divided into two categories: specific technologies, such as filtration and air cleaning devices, and more generic approaches to increasing building protection, such as building pressurisation strategies and isolation of areas of potential concern (e.g. mail rooms). These are employed primarily to limit dirt build-up on cooling coils and other wetted surfaces in order to reduce the potential for microbial growth and to maintain good heat transfer between the air and the coil surfaces. However, typical levels of filtration are not always very effective in removing biological particles of the sizes less than 1 μm (NIOSH, 2003). Some of the available retrofit technologies are given in Box 3.

Box 3 The retrofit technologies

- enhanced particle filtration
- gaseous air cleaning
- UV germicidal irradiation
- photocatalytic oxidation
- all filtration and air cleaning technologies
- work-area treatment
- system recommissioning
- envelope tightening
- building pressurisation
- relocation of air intakes
- shelter-in-place
- isolation of special spaces
- shutdown and purging
- automated HFAC response.

- **Rodent control**: rat infestation can be quite extensive. Surveyors and those carrying out initial site clearance are likely to come across colonies of rats in old abandoned structures, etc. Rat-proofing is very important in construction industry. Buildings should thus be kept in good repair. Concealed spaces and structural pockets, openings greater than 1.25 cm leading to voids and food spaces, gaps around penetrating fixtures (e.g. pipes or ducts passing through bulkheads or decks) regardless of location, should be obstructed with rat-proofing materials, and the insulation layer around pipes, where over 1.25 cm thick, should be protected against rat-gnawing.

Special systems to control emerging biohazards

Handling and containment of biohazardous materials

Currently, the biohazard facilities are being used not only to detect pathogenic microbes but also to do recombination deoxyribonucleic acid (DNA) tests with the development of biotechnology. In the biohazard facilities, handling of materials and the equipment should be based on risk assessment, and risk avoidance should be the guiding principle (Sherlock and Morrey, 2002).

The Commonwealth Gene Technology Act 2000 (OGTA) took effect on 21 June 2001. The OGT report is concerned with genetically modified (GM) materials and biotechnology. It requires specific procedures and strict material audit tracking, storage requirements, security and import requirements of material. This is particularly applicable to electronic related systems. It should also manage the release of GM material into the environment. Office of the Gene Technology presently uses third-party certifiers to certify facilities and processes (IEEE, 2009).

There are Standards for containment requirements categorised based on the types of biohazardous substances. These facilities are defined under Physical Containment: PC1, PC2, PC3 and PC4. They should be equipped with personal safety equipment, negative room pressure, non-potable water management, air filtering, surface treatments, safety showers, door access, and non-dust settling and fully cleanable rooms. Standards also cover fume cupboards.

For the containment of the biohazardous materials, the following guidelines should be followed as generalised for PC2, which are:

- Surfaces where they occur or likely to occur must be smooth, impermeable, cleanable and resistant to damage by cleaning/disinfection agents.
- Open spaces under benches accessible for cleaning.
- Hands-free wash basin (or disinfectant dispenser).
- Storage area for protective clothing.
- Eyewash equipment (or sterile eye-irrigation packs).
- Supply of disinfectants – labelled and dated.
- Procedures for decontaminating spills.
- Containment facilities for aerosols (Class I or II biosafety cabinet).
- Personnel trained (to keep records).
- Annual inspection.

Decontamination

Decontamination is required where the biohazardous materials have been handled or spilt around:

- Provision shall be made to decontaminate the biohazard areas or materials and the room independently with formaldehyde gas.

- Decontamination of the safety cabinet(s) shall be performed in accordance with the requirements of national specifications; Ministries of Environment and Health have specified in many countries.
- Decontamination of the room will require a closed-off damper in the discharge from the exhaust filter and means of closing the room replacement air aperture.
- A pest/vector control programme should be in place.

Field kits

Several kits are available for field use in case of emergencies, to protect against biohazards. A 19-piece Bloodborne Pathogen/Bodily Fluid Spill Kit has been designed to guard during biohazard clean-up and is approved by OSHA or UK Health and Safety Executive (HSE). The kit contains: (3) antiseptic cleansing wipes (sting free); (1) disposable gown (full sleeves); (1) disposable bonnet and several bags and wipes. A 176-piece first aid kit has been specifically designed for the contractor; it serves up to 25 people and contains a wide variety of items that are designed to handle first aid emergencies such as burns, major bleeding, minor cuts, eye injuries, etc.

Good working practices

General

- All procedures should be followed so as to minimise the hazards.
- Recommended dress code should be followed while at work.
- Design and operate processes and activities to minimise emission, release and spread of substances hazardous to health.
- Take into account all relevant routes of exposure – inhalation, skin and ingestion – when developing control measures.
- Control exposure by measures that are proportional to the health risk.
- Choose the most effective and reliable control options that minimise the escape and spread of substances hazardous to health.
- Where adequate control of exposure cannot be achieved by other means, provide, in combination with other control measures, suitable PPE.
- Check and review regularly all elements of control measures for their continuing effectiveness.
- Inform and train all employees on the hazards and risks from substances with which they work, and the use of control measures to minimise the risks.
- Ensure that the introduction of measures to control exposure does not increase the overall risk to health and safety.

Specific to biohazards (also applicable to laboratories)

- PPE must be stored in a well-defined place, checked and cleaned at suitable intervals, and when discovered to be defective, repaired or

ICE manual of health and safety in construction © 2010 Institution of Civil Engineers

replaced before further use. Military mask filters typically contain two types of media: an activated carbon media for gas/vapour adsorption and a HEPA media to protect against aerosol threats such as agents of biological origin (e.g. bacteria, viruses and toxins). It is capable of removing at least 99.97% of all airborne particulate hazards in the form of aerosols. PPE which may be contaminated by biological agents must be removed on leaving the working area, kept apart from uncontaminated clothing, decontaminated and cleaned or, if necessary, destroyed.

- Eating, chewing, drinking, taking medication, smoking, storing food and applying cosmetics is prohibited in the work areas.

- Persons working in biological laboratories should adhere to a strict code of personal hygiene; hands must be decontaminated immediately when contamination is suspected, after handling infective materials and before leaving the site. When gloves are worn, these should be washed or preferably changed before handling items likely to be touched by others not wearing gloves. Examples are – telephones, paperwork, computer keyboards and, where practicable, equipment controls should be protected by a removable flexible cover that can be disinfected.

- Bench tops and work surfaces should be regularly decontaminated according to the pattern of work and should be cleaned after use.

- Used materials awaiting disinfection should be stored safely.

- There must be safe storage of biological materials. All waste material, if not to be incinerated, should be disposed of safely by other appropriate means.

- Accidents and incidents must be immediately reported to and recorded by the person responsible for the work or some other delegated person.

- Effective disinfectants should be made available for immediate use in the event of spillage.

- The place where biological materials are stored should be easy to clean. Bench surfaces should be impervious to water and resistant to acids, alkalis, solvents and disinfectants.

- All infectious and toxic materials should be correctly labelled and storage areas should show appropriate warning notices.

- Personnel must receive suitable and sufficient information, instruction and training in the procedures to be conducted in case of spillage and emergencies.

- Training received must be recorded on a proforma and signed by the trainee; this should also include details of relevant documents read.

- Workers must ensure that other persons, e.g. cleaners, maintenance personnel, visitors, are not exposed to biological hazards.

Common disinfectants against biological agents

There is no universal disinfectant for biohazards. Disinfectants may also deteriorate on standing or be inactivated by detergents, organic matter, etc. Some common types are:

a) **Hypochlorite solutions**: commonly recommended concentrations – 1000 ppm for surface decontamination, 2500 ppm for discard containers, 10 000 ppm for spillages; these are active against bacteria (including spores) and viruses but have limited activity against fungi and tuberculosis bacilli are compatible with anionic/non-ionic detergents but corrode many metals, damage rubber and are inactivated by organic materials and so need frequent changing.

b) **Chlorine releasing granules**: contain sodium dichloroisocyanurate (NaDCC) and may also contain absorbent powders. They have a relatively long shelf life and are useful for spillages.

c) **Clear soluble phenolics**: active against vegetative bacteria (including tuberculosis bacilli) but not active against spores and have a limited effect on fungi; not active against many viruses. Compatible with anionic/non-ionic detergents and metals, and inactivated by rubber and some plastics.

d) **Glutaraldehyde**: has a similar range of activity to hypochlorites but does not corrode metal; does not readily penetrate organic matter and is relatively unstable once inactivated; is also a potent allergic sensitiser.

e) **Alcohols**: are normally used as 60–80% v/v solutions in water; are active against protozoa and many viruses and vegetative bacteria (but not tuberculosis bacilli); can be used as a disinfectant skin rub (often with addition of 5% w/v chlorhexidine); do not readily penetrate organic matter; are flammable.

f) **Virkon**: active against many organisms; relatively non-toxic and non-corrosive.

g) **Autoclave**: most biological agents (including bacterial and fungal spores) can be rendered non-viable by exposure to steam under pressure (1 hour at 121°C (1 bar)). The hazard from adequately treated material should be very low to none.

Spillage management

There should be contingency plans which must include spillage control, room evacuation, fumigation and decontamination and, if there is a risk of infection, first aid and medical treatment (prophylaxis) and health surveillance and counselling for exposed people. Equipment and materials required for dealing with spillages must be readily available and all workers must know the procedures (Sridhar *et al.*, 2009):

- Only people essential for carrying out repairs and other essential work may be permitted in the affected area and they must be provided with appropriate PPE and other necessary equipment.

- Employees (or their representatives) must be informed as soon as practicable after an incident or accident that has (or may have) released a biological agent.

- Employees have a responsibility for reporting such incidents or accidents immediately. If there is a risk of airborne infection, the site of incident must be evacuated as quickly as possible; fumigation may be required before reoccupation. Reoccupation is only after certification.

- Spillages should be contained and covered with disinfectant (powder or granules as appropriate) or absorbent paper/cloth soaked in disinfectant; the disinfectant should be allowed to act for at least 15 minutes.

- The debris should be swept gently into a dustpan using a piece of board or stiff card; any residual pieces of glass, etc. should be picked up with forceps or swabs; debris should be put in a suitable container for disposal by a safe route; additional disinfectant should be applied to contaminated surfaces.

- People attending casualties should avoid becoming contaminated themselves by observing personal hygiene and stipulated medical procedures; soap and water or a suitable detergent should be used.

- If there is a health risk to the community in the vicinity, there must also be contingency plans, made in consultation with local community physicians, to limit spread.

Biowaste management

Management of biowaste is very important at the site of generation. Every country has regulations under hazardous waste management. However, the basic requirements for safe handling are:

- Waste must be contained in a biohazard box or bag which is autoclavable with colour code, preferably yellow.

- Liquid wastes must be inactivated either chemically (e.g. bleach) or autoclaved.

- The solids must be autoclaved and then transferred to a different bag (colour coded red–white) to indicate that the waste has been deactivated; and special treatment should be given for radioactive waste and waste of a biosafety level 3 and higher.

The conventional disposal technologies are hazardous landfills, incinerators with control of emissions and ash, bioreactors and immobilisation/vitrification, depending on the degree of hazard the waste may pose to the community, the handler or the environment.

Monitoring and assessment of biohazards

Risk assessment

Since the Health and Safety at Work Act of 1974, there has been a requirement on all employers to assess the risks that would arise from their organisations' activities to themselves, their employees and those who might be affected. Biorisk assessment should take into cognisance the pathogenicity/infectivity, virulence/lethality, infective dose, therapy/prophylaxes, epidemic potential, resistance survival in the environment, geographic spread (endemic) and mode of transmission of the infective agent.

The biohazard risk may be assessed based on the hazard group to which they belong:

- Hazard Group 1: a biological agent unlikely to cause human disease.

- Hazard Group 2: a biological agent that can cause human disease and may be a hazard to employees; it is unlikely to spread to the community and there is usually effective prophylaxis and effective treatment available.

- Hazard Group 3: a biological agent that can cause severe human disease and presents a serious hazard to employees; it may present

a risk of spreading to the community, but there may be prophylaxis or treatment available.

- Hazard Group 4: a biological agent that causes severe human disease and is a serious hazard to employees; it is likely to spread to the community and there is usually no effective prophylaxis or treatment available.

A hazard has the potential to lead to an incident. An incident is an event which may be sudden or unexpected and uncontrolled, that has, or could have, resulted in an adverse outcome or consequences. Risk assessment aims to reduce the number/impact of incidents occurring, by determining what can happen, why, how and what can be done to prevent an incident (**Table 2**). Risk management is about utilising risk assessment, best practice and latest technology, to create an environment in which good clinical care and business support services can flourish with little or no disruption created by injury, ill health or other losses.

i) Identify all the hazards associated with the work.
ii) Identify all people who might be affected (including visitors; clients; public; new, young and inexperienced workers).
iii) Analyse the consequences of hazards causing an incident.
iv) Analyse the likelihood of the hazards causing an incident after considering existing control measures.
v) Evaluate the risks by applying numeric values to the consequence and likelihood then multiplying them together.
vi) Register, rank and prioritise the risks for action.
vii) Decide on whether the risk can be avoided, transferred, reduced or simply accepted as it is.
viii) Decide on reasonably practicable risk treatments if risk not to be accepted or transferred.
ix) Implement the risk treatments including any necessary training and safe systems of work.
x) Provide any essential and appropriate PPE for risks that cannot be controlled otherwise.
xi) Ensure all stakeholders are informed of the results and outcomes of the risk assessment.
xii) Review the risk assessment as the situation changes and periodically as a routine.
xiii) Implement any necessary adjustments to risk assessment and evaluate until it is reduced to the lowest or tolerable level.

Environmental impact assessment (EIA)

Under European and UK law, major infrastructure projects such as construction works must provide an environmental statement (ES). This is a full analysis of significant potential environmental effects of the construction and operation of the project at a local, regional and wider scale. It is also mandatory in many countries for any development projects. The EIA process consists of five principal steps:

1. **The scoping** – an initial and outline assessment to identify the range of environmental impacts that the scheme may

give rise to during construction and operation. This determines the specialist areas to be examined in detail.

2. **Baseline studies** – this establishes the environmental conditions which will exist when the scheme is implemented. This is done by gathering data, from official sources and by survey, about the existing environmental conditions and considering the nature of the scheme. Changes in environmental conditions have to be extrapolated to future years to establish the conditions that will exist when the scheme is constructed and operated.

3. **Assessment** – the analysis to identify and quantify, where appropriate, the constructional and operational impacts. Involves consulting bodies charged with statutory duties for protecting the environment holistically including biophysical, ecology and biodiversity, water, air and soil, geology and hydrogeology, socio-cultural, heritage, health, etc.

4. **Mitigation options** – the examination of ways to eliminate or reduce the significant or main adverse environmental effects to an acceptable level. These measures then become an intrinsic part of the proposed scheme.

5. **Production of the environmental statement** – including the non-technical executive summary.

As the baseline data gathering and scheme design progress, it is desirable for environmental issues to be addressed in the developing design, well in advance of fully defining the actual impacts and characterise mitigation measures.

Roles and responsibilities of stakeholders

i) Government

- Play an important role as legislators and regulators; the World Bank and development agencies can influence labour standards and working conditions on construction sites in many parts of the world; their procurement policies and conditions of tender should set exemplary standards.

- Develop a coherent legislative and policy framework on occupational health and safety in the sector; the national policy must include a system for promotion and enforcement of the regulations.

- Provide information, training and guidance on hazards and their prevention; negligent employers must be shown that they will face stiff fines, high compensation claims, social stigma and loss of licence or liberty.

ii) Employer

- Must be committed and adhere to labour standards; insist that these are respected by all subcontractors and suppliers; these labour standards are to be based on International Labour Organisation Conventions, including such fundamental human rights as freedom of association, the right to organise and the right to collective bargaining.

- Should be involved in building workforce skills in health and safety; several Construction Industry Training Boards have introduced mandatory training on health and safety; provide skills certification and recognition of prior learning, boosting quality and productivity as well as reducing injuries and ill health.

- Implement compulsory employers' liability insurance to cover all workers on site.

- Encourage worker participation in Company health and safety policies and systems for risk management; downsizing and outsourcing have created a construction industry dominated by precarious, informal contractual conditions, by sub-contracting and by bogus self-employment which has a direct and negative impact on health and safety.

- Ensure appropriate standards to prevent blood borne infections (eLCOSH 2005):

  - An employer to instruct each employee to recognise and avoid unsafe conditions. The employer would be required

Likelihood							Action
5 Almost certain	5	10	15	20	25	25	Prohibited Stop immediately Inform authorities
4 Likely	4	8	12	16	20	16–20	Very high priority
3 Possible	3	6	9	12	15	12–15	High priority
2 Unlikely	2	4	6	8	10	6–10	Medium priority
1 Rare	1	2	3	4	5	1–5	Low priority
Consequence	Insignificant	Minor	Moderate	Major	Catastrophic		

Table 2 Risk evaluation matrix
Data taken from Barnsley Primary Care Trust (2002)

to train workers in the hazards of bloodborne pathogens (Regulatory Standard 29 CFR 1926.21(b) (2)).

- Ensure collection and separation of sharps and other waste that may be contaminated (Regulatory Standard 29 CFR 1926.25).
- Ensure use of appropriate PPE. The contractor is required to provide gloves, boots, overalls and eye/face protection where appropriate (Regulatory Standard 29 CFR 1926.28).

In the UK, advice is available from HSE's biosafety website on controlling the risks of infection at work:

- environments should be cleaned and maintained
- encourage hand hygiene
- clean and remove spillages of blood and other body fluids
- clean care equipment
- minimise the risks of exposure to bloodborne infections
- provide PPE
- dispose of healthcare waste, including sharps
- procedures for handling and laundering work clothes
- management role in infection prevention and control.

iii) Clients' and contractors' associations

- Safety, health and welfare provisions are included as mandatory components in tender documents and all contractors should include them in their cost estimates.
- All management and supervisory staff on their sites have demonstrable competence in occupational health and safety and in management and supervisory skills.
- All workers should have a demonstrable skill level, incorporating occupational health and safety.
- All contractors respect labour standards.
- Structures and resources are in place to implement policy and comply with the law.
- There is proper communication and coordination between contractors and the participation of workers, including induction training.
- Occupational health and safety targets should be audited for each contractor on site; previous health and safety performance should be included in the selection criteria for tenders and all bids should present a detailed health and safety plan before work starts.

Selected case studies

Malaria in Mangalore, India (*The Hindu*, 2004)

In Mangalore, south India, apathy among people and the officials concerned was primarily responsible for the rise in the number of malaria cases. While 2609 cases of malaria were recorded between January and June 2003 in the city, the number rose to 5440 during 2004. The number of malaria cases reported in June alone was 1858. The main problem encountered was the reluctance of the people and those in the construction industry to follow preventive measures. While people were averse to using mosquito nets, those in the construction industry, hotels, and apartments did not bother to maintain cleanliness in their surroundings.

The Malaria Control Cell had surveyed 16 654 houses in the city, released guppy fish in 1699 wells, and persuaded people in 90 per cent of the houses to cover the overhead tanks. Under the Active Surveillance Programme, the cell had tested 7896 people for the disease. Of them, 174 people had been found to be carrying malaria parasite.

It had been proven world over that there was a direct correlation between a spurt in construction activities and an increase in the number of malaria cases. Malaria came back to haunt any developing city once every 10 to 15 years when construction activity witnessed a boom. The construction and hotel industry as well the managers of apartments, hostels, and so on have important roles in controlling the spread of the disease.

Construction industry in Nigeria

Nigeria, with a population of over 140 million, has a lot of construction activity going on in all the major towns and cities. The construction activities are mainly in the areas of buildings, airports, roads, estate developments, and healthcare facilities. A field survey was conducted in 18 such construction companies ranging from small scale, medium scale and large size. Site/Project Managers or Managing Directors were involved in the data collection.

- **Health and welfare**: some companies provide additional cash incentives rather than providing health facilities. Workers patronise private food vendors for their food needs and very often they are substandard with respect to nutritional quality and hygiene.
- **Biological hazards:** the knowledge is poor and some of them did not distinguish between biological hazards and environmental hazards. Larger companies with more than 100 workers have some programmes to address the biohazards. About 85% of the sampled populations confessed that they experienced some biohazards during their work. The prevalence of exposure to biohazards was more among workers in companies who were sub-contracted from larger companies. Airborne infections were more common. Respiratory diseases, musculoskeletal disorders of the back and other infections due to exposure to dusts, fumes, gases, parasites, etc. were also reported (Pattron, 2009).
- **Control of biological hazards**: main contractors have better facilities and equipment to cope with biohazards as they follow some standards.
- **Occupational health and safety**: multinational companies (those handling rehabilitation of the sewage treatment plants) usually provide PPE and an occupational health nurse visits the site twice a week and attends to any emergencies. Some noted hospitals are

also attached to the companies as retainers. Some identified first aiders provide first aid when in need and they are trained.

■ **Prevention of hazards**: many companies are reluctant to install any preventive or control devices for control of biohazards. This is mostly due to cost implication. Toilet facilities, change rooms and showers are only provided by the larger companies. Workers in smaller companies use open areas and sorrounding bush for such activities. A company reported a snake bite of a worker during such practices.

All the companies had appropriate signage and symbols and used them in high-risk zones within the construction sites. Usually, at the commencement of any project, the workers are given precautionary briefings. A lot more attention is given to safety than the health of workers.

New BSL-3 laboratory: in a newly constructed 'state-of-the-art' laboratory facility in USA (Global Biohazard Technologies, 2009)

This laboratory had been built and GBT was requested to review the Standard Operating Procedures (SOPs) and certify the laboratory to be in compliance prior to start-up. The laboratory was part of a new, state-of-the-art laboratory facility which had been in operation for several years. The laboratory was designed to BSL-3 requirements, but was being operated at BSL-2. During the evaluation of the facility it was found that although the two exhaust fans were each running at approximately 75% of their capacity there was virtually no directional air flow. A close evaluation of the laboratory demonstrated that none of the room penetrations were properly sealed, but the doors were virtually air tight. In addition, there was a canopy hood over the autoclave door on the containment side which was directly connected to the canopy on the non-contained side allowing air to be pulled from outside of containment through the canopy hoods. Remediation of the penetration sealing problems, removal of the door gaskets and removal of the containment side canopy hood should improve the directional air flow in the laboratory. Failure to have a third-party review of the design concepts resulted in delay of several months to the start-up time and thousands of dollars for remediation.

Public health laboratory

A public health laboratory was designed and built as part of a renovation project in an existing building without consultation with a knowledgeable biocontainment professional. Following design and construction by a reputable local A&E firm, GBT managing partners were requested to visit the lab for certification. It was found that although there was directional air flow from the corridor into the anteroom, there was virtually no differential between the anteroom and the laboratory. In addition it was noted that the ceiling was tiled with acoustical ceiling tiles and that the flooring was the tile floor that had been in the room prior to renovation. The biosafety cabinet was improperly installed and a number of other problems were uncovered during the evaluation.

The laboratory was eventually opened following more than a year of further renovation and at an extremely high cost.

A university research facility

A university completed the construction of a biocontainment laboratory facility with a suite of BSL-3 research laboratories and a BSL-3 animal facility with an aerosol testing laboratory and requested that GBT professional biosafety personnel visit the site and certify the laboratory and animal facility. The facility had been commissioned and GBT personnel had been involved with the review of the safety features of the laboratories and animal rooms from the onset of the design process. In addition, GBT personnel had visited the facility numerous times during the construction phase and had provided guidance on heating, ventialation and air-conditioning (HVAC) systems and on the sealing of penetrations and final finishing of the facility. Although some minor problems were noted, the facility was capable of being certified very soon after completion. GBT personnel, as part of the certification process, reviewed the safety SOPs for the facility and assisted the health and safety personnel of the university in developing appropriate, site-specific SOPs which allowed for a safe and efficient operation of the facility.

Summary of main points

Biohazards are closely interrelated to environmental and occupational health risks. They can be prevented or significantly mitigated by activities in various sectors in addition to health. Infrastructure, energy, and agriculture are primary. The infrastructure sector, with the construction industry in particular, has huge potential to improve health outcomes and save lives: housing, water, sanitation, drainage, transportation, urban development and energy projects could avert a large percentage of the deaths and disability-adjusted life year (DALY) loss.

■ Many biological agents are communicated by way of air, such as exhaled bacteria or toxins of mouldy grains. The production of aerosols and dusts should be avoided in the manufacturing process, during cleaning and/or maintenance.

■ Good housekeeping, hygienic working procedures and use of relevant warning signs are key elements of safe and healthy working conditions.

■ Many microorganisms have developed mechanisms to survive or resist heat, dehydration or radiation, for example by producing spores. The workplace must develop decontamination measures for waste, equipment and clothing, and appropriate hygienic measures for workers, as well as proper instructions for safe disposal of waste, emergency procedures and first aid.

■ All construction workers, particularly managers and supervisors have in place an effective safety and health programme.

■ All hazards occurring are audited regularly and solutions followed. Increase the level of training pre-employment and on-job for all construction workers including employees and supervisors. These measures would reduce, prevent or eliminate the biohazards and make the construction industry safer.

References

Atsui, K. The Present State of Biohazard Guide System and Facilities. *Journal of Japan Air Cleaning Association* 2007, **4**(5), 297–303.

Barnsley Primary Care Trust. *What is Risk Assessment?*, 2002. Available online at: http://www.barnsleytrust.org

Biohazard Cleaning. *What is a Biohazard?*, 2008. Available online at: http://www.biohazardcleaning.co.uk/links/html

Biohazard UK Ltd 2008. Available online at: http://www.biohazarduk.com

Brunekreef, B. and Holgate, S. T. Air Pollution and Health. *Lancet* 2002, **360**, 1233–1242.

Bupa. *Insect Bites and Stings*, 2009, Bupa's Health Information Team, July. Available online at: http://www.bupa.co.uk/individuals

CDC. *Bioterrorism Overview, Centers for Disease Control and Prevention*, 2008, 12 February. Available online at: http://www.bt.cdc.gov/bioterrorism/overview.asp

City of London. *Health and Safety – City of London Display Screen Equipment Project Survey*, 2009. Available online at: http://www.cityoflondon.gov.uk

Coker, A.D. and Olutage, F.A. *Combating the Guinea worm scourge in Nigeria: An Engineering approach in traditional and modern health systems in Nigeria*. Africa World Press Inc. Trenton, N.J. 2006, 417–427.

eLCOSH. Exposure to Biohazards. *Professional Safety Magazine* 2005, **50**(8).

Global Biohazard Technologies. *Case Studies*, 2009. Available online at: http://www.globalbiohazardtechnologies.com

Goetsch, D. L. *Occupational Safety and Health*, 3rd edition, 1999, Upper Saddle River, NJ: Prentice Hall.

Harborne, D. J. Emergency Treatment of Adder Bites: Case Reports and Literature Review. *Archives of Emerging Medicine* 1993, **10**(3), 239–243.

IEEE. *Biohazards in Engineering, Global History Network*, 2009. Available online at: http://www.ieeeghn.org

NIOSH. *Guidance for Protecting Building Environments from Airborne Chemical, Biological, or Radiological Attacks*, 2002, National Institute for Occupational Safety and Health, DHHS (NIOSH) Publication No. 2002-139.

NIOSH. *Guidance for Filtration and Air-cleaning Systems to Protect Building Environments from Airborne Chemical, Biological, or Radiological Attacks*, 2003, National Institute for Occupational Safety and Health, DHHS (NIOSH) Publication No. 2003-136.

NIOSH. *Hazards to Outside Workers, NIOSH Safety and Health Topic*, 2008. Available online at: http://www.niosh.org

Pattron, D. D. *Pocket Guide: Worker Safety Series – Construction*, 2005, Washington, DC: OSHA.

Pattron, D. D. Musculoskeletal Disorders – A Major Source of Workplace Injuries: Causes & Recommendations for Prevention. *Curepe: The Health Safety and Environment Quarterly* 2008, 1, 46–49.

Pattron, D. D. *Potential Hazards & Proactive Solutions in the Construction Industry – Construction Industry, Idea Marketers*, 2009. Available online at: http://www.onlinemarketers.com

Persily, A., Chapman, R. E., Emmerich, S. J., Dols, W. S., Davis, H., Lavappa, P. and Rushing, A. Building Retrofits for Increased Protection against Airborne Chemical and Biological Releases NISTIR 7379, 2007, National Institute of Standards and Technology, Technology Administration, US Department of Commerce, pp. 1–179.

Schwartz, J. Air Pollution and Hospital Admission for Heart Disease in Eight US Counties. *Epidemiology* 1999, **10**, 17–22.

Sherlock, R. and Morrey, J. D. *Ethical Issues in Biotechnology*, 2002, Lanham, MD: Bowman and Littlefield.

Sridhar, M.K.C, Wahab, W.B., Agbola, S.B. and Bacliane, A. *Health Care Wastes Management. A handbook for developing countries*. University Press; Ibadan, Nigeria, 2009, 1–244.

The Hindu. People's Apathy Aiding Spread of Malaria: Expert, 2004, The Hindu On-line Edition, July. Available online at: http://www.hindu.com/2004/07/04/01hdline.htm

Unionsafe. Hazards in the Work Place Fact Sheet – Health hazards in construction sites, *Principles of Biosafety* 2009, 1–3. Available online at: http://www.unionsafe.labor.net.au

Referenced legislation and standards

Control of Substances Hazardous to Health Regulations 1994. Statutory instruments 1994 3246. London: HMSO.

Council Directive 90/679/EEC of 26 November 1990 on the protection of workers from risks related to exposure to biological agents at work. *Official Journal of the European Union* L374, 1–12.

Health and Safety at Work etc Act 1974, Elizabeth II 1974, Chapter 37, Update Date: 199206 AN – Accession Number: 00580, 1974.

Management of Health and Safety at Work Regulations 1999. Statutory instruments 1999 324. London: HMSO.

Regulations (Standards – 29 CFR) Part 1926 Safety and Health Regulations for Construction. Occupational Safety and Health Administration (OSHA), United States Department of Labor. Available online at: http://www.osha.gov/pls/oshaweb/owasrch.search_form?p_doc_type=STANDARDS&p_toc_level=1&p_keyvalue=1926

Further reading

Hughes, P. *Introduction to Health and Safety in Construction*, 3rd Edition, 2008, Oxford: Butterworth Heinemann.

Reese, C. D. and Eidson, J. V. *Handbook of OSHA Construction Safety and Health*, 2006, Broca Raton, FL: CRC Press.

Stellman, J. M. *Encyclopaedia of Occupational Health and Safety: The body, health care*, 2009. Vienna: International Labour Organisation, International Labour Office. Available online at: http://www.ilo.org

Websites

Biosafety Information (UK) http://www.hse.gov.uk/biosafety/information.htm

Centers for Disease Control (USA) http://www.cdc.gov/

Health and Safety Executive, Infections at Work and GMO http://www.hse.gov.uk/biosafety/index.htm

Institution of Civil Engineers (ICE) Health and safety http://www.ice.org.uk/knowledge/specialist_health.asp

Chapter 13

Controlling exposure to physical hazards

Theo C. Haupt Building Construction Science, Mississippi State University, USA

doi: 10:10.1680/mohs.40564.0151

Physical hazards occur on all construction sites and threaten the overall health and safety of construction workers. These hazards are the most common type of occupational hazard on construction sites. In 2007 in the United Kingdom alone just less than 100 000 construction workers suffered from an illness which was either caused or made worse as a result of working in the construction industry. Physical hazards include unsafe conditions that can cause injury, illness or death. Exposures and the consequences of exposure to physical hazards can be mitigated and controlled by construction employers through hazard identification and risk assessment techniques, monitoring and management of exposures. This chapter discusses several physical hazards and their health effects. Several practical preventative interventions and mitigating strategies are outlined.

CONTENTS

Box 1	Key learning points

- Understanding the nature of physical hazards.
- Awareness of the consequences of exposure to physical hazards.
- Recognising the most dominant physical hazards.
- Assessing the risk of exposure to physical hazards.
- Applying the hierarchy of control.
- Identifying the physical health effects of exposure.
- Understanding the contribution of designers in mitigating exposures to physical hazards.
- Identifying management and environmental issues.

Introduction

Construction sites are notorious for being fraught with physical hazards, largely as a result of the nature of construction activities themselves as well as the management of these activities and the physical environments in which they occur. Exposure to these hazards continues almost unabated despite increasingly restrictive legislative frameworks in most countries. For example, in 2007 in the United Kingdom (UK) alone just less than 100 000 construction workers suffered from an illness which was either caused or made worse as a result of working in the construction industry

The nature of physical hazards

A physical hazard is any hazard that threatens the physical health and safety of a construction worker. It is a factor within the working environment that is harmful to a worker's body

without him or her necessarily touching it. Physical hazards are the most common type of occupational hazard on construction sites. They include unsafe conditions that can cause injury, illness or death. Examples of physical hazards include:

- Vibration due to, for example, using a concrete vibrator or plate compactor.
- Electrical hazards such as frayed cords, missing ground pins and improper wiring.
- Unguarded machinery and moving machinery parts where guards have been removed, or moving parts that a worker can accidentally touch such as conveyor belts to move bricks.
- Constant loud noise without protection from, for example, power tools exceeding minimum noise levels of 85 dBA.
- High exposure to sunlight and/or ultraviolet rays, heat or cold.
- Exposure to air pollutants such as organic solvents from paints, respirable dust and fumes from diesel engines that exceed workplace exposure limits (WELs) (also known as occupational exposure limits (OELs) in some countries).
- Working from and at heights, including ladders, scaffolds, roofs or any raised work area.
- Working with mobile equipment such as cranes and earthmoving equipment.
- Spills on floors or tripping hazards, such as cluttered working areas or electric leads running across the floor.
- Working in awkward positions such as reaching overhead.

Fortunately, most physical hazards are easily identifiable. However, they are often ignored because of their widespread pervasiveness on construction sites, lack of knowledge about the hazards on the part of workers, resistance by management to making the necessary improvements in terms of time or funding, or delays in making the required changes to mitigate the hazards effectively.

The consequences of exposure to physical hazards

Many hazards such as an incorrectly erected scaffold are immediate threats to the safety of construction workers and, therefore, require immediate attention if either the collapse of the scaffold or the risk of workers, material and equipment falling is to be prevented. Other hazards, such as ergonomic hazards which expose workers to working in awkward positions for extended periods of time, can produce health problems such as lower back pain over time. However, the full extent of the problem may only manifest itself after many months or years of exposure. It is imperative that both immediate and longer term threats are mitigated if the health and safety of workers is to be maintained.

Hazard identification and risk assessment (HIRA)

A hazard refers to the likelihood of something causing harm while a risk refers to the likelihood that harm will occur. Once risks have been identified, focus can be placed on high risk processes and the mitigation of exposure to them. The purpose of the HIRA is to assist in eliminating or mitigating hazards, which if not eliminated or mitigated could require specialised training of workers, environmental monitoring, medical surveillance, biological monitoring, special personal protective equipment (PPE), or some combination of all of these. A HIRA is required to ensure the reduction or elimination of risk to those who will be involved in any stage of the construction process, or during the maintenance of the structure or services once complete, for the entire life of the structure or services. The process does not need to be absolutely precise, but should deal with known hazards, issues or risks where possible or reasonable. Exposure to physical hazards ought to be considered during, for example:

- client meetings
- site inspections and discussions
- preparation of the concept design
- determination of project duration
- preparation of project documentation
- pre-tender meeting
- evaluation of tenders
- pre-qualification of contractors
- preparation of detailed design
- site handover

- site meetings
- preparation of working drawings
- conducting constructability reviews
- coordination of the design.

Additionally, consideration for physical hazards should be given when, for example:

- compiling project specifications with reference to the composition, mass, surface area, shape and texture of construction materials;
- working on the general design, detailing, plan layout and elevations;
- deciding on the best methods of fixing building elements such as curtain walling;
- deciding the type of structural frame and selection of internal and external finishes; and
- determining the final position of building components.

While risk in construction is generally considered to comprise two primary elements, namely the likely severity, consequences or impact of harm or ill health, and the likelihood or probability that harm or ill health could occur, it might be necessary to also consider the number of persons that could possibly be exposed and the duration of that exposure. Given that risk assessments should always be kept simple, these additional considerations could make the assessment more complex.

The steps involved in any risk assessment may vary slightly depending on the type of risk assessment used (see Chapter 6 *Establishing operational control processes*, Chapter 8 *Assessing health issues in construction* and Chapter 9 *Assessing safety issues in construction*). However, the risks need to be determined for each of the various construction activities. Thought must be given to the method of construction, processes to be followed, and materials, plant and equipment used in order to limit exposure. Where necessary, Material Safety Data Sheets (MSDSs) should be obtained from suppliers to determine short or long term health risk exposure to workers, and also to establish whether alternative materials should rather be used. For example, where solvent based paints have been specified for use in a confined space and ventilation is likely to be problematic, a water based paint should be considered, as PPE should only be used as a last resort, and also where no other safer option is available.

Further, previous knowledge is useful where work activities are similar – although no two projects are exactly alike. For example, the placing of storm water culverts in position invariably involves the same work activities. However, environmental factors such as soil type, ground conditions and surrounding traffic densities and flows may alter the risks significantly. Similarly, the installation of roof sheeting will depend on the pitch of the roof, the length of the sheets, their profile and the weather conditions. Since workers would need to be protected from falling, attachments for fall arrest lines and/or use of safety nets and air bags may be required.

However, assessment of the risks is not enough. The risk exposure needs to be quantified. This quantification is necessary in order to make an informed decision with respect to deciding whether a design, detail, method of fixing or specification is appropriate or needs to be altered. Based upon the 'reasonably practicable' principle, a decision can be made to amend or leave any of these unchanged. Furthermore, a precise estimate is not required, as doing so would be too time consuming, and often there is a lack of data available. What is important is not so much the method, but the decisions that flow from the risk assessment process. Once the risk has been quantified, the hierarchy of control is to be followed, again as far as is 'reasonably practicable', namely:

- elimination of the hazard;

- substitution of the construction method or material;

- containment of the exposure;

- introduction of an engineering intervention;

- use of safe working or operating procedures; and

- proper use of PPE.

The measures that minimise the risk and other less effective measures require more frequent reviews of the hazards and systems of work (Queensland Government, 2007).

Examples of dominant physical hazards

Slips, trips and falls from height and on the same level

Several studies suggest that working at height is an historically high risk area across many construction activities on sites. For example, Bentley et al. (2006) citing Jeong (1998), Workers' Compensation Board of British Columbia (1997), NIOSH (2000), and Health and Safety Commission (2005) report that:

- Falls from height in South Korea comprised 42% of fatal construction accidents and falls on the same level 7%.

- In British Columbia, falls from height accounted for 20% of disability claims with 17 workers being killed.

- In the USA 49.9% of deaths were due to falls from height.

- In the UK 30% of all incidents resulting in major injury to construction workers were due to falls from height.

Falls occur as a result of exposure to uneven ground surfaces, cluttered work environment due to poor housekeeping and slippery conditions. The consequences of slips, trips and falls range from minor injury requiring first aid treatment to serious injury

resulting in either temporary or permanent disablement due to bone fractures and, in the worst cases, death. When a worker falls it is possible that he or she will suffer a blow to the head or body resulting in the brain temporarily not being able to function normally due to it being shaken inside the skull. This concussion may result in the worker becoming lightheaded or passing out. It is also possible that concussion may cause more serious problems and might require surgery or lead to difficulties with movement, learning or speaking. Some symptoms of post-concussive syndrome include:

- changes in ability to think, concentrate or remember;

- headaches or blurred vision; and

- dizziness, light-headedness or unsteadiness that makes standing or walking difficult.

Therfore, any worker who falls should be examined by a medical doctor.

Considering that not everyone is comfortable with working at heights, it is important that the pre-employment medical examination includes questions that will highlight the possibility of having a fear of heights or working at heights. This situational phobia will result in the affected workers possibly losing control, feeling physically stressed or afraid, fainting or having a panic attack. Obvious symptoms are rapid breathing or hyperventilation, irregular or racing heartbeat, and feeling dizzy, sweaty or shaky. Treatment of a phobia of heights includes counselling and cognitive-behaviour therapy that requires systematic desensitisation, during which workers will go through a series of steps that bring them closer to the situation or activity. In some cases, doctors will prescribe some form of medication to ease the symptoms of panic and anxiety.

Responses to the questions listed in **Table 1** will assist in identifying and controlling slips, trips and falls from height and on the same level, and probably other exposures.

It is important to ensure that workers have received prior training in the proper use and maintenance of scaffolds and ladders to avoid the possibility of injuring themselves. For further information see Chapter 14 *Working at height and roofwork*.

Lifting of materials

Frequent manual material lifting that includes, for example, regular lifting of heavy concrete blocks at or above shoulder level results in overexertion injuries. Standard-weight concrete blocks typically weigh approximately 16 kg each with physical dimensions of $0.2 \times 0.2 \times 0.4$ m. Cement is commonly supplied in pockets or sacks weighing from 25 kg in the UK to 50 kg elsewhere that are often lifted to a height appropriate for the contents to be tipped into a concrete mixer or wheelbarrow. Spinal compression varies between in excess of 3400 N and 6400 N (Haines et al., 1997). Lifting injuries are typically to the lower back region, forearms and shoulders. Further, the associated medical costs are extremely high accompanied by lost workdays resulting in overall reduced productivity. The

Fall protection	Yes	No	Action
Have all exposures to falls been assessed for possible methods of prevention?			
Are appropriate guardrails, toeboards/kickboards or catch nets or platforms in at every exposed edge where falls could occur?			
Are platforms being used as first choice for working places?			

Safe access	Yes	No	Action
Can all work areas be accessed safely, with all access equipment and gangways, and other walking surfaces level, clear and without any obstructions?			
Are all structures being worked on stable, safe and within permissible loading capacities?			
Is the site tidy and uncluttered to prevent accidental slipping?			
Are all protruding nails in timber (foot rippers) removed or hammered down flat?			

Scaffolding	Yes	No	Action
Can all platform levels be accessed safely?			
Are ladders integrated into the scaffolds?			
Are all working platforms fully and properly boarded out?			
Are all boards properly secured to prevent movement?			
Are platforms free of unnecessary waste material?			

Ladders	Yes	No	Action
Have other forms of access been considered such as a scaffold?			
Are ladders used only as means of access and only for short periods of time?			
Are there firm site rules to prevent workers from standing or working on the top third of stepladders?			

General housekeeping and material handling	Yes	No	Action
Has the project site layout been well-planned?			
Is the site constantly being kept clean and uncluttered?			
Are sufficient waste containers placed in strategic positions?			
Are unused materials properly stacked and stable?			
Has all used timber been picked up and any protruding nails been removed or hammered down flat?			
Are passageways always kept clear and accessible?			

Roofing	Yes	No	Action
Is edge protection provided to prevent workers, materials and equipment falling from sloping or flat roofs?			
Are crawling ladders or crawling boards used where the pitch of roofs is greater than 15 degrees?			
Do the roof battens provide a safe handhold and foothold?			

Table 1 Checklist to prevent slips, trips and falls

task of lifting materials is exacerbated by not only the weight or mass of the materials themselves but also their composition, regularity and uniformity of their shape and size, centre of gravity, surface area, packaging and surface texture. Other factors affecting manual handling of material include the layout of the construction site, location and manner of the material storage, management and logistical planning of the material supply, lifting height, rate of lifting (such as 30 lifts per hour) and whether the work is being done to the exterior or interior of the structure (Haines et al., 1997).

ICE manual of health and safety in construction © 2010 Institution of Civil Engineers

Interventions are needed to overcome the negative effects material handling require; for example, proper training in lifting techniques and safer work practices, reducing the mass and size of the material including packaging, improved planning, organisation and layout of construction activities, and better management of material deliveries and storage locations. Training construction workers in proper lifting and manual handling techniques will also contribute to avoiding the negative health consequences of poor manual handling of construction material.

Excavations

Excavations potentially result in damage to plant and equipment, serious injuries, exposure to disease and multiple fatalities. Health hazards present in the form of poor and contaminated soil conditions, contaminated groundwater, working in water and the presence of toxins in the air in the confined spaces of excavations.

Several factors need to be considered to prevent exposure to these hazards such as, for example:

- the depth and width of the excavations;
- the method of excavation – whether mechanically or by hand;
- the type of soil;
- the level of the groundwater table and the presence of surface water;
- detected or undetected spills and leaks;
- previous use of the site and surrounding area;
- presence of oxygen deficient atmosphere and/or toxic gases; and
- presence and concentrations of pollutants.

During excavation operations, construction workers can potentially come into contact with waterborne and water based diseases in ground- and surface water. These include the presence of human or animal faeces in water, urine infected by bacteria, pathogenic viruses and the presence of parasites. Soils could be contaminated that require special precautions with respect to excavation, handling, stockpiling, transportation, deposition and disposal. Low levels of oxygen, inert atmosphere, and/or presence of toxic gases such as carbon monoxide may result in asphyxiation. Workers may then become dizzy, disoriented and black out. When the sides of excavations collapse and workers are trapped they might experience compressive asphyxia due to their chest or abdomen being compressed. Exposure may be mitigated by means of pre-construction surveys, inspecting excavations regularly, remediating contaminated soil promptly, and monitoring air quality continuously during excavation operations. The oxygen content should not be less than 19.5% or more than 23.5% under normal atmospheric pressure. Where workers are required to physically work in water, they should wear proper personal protective clothing in the presence of an efficient de-watering system.

Further information on these hazards can be found in Chapter 15 *Excavations and piling*, Chapter 16 *Confined spaces* and Chapter 21 *Working on, in, over or near water*.

Tunnelling

Tunnelling activities involve high manageable risks. From a design point of view tunnel design is different. For example, it is difficult to determine with absolute accuracy geological properties and their variability along the tunnel. Some of the hazards that construction workers face during tunnelling operations include:

- changing conditions;
- limited space;
- access;
- air contamination;
- underground fires;
- sudden flooding;
- explosions;
- tunnel collapse;
- moving heavy equipment;
- noise;
- compressed air use;
- wet and uneven slippery surfaces;
- falling objects;
- overhead seepage;
- ground gas and water inrush;
- contaminated groundwater;
- reduced visibility;
- vibration;
- heat and humidity;
- hazardous substances; and
- power failure (Queensland Government, 2007).

Compressed air tunnelling usually occurs when working in very soft and extremely wet conditions and where other forms of preventing excessive ingress of water or tunnel collapse are impractical. Workers under these conditions are faced with a pressurised atmosphere and might suffer decompression sickness as a result. Decompression sickness is also known as caisson disease and results from inadequate decompression after working in a pressurised environment for some time. Compressed air can cause bubbles of air in the bloodstream. In the case of high levels of bubbles, complex reactions occur in the body that may result in numbness, paralysis, higher cerebral function disorders, congestive lung symptoms and circulatory shock. Symptoms include unusual tiredness and fatigue, shortness of breath, pain in joints and muscles, itchy skin, dizziness, vertigo, numbness, tingling and paralysis. Treatment of decompression sickness is recompression and increasing oxygen levels. The prevention of injuries as a result of working in a pressurised environment is possible through ensuring the use of the lowest possible air pressure to get the work done and training of workers in safe work procedures and hazard recognition.

Physical health effects of exposure to physical hazards

Operators of construction equipment experience health hazards such as whole body vibration, awkward postural requirements including static sitting, dust, noise, psychosocial factors, dust, diesel exhaust, asphalt and/or welding fumes, noise, temperature extremes, time pressure, and shift work. Exposure by workers over time to these unmitigated physical hazards results in several physical health effects. These include various cancers, musculoskeletal disorders, communicable disease infections, gastro-intestinal tract (GIT) infections and respiratory disorders. Hypertension also frequently presents itself in construction workers. These health effects result in various disabilities, reduction in work productivity, absenteeism, increase in medical costs and health care benefits, and ultimately increased construction costs. Examples of the health effects include the following.

Cancer

As a result of its physicochemical properties that include resistance to electricity, chemicals, heat, and fire and is relatively low cost, asbestos was widely used in construction until recently in the form of asbestos-cement fireproof panels, roof sheeting and insulation. Lengthy exposure to asbestos and ingestion or inhalation of asbestos fibres may cause asbestosis (evolutive lung fibrosis), pleural fibrosis and calcification, lung cancer and mesothelioma. The risk of infection is proportional to the duration and intensity of exposure (Rom et al., 2001). Early diagnosis of these cancers is often difficult and delayed both for the non-specific clinical manifestations of the disease and morphological variability (Attanoos and Gibbs, 1997). Persons previously exposed to asbestos on construction sites remain at risk since asbestos induced disease, especially cancer, may develop up to 40 years after exposure (Vogelzang, 2002). Exposure is likely during the removal of existing asbestos during repair, refurbishment and maintenance activities. Large proportions of asbestos used in the past are still present in the general environment and cause the release of fibres into the air because of aging and disintegration. Careful medical surveillance as part of a detailed record of occupational history might promote the possibility of earlier detection and diagnosis which might lead to prolonged survival and improved quality of life of infected workers.

Most countries have introduced highly prescriptive legislation to deal with the use of and exposure to asbestos such as OSHA 29 CFR 1926.110 in the USA and the Control of Asbestos Regulations 2006 in the UK. Asbestos work must be conducted in a regulated area with only trained and authorised workers, wearing respirators and other protective clothing. The removal of asbestos must be undertaken by a licensed contractor and requires special treatment to prevent the release of fibres into the air and improper disposal of waste.

Musculoskeletal and sensory effects

Repeated motions performed in the course of normal work or daily activities such as continuously working in static positions such as kneeling for long periods of time have been found to be a major contributor to the manifestation of repetitive motion disorders (RMDs) such as meniscus lesions, bursitis, tendonitis, ganglion cyst, trigger finger and osteoarthritis of the knees (Schildge et al., 1997). Other contributions include repetitive and heavy lifting, bending and twisting, repeating an action too frequently, uncomfortable working position, exerting more force than required, and working too long without breaks. The HSE estimates 11.6 million working days a year are lost in the UK to work related MSDs. In the USA between 25% and 33.3% of all lost workday injuries were sprains and strains. About 25% were due to overexertion, 15% were injuries due to lifting and about 23% were back injuries. Residential construction contributed the highest rate of carpal tunnel syndrome (CTS) while roofing and sheet metal work caused the highest rate of tendonitis. CTS manifests as tingling, numbness, weakness or pain in the fingers or hand due to pressure on the median nerve in the wrist. Contributory causes include making the same hand and/or wrist movements again and again as well as wrist injuries and bone spurs. Typically, carpenters, plasterers and painters suffer with CTS. Treatment will include the wearing of a wrist splint and, in the worst chronic cases, surgery. Back injuries are presented most commonly among bricklayers and floor layers (Schneider, 1997). Common causes of lower back pain include:

- lumbar strain where ligaments, tendons and/or muscles of the lower back are stretched;
- nerve irritation of the lumbar spine;
- lumber radiculopathy caused by damage to the discs between the vertebrae;
- bone encroachment where the space for the adjacent spinal cord and nerves is reduced; and
- bone and joint conditions that arise from injury such as fractures.

Slipped or herniated discs, where the tissue that separates the vertebral bones of the spinal column has been ruptured, manifest as a result of repetitive body movements during construction activities. Workers experience pressure against one or more spinal nerves causing a shooting pain known as sciatica, weakness or numbness in the area served by these nerves. Treatment is in the form of anti-inflammatory and muscle-relaxant medications, exercises, and physical therapy. In severe cases surgery may be required. Lappalainen et al. (1997) developed an observation chart that could be used to identify musculoskeletal risks during construction activities. By using the observations recorded on the chart (**Table 2**) the

Body region	Type of load	Descriptive figure	Criteria for harmful load	Criteria for a very harmful load
1. Cervical spine	Bending or twisting		a. Head bent backwards or extremely forward b. Head turned or twisted extremely to the side	a. Posture both bent and twisted
2. Upper extremities	Holding in raised position		a. Upper extremities over the shoulder level b. Shoulders hunched or raised more than 30 degrees	a. Both hands are held high b. Posture lasts a long time c. Movement is quickly repetitive
3. Wrists or elbows/ forearm	Bending/broad grip or repetitive moment (twisting or bending)		a. Wrist in nearly maximal flexion b. Too broad a grip c. Repetitive movement >10 times/min (light load) or >1/min (heavy load)	a. Repetitive movement >10/min and heavy load
4. Back	Bending or back twisted/ bent		a. Bent forward >20 degrees b. Bent backward c. Shoulders clearly twisted or bent sideways in respect to the waist	a. Stooped or twisted/bent posture sustained for a long period b. Use of afore-mentioned posture while carrying a load
5. Back/legs	Lifting or use of great strength (carrying, pulling, pushing)		a. Load >15 kg/distance 45 cm b. Load >25 kg/distance 30 cm c. Strong effort or heavy load	a. Load heavy and difficult to handle b. Lifting distance great c. Handling of load exposing the worker to sudden or tearing movement
6. Knees	On knees or squatting		a. Load on knees or dependent on knees b. Buttocks almost touching heels	a. Long duration or poor or unsuitable base (e.g., sharp corners)

Table 2 Interpretation criteria for the observations
Reproduced, with permission, from Lappalainen et al. 1997.

degree of musculoskeletal well-being (DMW) can be calculated as a percentage as follows:

$$DMW \% = S/(S + H + 2VH) \times 100$$

where S = the observations representing suitable workloads, H = observations representing harmful workloads, and VH = observations representing very harmful workloads.

Observers do systematic walk-throughs of the construction site and observe workers for about one minute. They evaluate the work activity with respect to all six regions of the body using a scale, namely suitable, harmful or very harmful. At the same time observers, by using a sample list, check that workers have the proper tools and equipment needed to ensure a high quality product. They record the musculoskeletal load and the preventive measures such as, for example, missing tools or equipment on the survey form. In the 'Comments' column the missing tools and equipment are recorded and the 'very harmful' work tasks are listed.

Whole body and hand–arm vibrations occur during the use of hydraulic and power tools – such as: concrete and paving breakers; pokers and compactors; sanders, grinders and disc cutters; hammer drills; chipping and jack hammers; chainsaws; scabblers or needle guns – and operating mechanical plant and equipment such as rollers, dump trucks and excavators. Hand–

arm vibration syndrome, for example, is the most compensated disease in the UK according to the HSE (2001). Studies by Bovenzi and Hulshof (1999) found that whole body vibration was associated with an increased risk for lower back pain, sciatic pain, and degenerative changes in the spinal system, including lumbar intervertebral disc disorders. These ailments manifest largely in drivers of trucks, cranes and earthmoving equipment such as front-end loaders and bulldozers. If these exposures and consequences are to be mitigated then it is important to quantify vibration and postural requirements on construction sites. In practice this is difficult to do.

Similarly, hand–arm vibration syndrome (HAVS) can result in several types of disorders. There is strong evidence of a positive association between high level vibration exposure and the vascular symptoms of HAVS (Falkiner, 2003). Vascular disorders relating to damage of the vessels that circulate blood manifest with symptoms such as whiteness of the fingers accompanied by numbness and 'pins-and-needles', and loss of finger dexterity. Neurological disorders resulting from damage to the nervous system are characterised by numbness, tingling sensations and loss of hand-grip strength. Additional symptoms include pain or stiffness in the hands and lower arms, and a decline in strength and dexterity (Edwards and Holt, 2006). In order to mitigate exposure to excessive hand–arm vibration the recommended control measure after assessment of the risk of exposure is continual monitoring. The consequence of such monitoring and management of exposure might be consideration for shorter safe trigger-times when using hand-held construction tools. These reduced trigger-times might require multiple operator rotas in situations where tool use is required throughout an entire working day (Edwards and Holt, 2006). Workers and operators can also be trained to visually protect their hands from injury and to regularly inspect their hands for injury.

Skin disease

In construction, occupational contact dermatitis is caused by exposure to nickel, chromium, cobalt and their compounds, latex and rubber chemicals, epoxy resins, and various paint and wood preservatives. Symptoms are red, sore, itchy, scaly and blistered skin. In severe cases the skin can crack and bleed and it could spread all over the body. Construction workers have a substantial risk of developing contact dermatitis caused by cement with its main allergen potassium dichromate. In severe cases workers leave the industry. A study by Bock et al., (2003) found that construction workers developed skin problems after median periods of exposure ranging between five and 12 years, depending on the occupational group, with hands being the most frequent location of occupational skin disease followed by the face and legs. The most likely construction workers to develop irritant or allergic contact dermatitis include tilers, terrazzo workers, painters, plasterers, bricklayers, plumbers and machine tool operators, and general workers who mix cement mortar (Athavale et al., 2007). The prevalence of potassium dichromate has declined in Europe, in particular following a concerted effort to reduce chromium VI levels. The addition of ferrous sulphate to cement in recent years has reduced the amount of chromium in cement. Allergic contact dermatitis occurs more frequently than irritant contact dermatitis. Other allergens found in construction that result in sensitisation include epoxy resin, cobalt chloride, thiuram mix, nickel sulphide, p-phenylenediadiamine and colophony.

Interventions include the application of the hierarchy of control and substitution of materials containing the substances that cause skin disorders. MSDSs have to be carefully scrutinised to determine the composition of materials. Where exposure cannot be avoided, workers need to be trained in the proper use of the materials and provided with the necessary protective clothing and equipment. They also need to be trained in the proper use and maintenance of both protective clothing and equipment.

Exposure to the sun and its damaging ultraviolet (UV) rays results in sunburn, blistering, aging of the skin and skin cancer. UV rays damage DNA, the genetic material that makes up genes which control the growth and overall health of skin cells. If the genetic damage is severe, a normal skin cell may begin to grow in the uncontrolled, disorderly way of cancer cells. Basal cell carcinomas usually occur on the face, neck, V-shaped area of the chest, and upper back and less often on the top sides of the arms and hands. However, squamous cell carcinomas also appear most often on the face and neck, V-shaped area of the chest, and upper back but are more likely than basal cells carcinomas to form on the top of the arms and hands. Melanoma is a serious form of skin cancer that starts in the pigment-producing cells called melanocytes. These cells become abnormal, grow uncontrollably and aggressively invade surrounding tissues. Surgical removal or excision is the most effective treatment. Prevention is by training workers to reduce exposure to the sun during the middle of the day, wear protective clothing, using sunscreen daily and examining the skin regularly.

Circulatory and GIT diseases

Exposure to lead and lead based paint presents similar challenges as asbestos. In particular, lead based paint that is flaking off is hazardous. When surfaces painted with lead are scraped, sanded, burned or heated to remove the paint, lead dust could form and be inhaled. Similarly, when lead pipes are cut or torched, dust could form. Lead chips and dust could settle on surfaces and objects that workers come into contact with. Should the area be vacuumed, swept or walked through, the resultant dust may be inhaled. Effects of lead exposure include lung infections, high blood pressure, digestive problems, nerve disorders, memory and concentration problems, and muscle and joint pain (Manuel, 2003).

Prescriptive legislation such as, for example, OSHA 29 CFR 1926.62 requires the implementation of specific procedures to protect workers from exposure to potentially harmful quantities of airborne lead dust. Workers may also be required to wear special respirators and protective clothing to protect them from exposure. Lead paint debris has to be disposed of taking special precautions.

Respiratory diseases

Chronic exposure to air pollutants in the form of dust such as timber sawdust and cement, epoxy resins, dust from latex rubber, organic solvents used in paints and cleaning aids, and fumes from diesel that contain benzene and petrol engines cause irritation of the eyes, nose and throat in addition to pulmonary function impairment and is considered a human carcinogen. Breathing these particles may cause allergic respiratory symptoms, mucosal and non-allergic respiratory symptoms, and cancer. A study comparing carpenters with painters in Finland found that painters reported more asthma-like, rhinitis, laryngeal and eye symptoms than carpenters. Painters with painting experience ranging between one and ten years had a threefold risk of asthma compared with carpenters. Chronic bronchitis was linked to the painting trade and the duration of exposure. The results indicate a higher risk for respiratory symptoms and chronic bronchitis among construction painters than among carpenters (Kaukiainen et al., 2005).

Control measures for exposure to dust should include:

■ Hosing down dust with water at the point of dust generation. Water can be used through non-electric cutting or grinding tools to reduce the dust that is generated.

- Using a dust control system.

- Encouraging good work practices to minimise exposures to nearby workers or the public.

- Dust levels in the air should be monitored by a competent person. Using a suitable type of respirator until adequate dust controls are put in place after workers have been trained in their proper use and maintenance.

Medical or health surveillance programmes that include chest X-rays on workers most likely to be exposed will enhance the prospects of early detection.

Also see Chapter 11 *Controlling exposure to chemical hazards*.

Designing for health

Considering the impact of designers on construction health and safety, there has been a growth of interest in designing for health. Construction activities by their nature require bending, working in awkward or cramped positions, reaching away from the body and overhead, repetitive movements, handling heavy material, plant and equipment, use of body force, exposure to vibration and noise, and climbing and descending (Schneider and Susi, 1994). Ergonomics should be considered a fundamental knowledge in designing things and systems (Lee, 2005). Designers influence ergonomics given that they evolve the concept, execute the detailed design, provide details and specify materials. These dictate the materials, construction methods and processes used on project sites (Smallwood, 1996). According to Smallwood (1997) the five most frequently cited ergonomic problems reported by building contractors in South Africa are daily repetitive movements: climbing and descending, use of body force, bending, twisting the back, and reaching overhead. Workers, on the other hand, reported that they encountered on a daily basis bending or twisting the back, repetitive movements, reaching away from the body, reaching overhead, and climbing and descending in order of daily frequency.

Bricklayers, for example, experience one-handed repetitive lifting of bricks which can weigh between 1 kg and 6 kg, and working with mortar in awkward positions for the lower back. Their assistants typically manually transport both bricks and mortar, involving manual lifting and carrying materials, and also pushing and pulling wheelbarrows. A study by Van Der Molen et al. (2004) found that it was likely that a reduction of frequent trunk flexion and of manual materials handling would decrease the occurrence of back complaints in bricklayers while a reduction in pushing and pulling wheelbarrows and material handling above shoulder height would decrease shoulder complaints. Lifting devices were designed to adjust the working height of materials and the transport of materials was mechanised (refer to **Figure 1**). There was significant reduction in the demands on the lower back of bricklayers and frequency and duration of trunk postures of their assistants. The stacking of bricks themselves was redesigned. By modifying scaffolding used

Figure 1 Example of mechanical lifting device for bricks
© Theo Haupt, 2007

by bricklayers the adjustment of working height is possible with positive health effects.

Ergonomic interventions with the requisite direct worker participation and management support are possible by, for example:

- Reducing extreme joint movements by minimising working conditions that require unusual posture such as twisting the spine, reaching above the head and leaning to one side.

- Reducing excessive force through opting for mechanical assists rather than muscle.

- Reducing highly repetitive and monotonous tasks by providing workers with a larger and more varied number of construction activities to perform.

The thrust of designing for health is the reduction of the risk of injury, illness and fatalities by integrating decisions affecting health in all stages of the design process. Designers can use both their knowledge and influence to design-in health and safety features that will improve the actual construction of the facility itself and its maintenance after completion as part of 'making the job fit the worker'. The study of human body measurements to assist in the understanding of human physical variations is known as anthropometry. Designers could, for instance, design parapet walls at guardrail height to protect against falls from heights, design attachment points into concrete to provide convenient and available tie-off points for personal fall arrest systems such as in window openings in high rise buildings (refer to **Figure 2**), specify modules to be prefabricated on the ground or in controlled factory environments to limit work at height and material handling of individual components such as staircases installed mechanically to prevent falls through openings in floors, design trenches at minimum feasible depths, design permanent staircases and walkways to be constructed early in the construction phase so that the use of temporary scaffolding is minimised, and substitute materials such as toxic materials in coatings and adhesives to reduce risk through design and specification. These examples demonstrate the positive impact of ergonomic friendly designs on the actual construction process,

post-construction maintenance process and daily end use of the constructed facility. Designers should, therefore, give ergonomic considerations equal weighting with other design considerations.

Management and environmental issues

Threats to the health and overall well-being of construction workers are also attributable to several management and environmental issues. It is not unusual for management to expect workers on construction sites to work for extended working hours or long shifts in order to increase productivity when falling behind on the project schedule, to avoid penalties or accelerate or compress the schedule to achieve an earlier completion date and a possible bonus from the Client. Unfortunately the negative health impacts of these strategies on workers are often overlooked.

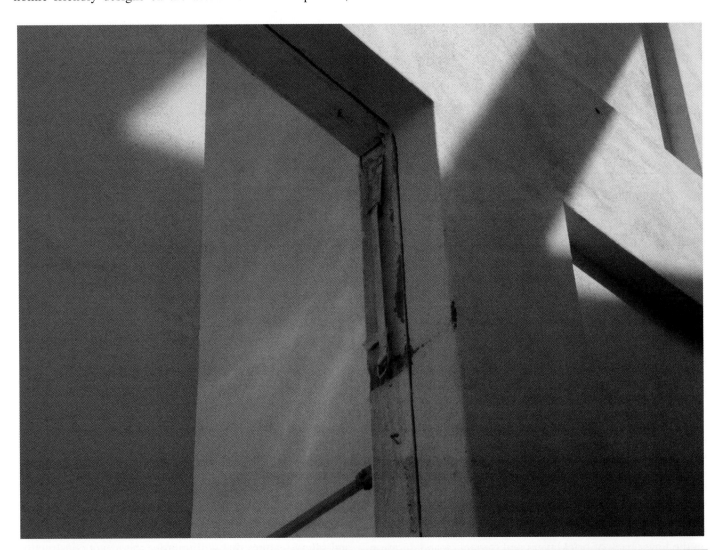

Figure 2 Example of designed-in attachment for personal fall arrest system
© Theo Haupt, 2008

ICE manual of health and safety in construction © 2010 Institution of Civil Engineers

Sleep deprivation and fatigue

Working long hours and shifts results in sleep deprivation and mental fatigue. Lack of sleep or sleep deprivation can have potentially serious health effects on workers in the form of physical and mental impairments. These impairments include reduction of their ability to think lucidly and concentrate on the task at hand, handle stress, maintain healthy immune systems and moderate their emotions. Without adequate rest, the ability of the brain to function properly quickly deteriorates. As a result the brain has to work harder to counteract the effects of sleep deprivation. Other typical effects of sleep deprivation include:

- depression
- heart disease
- hypertension
- irritability
- slower reaction times
- slurred speech
- tremors.

After having worked unduly long hours or multiple shifts it is likely that construction workers will experience weakness and feelings of tiredness accompanied by sore muscles. As a result of overwork and lack of sleep, workers may become fatigued where they have feelings of tiredness, exhaustion and lack of energy.

The overall consequences of lack of sleep and mental fatigue can be catastrophic on a construction site and those workers who operate plant and equipment or drive are particularly susceptible. Exposure to the consequences of overtime and shift work can be avoided by proper project planning and consideration of workers.

Noise

As a result of working in an environment where the typical noise levels regularly exceed the permissible minimum in the UK of 80 dBA, construction workers are increasingly presenting with noise induced hearing loss. Continued or repeated exposures to high intensity sounds can cause acoustic trauma to the ear, resulting in hearing loss, ringing in the ears (tinnitus) and occasional dizziness (vertigo), as well as non-auditory effects, such as increases in heart rate and blood pressure. There is a direct correlation between the duration of exposure to loud noise and hearing loss suggesting that the longer the exposure, the greater the damage. In the USA, for unprotected ears the allowed exposure time decreases by one-half for each 5 dB increase in the average noise level. For instance, exposure is limited to 8 hours at 90 dB, 4 hours at 95 dB and 2 hours at 100 dB. The highest permissible noise exposure for the unprotected ear is 115 dB for 15 minutes per day. Any noise above 140 dB is prohibited. When noise measurements indicate that hearing protectors are

needed, the employer must offer at least one type of earplug and one type of earmuff without cost to workers. Earplugs and earmuffs potentially reduce noise levels to the ear by 15 to 30 dB. Further, use of earplugs and muffs together usually adds 10 to 15 dB more protection than if either were used separately. If the annual hearing test reveals a hearing loss of 10 dB or more in the higher sound frequencies in either ear, the worker must be informed. Additionally, construction workers must wear hearing protectors when noise averages more than 80 dB for an 8 hour day. As a result of reduced hearing, workers are exposed to unnecessary hazards presented by moving plant, heavy equipment and vehicles. Workers are also likely to misunderstand work instructions given them as a result of not hearing them properly. Exposure to noise can be controlled by reducing the operating noise levels of noise-generating plant and equipment through muffler systems. Measuring noise exposure is done with noise dosimeters or integrating sound level meters. The integrating sound level meter is a hand-held instrument, while the noise dosimeter is a small device worn by the worker whose exposure is being measured. Each time noise measurements are made, the equipment should be checked or calibrated. Other mitigating interventions include avoiding noisy processes, scheduling noisy activities for when fewer workers are present and isolating noisy processes by placing temporary sound barriers or screens. Requiring workers to undergo regular double baseline audiometric examinations and tests as part of a managed medical surveillance programme will highlight any reduction in hearing and allow measures to be taken to reduce the likelihood of further hearing loss.

Extreme temperatures

Construction work, by its nature, is impacted by environmental influences such as changing weather conditions and temperatures. Decrease in the internal temperature of the body constitutes a threat to life. When construction workers work in situations where their bodies are unable to warm themselves or maintain normal body temperature, it is likely that they will experience cold stress disorders. These disorders could lead to tissue damage and possibly even death. The amount of cold stress that workers may be exposed to is influenced by the presence of dampness, contact with cold water or surfaces, lowering of body temperature due to condensation caused by high humidity at low temperatures, and wind velocity. Cold stress is also influenced by construction workers working in clothing that may be wet or inadequate to cope with the extreme cold, being exhausted or fatigued, suffering from chronic illnesses such as diabetes, heart disease, vascular and thyroid problems, and using medication. Working in a cold and wet environment for long periods of time can cause trench foot, where the feet of workers experience tingling and itchy sensations, burning, pain and even swelling. Other common health consequences of exposure to extreme cold include: chilblain which presents as red swollen and tender skin often hot to the touch and itchy; frostbite where the skin tissue actually freezes leading to cellular hydration; and general hypothermia as the body temperature drops below 36^0 C causing normal muscle and brain functions to be impaired. Typical warning signs of hypothermia include nausea, dizziness, fatigue, euphoria, irritability, pain in feet, hands or ears, as well as severe shivering. Exposure to extreme cold working conditions may be controlled by:

■ monitoring exposure especially when the work environment temperature drops below $45^0 F$ or $-43^0 C$;

■ introducing engineering controls to reduce risk of exposure through, for example, wind breaks or shields or thermal insulating material on plant and equipment handles and surfaces;

■ job rotation and rest periods;

■ training and empowering workers to adopt safe work procedures to overcome the effects of cold temperature extremes;

■ wearing personal protective thermal clothing in layers to insulate against the extreme cold conditions using ISO 11079:1993; and

■ developing emergency first aid procedures.

Exposure to high temperatures in the working environment results in heat stress or heat stroke. The consequences may be minor but also catastrophic in extreme circumstances. Symptoms of heat stress include feelings of tiredness, increased irritability, clammy skin, confusion and disorientation, feelings of light-headedness, reduced ability to focus and concentrate, and muscular discomfort and cramping. On the other hand, symptoms of heat stroke include lack of sweating or perspiration, confusion and disorientation, raised body temperature, hot and dry skin, and loss of consciousness and convulsions. Exposure to extreme hot working conditions may be controlled in the same way as for extreme cold by also, for example:

■ providing adequate ventilation;

■ introducing a monitoring regime;

■ job rotation and rest periods;

■ providing appropriate personal protective clothing such as chiller vests and hats;

■ providing adequate supply of cool drinking water; and

■ training and empowering workers to adopt safe work procedures to overcome the effects of heat exposure.

Radiation

When construction workers are exposed to radioactive materials and radiation in the work environment without proper protection they will suffer damage to their bodies as a result of ionising radiation. In addition to exposure from external sources, radiation exposure can occur internally from ingesting, inhaling, injecting or absorbing radioactive materials. Both external and internal sources may irradiate the whole body or a portion of the body. The most common forms of ionising radiation are alpha and beta particles or gamma and X-rays. High levels of radiation lead to somatic changes that result directly from the radiation dose and manifest in the form of malignancy

(Eichholz, 1997). High levels of exposure may also lead to genetic changes that result in changes in the genes or chromosomes of reproductive cells. These changes are handed down to descendants of the affected worker in the form of gene mutations, chromosome aberrations and changes in the number of chromosomes themselves. The health effects of exposure may either be stochastic. which is associated with long term chronic low-level exposure to radiation, or non-stochastic as a result of high level exposure. Stochastic effects generally result in cancer, and changes in DNA resulting in mutations of cells. On the other hand, non-stochastic effects may not be cancerous. Rather they include burns and radiation sickness, also known as radiation poisoning, resulting in premature aging or possibly death. The initial signs and symptoms of treatable radiation sickness are usually nausea and vomiting. The amount of time between exposure and the onset of these symptoms is, in fact, a relatively reliable indicator of how much radiation a person has absorbed. After the initial onset of signs and symptoms, a person with radiation sickness then experiences a brief period of a few days with no apparent illness. This period precedes the onset of signs and symptoms indicating a more serious illness. Tell-tale signs include nausea, weakness, hair loss, vomiting blood, bloody stools, diarrhoea, high fever, skin burns or diminished organ function. Prevention or mitigation of the effects of exposure may be achieved by reducing the time of exposure to radiation sources, increase the distance from radiation sources and making use of shielding to reduce the magnitude of exposure. Shielding can include heavy clothing and the use of lead. All individuals who, in the course of their employment, are likely to receive a dose of more than 100 millirem in a year, must receive adequate training to protect themselves against radiation. The treatment goals for radiation sickness are to prevent further radioactive contamination, treat damaged organs, reduce symptoms and manage pain.

Electro-magnetic fields

Electric and magnetic fields (EMFs) are areas of energy that surround any electrical device that in the case of construction includes power lines which always emit magnetic fields and electrical wiring. Any device connected to an electrical outlet, when the device is switched on and a current is flowing, will have an associated magnetic field that is proportional to the current drawn from the source to which it is connected. The strength of a magnetic field decreases rapidly with increased distance from the source. Electromagnetic fields are not shielded by most common materials, and pass easily through them. The risk of exposure to EMFs is highest among electricians, telephone linemen and electric power workers. Workers who maintain transmission and distribution lines may be exposed to very large electric and magnetic fields. Within electric power generating stations and substations, electric fields in excess of 25 kV/m and magnetic fields in excess of 2 mT may be found. Welders can be subjected to magnetic field exposures as high as 130 mT. Although there has been no conclusive scientific evidence to suggest that exposure to EMFs poses a direct threat to human health, some studies suggest that exposure to EMFs may contribute to Alzheimer's disease and is carcinogenic (Scientific Committee on Emerging and Newly Identified Health Risks, 2009). Electromagnetic hypersensitivity is a condition reported by persons who have been exposed to EMS. Protection from 50/60 Hz electric field exposure can be relatively easily achieved using shielding materials. This is only necessary for workers in very high field areas. More commonly, where electric fields are very large, access of workers is restricted. Where magnetic fields are very strong the only practical protective method available is to limit the number of workers exposed and the duration of that exposure.

Summary of main points

Construction work and construction sites pose numerous physical hazards to all persons using the site and its access routes. It is important to assess and quantify risks to enable practical preventative interventions to be put in place to, as far as is *reasonably practicable*, mitigate the hazards. Workers should be continually assessed and monitored for exposure to physical hazards so that, if they are showing signs of being affected, steps can be taken to reduce any further effects.

References

Athavale, P., Shum, K., Chen, Y., Agius, R., Cherry, N. and Gawkrodger, D. Occupational Dermatitis Related to Chromium and Cobalt: Experience of Dermatologists (EPIDERM) and Occupational Physicians (OPRA) in the UK Over an 11-year Period (1993–2004). *British Journal of Dermatology* 2007, **157**, 518–522.

Attanoos, R. and Gibbs, A. Pathology of Malignant Mesothelioma. *Histopathology* 1997, **30**, 400–418.

Bentley, T. A., Hide, S., Tappin, D., Moore, D., Legg, S., Ashby, L. and Parker, R. Investigating Risk Factors for Slips, Trips and Falls in New Zealand Residential Construction Using Incident-Centred and Incident-independent Methods. *Ergonomics* 2006, **49**(1), 62–77.

Bock, M., Schmidt, A., Bruckner, T. and Diepgen, T. Occupational Skin Disease in the Construction Industry. *British Journal of Dermatology* 2003, **149**, 1165–1171.

Bovenzi, M. and Hulshof, C. An Updated Review of Epidemiologic Studies on the Relationship Between Exposure to Whole-body Vibration and Low Back Pain (1986–1997). *International Archives of Occupational and Environmental Health* 1999, **2**(6), 351–365.

Edwards, D. J. and Holt, G. D. Hand–arm Vibration Exposure from Construction Tools: Results of a Field Study. *Construction Management and Economics* 2006, **24**, 209–217.

Eichholz, G. G. *Environmental Aspects of Nuclear Power*, 1977, Ann Arbor, MI: Ann Arbor Science.

Falkiner, S. Diagnosis and Treatment of Hand–arm Vibration Syndrome and its Relationship to Carpal Tunnel Syndrome. *Australian Family Physician* 2003, **32**(8), 530–534.

Haines, H. M., Mansfield, S. J., Haslegrave, C. M. and Wilkinson, N. J. Handling Sacks of Materials. Abstracts from the 1st International Symposium on Ergonomics in Building and Construction, 1997, 30 June–2 July, Tampere, Finland. The Center to Protect Workers' Rights. Washington, DC, 112–114.

Health and Safety Commission. *Comprehensive Statistics: Construction*, 2005, Bootle: National Statistics, Bootle, Health and Safety Commission.

Health and Safety Executive (HSE). *Hand–arm Vibration Syndrome Statistics Summary*, November 2001.

Jeong, B. J. Occupational Deaths and Injuries in the Construction Industry. *Applied Ergonomics* 1998, **29**, 355–360.

Kaukiainen, A., Riala, R., Martikainen, R., Reijula, K., Riihimäki, H. and Tammilehto, L. Respiratory Symptoms and Diseases Among Construction Painters. *International Archives of Occupational and Environmental Health* 2005, **78**(6), 452–458.

Lappalainen, J., Kaukiainen, A. and Viljanen, M. Survey of Musculoskeletal Risks at Construction Sites – Description of the Observational Method and Instructions for Its Use. Abstracts from the 1st International Symposium on Ergonomics in Building and Construction, 1997, 30 June–2 July, Tampere, Finland. The Center to Protect Workers' Rights. Washington, DC, 101–103.

Lee, K. S. Ergonomics in Total Quality Management: How Can We Sell Ergonomics to Management? *Ergonomics* 2005, **48**(5), 547–558.

Manuel, J. S. Unbuilding for the Environment. *Environmental Health Perspectives* 2003, **111**(16), 880–887.

National Institute for Occupational Safety and Health (NIOSH). Worker Deaths by Falls: A Summary of Surveillance Findings and Investigative Case Reports, DHHS (NIOSH), Publication No. 2000–116, 2000, Cincinnati, OH: NIOSH.

Queensland Government. *Tunnelling Code of Practice 2007*, 2007, Queensland: Workplace Health and Safety Queensland, Department of Justice and Attorney-General.

Rom, W., Hammar, S., Rusch, V., Dodson, R. and Hoffman, S. Malignant Mesothelioma from Neighbourhood Exposure to Anthrophylite Asbestos. *American Journal of Industrial Medicine* 2001, **40**, 211–214.

Schildge, B., Wakula, J. and Rohmert, W. Ergonomic Analysis of Load on the Knees at Tile Setter's Work. Abstracts from the 1st International Symposium on Ergonomics in Building and Construction, 1997, 30 June–2 July, Tampere, Finland. The Center to Protect Workers' Rights. Washington, DC, 163–165.

Schneider, S. Musculoskeletal Injuries in Construction: Are They a Problem? Abstracts from the 1st International Symposium on Ergonomics in Building and Construction, 1997, 30 June–2 July, Tampere, Finland. The Center to Protect Workers' Rights. Washington, DC, 169–171.

Schneider, S. and Susi, P. Ergonomics and Construction: A Review of Potential Hazards in New Construction. *American Industrial Hygiene Association Journal* 1994, **55**, 635–649.

Scientific Committee on Emerging and Newly Identified Health Risks. *Health Effects of Exposure to EMF*, 2009, Directorate-General for Health and Consumers, European Commission.

Smallwood, J. J. The role of project managers in occupational health and safety, In Dias, L. M. A. and Coble, R. J. (Eds), *Implementation of Safety and Health in Construction Sites*, Proceedings of the First International Conference of CIB Working Commission W99, 1996, Lisbon, Portugal. Rotterdam: A. A. Balkema, 227–236.

Smallwood, J. J. Ergonomics in Construction. Abstracts from the 1st International Symposium on Ergonomics in Building and Construction, 1997, 30 June–2 July, Tampere, Finland. The Center to Protect Workers' Rights. Washington, DC, 184–187.

Van der Molen, H. F., Grouwstra, R., Kuijer, P. F. M., Sluiter, J. K. and Frings-Dresen, M. H. W. Efficacy of Adjusting Work Height and Mechanizing of Transport on Physical Work Demands and Local Discomfort in Construction Work. *Ergonomics* 2004, **47**(7), 772–783.

Vogelzang, N. Emerging Insights into the Biology and Therapy of Malignant Mesothelioma. *Seminars in Oncology* 2002, **29**, 35–42.

Workers' Compensation Board of British Columbia. *WorkSafe Construction Industry Focus Report*, 1997, May, A Publication of the Research and Evaluation and Engineering Sections, Prevention Division, Vancouver: WorkSafeBC.

Referenced legislation

Control of Asbestos Regulations 2006 Reprinted November 2006, January 2007 and March 2007 Statutory instruments 2739 2006, London: The Stationery Office.

ISO/TR-11079. *Evaluation of Cold Environments – Determination of Required Clothing Insulation (IREQ)*, 1993, Geneva: International Standards Organisation, Geneva. Occupational Safety and Health Administration (OSHA), United States Department of Labor. Available online at: http://www.osha.gov/pls/oshaweb/owasrch. search_form?p_doc_type=STANDARDSandp_toc_level=1andp_keyvalue=1926 Regulations (Standards – 29 CFR) Part 1926 Safety and Health Regulations for Construction.

Further reading

Flook, V. *A Comparison of Oxygen Decompression Tables for Use in Compressed Air Work*, 2003, HSE Research Report 126, London: HSE.

Slocombe, R., Buchanan, J. and Lamont, D. Engineering and Health in Compressed Air Work. Proceedings of the Second International Conference on Engineering and Health in Compressed Air Work, 2003, held at St Catherine's College, Oxford from 25 to 27 September 2002.

Websites

British Tunnelling Society http://www.britishtunnelling.org.uk/

Institution of Civil Engineers (ICE), Health and safety http://www.ice.org.uk/knowledge/specialist_health.asp

medicinenet.com http://www.medicinenet.com/

Sleep Deprivation.com http://www.sleep-deprivation.com/

Section 5: Safety hazards

Chapter 14

Working at height and roofwork

Peter Fewings School of Built and Natural Environment, University of the West of England, Bristol, UK

Working at height, even low height, is an inherently dangerous activity and falling from height in construction is still top of the list for fatalities, and produces the most major injuries after the slips and trips category. The Work at Height Regulations 2005 now regulate a coordinated approach to reducing risk which includes a hierarchy of avoiding it, preventing falls and mitigating the effect of them. This chapter looks at the range of situations that cause falls, the effective planning to reduce or to avoid these risks, the use of access equipment such as ladders, scaffolding, mobile elevated working platforms, access towers and rope access. It also looks at the particularly hazardous case of roofwork. It suggests safe working methods, but many of these will be helped by the development of a willing and motivated culture among the workforce you supervise. It discusses particular responsibilities for the Supervisor to plan, inspect and maintain working equipment, inform and train. It also discusses the requirements for others which you are expected to check.

doi: 10:10.1680/mohs.40564.0167

CONTENTS

Box 1 Key learning points

- Feel confident you know the range of requirements required for planning, appraising equipment and competence for working at height.
- Distinguish the different types of access related work equipment and the hazards associated with their use and how to apply fall protection.
- Understand the basic principles and equipment associated with industrial rope access and the need for rescue systems.
- Understand the priority measures applied to roofwork and its particular dangers.
- Be encouraged to look at alternatives to reduce work at height exposure and develop hazard awareness in the workforce.

Background and application

Working at height means raised working with an opportunity to fall. Falling from height accounted for 34 of 72 fatalities in construction in 2007/8 (HSE, 2009a) and is the second biggest cause of major injuries. Over half of the major injuries were from low falls. Good practice for working safely at height is a critical management issue as one in six sites inspected by Health and Safety Executive (HSE) construction inspectors were failing to address work at height risks on site (HSE, 2009b).

The Work at Height Regulations 2005 (WAHR 2005) were especially formulated in the United Kingdom (UK) to deal with the particular requirements to plan and regulate the risks for working at height. It was revised in UK in 2007 to apply to organised climbing and caving activities.) If you can find ways of working which are not exposed to falls this would be a creative approach to the problem, although construction has

an inherent need to work near exposed edges and to access the outside of buildings. Here the Regulations require a reduction of risk. The use of 'cherry pickers' to aid the erection of structural steel, where previously it was common to climb stanchions and run along unprotected lengths of steel to get access, would be a simple example of improvements which might be expected. There is a hierarchy of prevention, maintenance and inspection, discussed in this chapter.

First you need to reduce instances of working at height. You can reduce risk by protecting from, or mitigating the effect of falls where you have to work at height either by collective or personal provision. Collective provisions are preferred. You will need to maintain protected areas on the basis of responsible controlled access and regular monitoring. Competent design and inspection goes without saying. Irresponsible use can compromise equipment for others, such as removing barriers for receiving loads and not replacing them. This creates a false sense of security and all users are required to report dangerous equipment or usage.

It is permissible to work from an existing safe place of work without invoking the Regulations. For example, this may apply to the stripping out of fittings on an upper floor of a building being demolished if the windows are still in place and there is a staircase access. It would not be a workplace subject to falls from height until the windows are removed.

Culture

Climbing and mountaineering are inherently dangerous because they are based on height and yet proportionately they have been subject to fewer accidents. Those who do it have a very high personal standard of safety and a sharpened awareness of their critical responsibility for and dependency on others. As a climber you check your equipment meticulously, you are acutely aware of the consequences of the slightest mistakes and you often have well thought out team backup systems if something does go wrong. This should be your model for working at height.

On construction sites the same acute dangers are there, but workers arrive who are not conscious of the height risks of their regular workplace or become too complacent of their ability to avoid harm.

A safe culture really means cultivating a healthy attitude towards risk in health and safety, including height awareness. It is a narrow line between developing a neurosis and understanding instinctively a safe and healthy environment. A neurotic worker is programmed with a fear of their own inability to move away from danger, a safe worker is a knowledgeable worker with an ability for continuous improvement towards a more safe and healthy environment. It is this committed developmental approach which makes the culture and not a belief in a perfect environment. Many companies try to create a culture such as:

■ 'Don't walk by' approach. This is an emphasis on many eyes are better than one, making it more likely to spot substandard or compromised provision. They encourage all staff and operatives, external observers, senior managers, visitors and members of the public, to report contraventions and suggest improvements.

■ Behavioural approach. Companies work hard to amend dangerous or shoddy habits, training supervisors to ensure uptake of appropriate interventions. Such interventions should be those that are proven the most effective for long-term improvement in health and safety behaviour and attitude, and should enanate from the supervisors. Language and psychology are important.

■ Zero tolerance approach. Use of a traffic lights warning system of red for removal from site, amber for a warning and re-induction with financial penalty for the employer, and green for go-ahead. Poor general performance of a subcontractor endangers their future procurement.

Collective provisions are preferable as they provide better protection for unexpected events. Human nature is such that it avoids the use of personal protective equipment (PPE) when it is perceived there is no danger. Even prescriptive requirements for wearing such equipment at certain times or places may be ignored if the reason does not seem logical, i.e. that there is no *perceived* danger. This depends upon training and the development of a more cautious culture in construction which has traditionally not been risk averse.

Work at height avoidance

It is worth you considering other methods of constructing which cut down prolonged work at height such as more prefabrication at ground level and more completed components in situ whilst working off mobile platforms or fixing from within the building. An example of this would be the use of longer, structurally strong cladding panels which preclude the use of fixing rails, spanning whole bays of structural steel and cutting

down the need to have long periods working at height and less people doing that. An example of this is a steel framed industrial unit or hanger where units fixed from the inside are craned in whilst fixers working off the existing floor are protected by the hanging unit. Volumetric prefabrication may go further in the supply of repetitive units such as finished hotel or student rooms. Even a steel frame may be eliminated if the units are structural by using a stacking system around a central core. An example of this is the Unite system for student accommodation. It is also possible, in the case of timber frame structures, to prefabricate the whole finished sections of roof and lift as a single component onto the walls. This allows for just a few people to work at height whilst guiding the units into position and bolting them together. Other forms of prefabrication are preformed bridge structures, which are lifted into prepared bearing positions for the whole span. The fact is that creative ways can be designed in at an early pre-construction stage.

Definition and assessment of working at height

The definition of working at height *used to be* defined as work 2 m above ground, but many serious injuries have happened with falls below this height; although they are less likely to be fatal they are still disabling. *Now* working at height is defined in WAHR 2005 with two parts which are not so prescriptive:

(a) A place is at height if a person could be injured from falling from it, including a place at, or below ground level.

(b) Obtaining access to or egress from such place while at work, except by a staircase in a permanent workplace, where, if measures required by these Regulations were not taken, a person could fall a distance liable to cause personal injury.

This requires the employee and the employer supervisor to make a risk assessment of any place where you may work and then, if at height, to take note of the precautions that help to reduce or eliminate risk.

It is important to note that you can:

■ fall off something – a platform, roof, ladders, an edge or a rope

■ fall into something – an excavation or a hole

■ fall down something – a staircase (in a temporary workplace) or a steep slope

■ fall through something – a floor or a roof or a glass atrium

■ drop something on someone else.

Each of the above situations puts you or an object into a position of freefall with the risk of a hard landing to damage a vulnerable part of your body or to hit or trap someone else. In doing so you may also bring down other weights on top of you which may crush or injure you or others. It includes any height as well as low level step-up platforms (hop-ups). We will not be considering falling over something (slips and trips) as this is not the focus of WAHR.

Remember:

- Is it at height?
- Can I drop anything on someone?

Low falls are generally defined as falls below 2 m. In certain instances it is necessary to have edge protection on lower working platforms. HSE in its WAHR training presentation gives an example of a 1 m raised pedestrian walkway with frequent use, working with your back to the edge of the platform or working above hazardous surfaces which might cause impalement, as a place where a guard rail is necessary. In some instances such as a loading bay or working off a small kick stool a guard rail might be impracticable. In other cases, such as a 1m trestle above a grass surface, a serious injury might not be likely (HSE, 2005a).

Responsibilities

- Employers have duties to set up safe systems for their employees, supervisors or to ensure 'those who control the work of others' have a competent knowledge and plan/coordinate health and safety including work at height.

- Clients or building owners who contract others to do work at height have responsibility for competent appointment, taking professional advice, adequate resources and to ensure planning and coordination take place.

- Designers are required to minimise work at height in construction and use.

- The main contractor becomes responsible for coordination and checking work at height once the site starts.

- An employee has a duty to report any safety hazard and also use any PPE supplied properly and undergo suitable training and instruction. They should seek further instructions if they think their directed work method is unsafe after viewing the workplace,

- Self-employed have responsibility for themselves.

Risk assessment

The general principles of risk assessment are just as relevant in working at height as other work hazards. You need to think through the significant risks beforehand of your choice of access method and to:

- identify *all* those at risk of falls or being hit by them;
- organise in sufficient time resources to reduce risk of falls;
- ensure competence in use of equipment.

You need to hand over your method statement so that the principal contractor is aware and can assess danger to others. Many platforms at height are used by others. Designers need to be aware of the impact their design has on working at height and to reduce more dangerous activities such as working off fragile roofs, cleaning windows or cladding off ladders or over glass, changing lightbulbs at height and cleaning out overhanging gutters. You and the CDM coordinator could check this. The Construction (Design and Management) Regulations (CDM) 2007 require a coordinator to assess design health and safety; also see Chapter 3 *Responsibilities of key duty holders in construction design and management.*

Principles of working at height

The overriding principle when working at height is to do all that is reasonably practicable to stop people from falling and/or dropping objects on those below. This includes doing a risk assessment, pre-planning the work and developing a way of working to prevent or reduce risk to a reasonable level if it is not possible to do the same work safely at ground level. This gives a hierarchy:

AVOID WORKING AT HEIGHT IF POSSIBLE

⇩

IF UNAVOIDABLE **PREVENT FALLS** BY USE OF EQUIPMENT

⇩

MITIGATE THE FALL DISTANCE AND CONSEQUENCES

A collective passive provision is better that an active personal choice such as hooking on or wearing something.

Planning and choice

The second principle is in the sensible choice of system for working at height. *You should be aware of the hierarchy of control and should* consider how to reduce risk in this order:

- You can *prevent* falls with work constraints which may be collective by the use of barriers restricting entry to dangerous areas or individually by using harnesses or clip-in boots connected to the structure to constrain access to the edge or to suspend in a controlled way (work positioning).

- You can collectively *reduce the impact* of falls by using nets or cushions or barriers which either break falls or create physical edge protection which arrest slips such as handrails and toeboards.

- You can personally *protect* by offering fall restraint such as harnesses which break the fall. This is an active system and takes the form of equipment which needs to be issued, willingly and knowledgably worn and properly maintained so that it is failsafe in a case of falling.

- You can use equipment that does not arrest fall, such as ladders and hop-ups and minimise the risk of a fall through with good practice training, instruction and supervision.

Note that rope access is a permissible method of access you could use where working platforms are difficult, impracticable or dangerous to construct. Here certain specialist best practice applies and this is discussed later.

The above hierarchy prioritises the collective passive systems of protection you can use such as a barrier to prevent falling above a high level net or cushion which would catch someone falling and in addition you can issue PPE which needs instruction in use. Stand-alone active controls are lower in the hierarchy and depend on training and could have significant harmful consequences if your worker disregards or misunderstands safe working.

Other principles

A third principle is that you do *not* put in place protection which *jeopardises others more than it protects*. An example would be the use of two scaffolders working for two hours at the edge to erect a barrier in order to protect one person working there for an hour.

A fourth principle is that your fall protection systems should have a methodology for *rescue* which is thought through. You may use the emergency fire service or you may need help and equipment from a specialist team to rescue an injured person hanging from a rope and to reduce their exposure to suspension trauma.

A fifth principle is to regularly *inspect* working platforms and work protection systems by competent persons, keeping records such as the known service life of components, loading restrictions and tests. Any nets, ropes or soft landing must be installed correctly and maintained.

Box 2	Case Study 1
A site manager was fined £10 000 after a worker fell 2.5 m and was injured when a joist collapsed in a building being refurbished as a supermarket. The cited reasons were failure to manage risk, the lack of guardrails on some parts of the scaffold and the difficulty of access from the building to the scaffold. This indicates the interconnectedness of measures possibly creating unsafe evasive action with no fall protection (Griffiths, 2009).	

Work at Height Regulations 2005 (WAHR)

The WAHR 2005 encourages people not to work at height unless they have to and gives guidance when you do. The HSE did not produce an approved code of practice to give guidance for safe procedures because they believed that there was adequate guidance already. The regulations require the following actions and are furnished with six schedules to focus guidance for particular work at height equipment.

Table 1 indicates the different categories in WAHR 2005 to cover in making risk assessments. It is vital that you cover organisation and planning of the job, effective selection of the equipment, competence of the installers and users of the equipment and consider the requirements for the safe application of that equipment and there is a need for special consideration for work on fragile surfaces such as roofs where so many accidents have occurred.

Other regulations in WAHR deal with falling objects, exclusion from danger areas, inspection of work places and equipment. There is also a duty for every person to report any dangerous activity or defect to their superior and to use equipment as they have been trained and instructed. Inspection of lifting equipment and ropes for access comes under the Lifting Operations and Lifting Equipment Regulations (LOLER clauses 9 and 10).

Equipment used for working at height

Work equipment includes scaffold, access towers and machines for access as well as specialist equipment you might use such as ladders or harnesses. All sorts of situations put people at risk from falling and different equipment has been designed to provide safe access. Some jobs which need this equipment are roofing, erecting steel frames, erecting scaffold, accessing windows to fix or clean, erecting cladding, painting or working at soffit level, fixing services or equipment in ceiling spaces, erecting false work and concreting walls and floors, operating and accessing tower cranes. These types of jobs need access equipment.

Mobile elevating work platforms (MEWPs)

MEWPs are a popular way to access work at height. They can be quite small for single use internally to go through doors and work in ceiling spaces, or very large working four or five storeys into the air for MEWPs external cladding. Each job needs planning, competent workers and additional provisions to stop them toppling over or dropping things on others. MEWPs come in several forms as:

- individual platforms for short term access, often maintenance, nick named cherry pickers;

- moveable work platforms for continuous working often in the form of a scissor lift. This enables an alternative to scaffold and can also hold a small amount of material such as cladding sheets;

- vehicle mounted boom such as those used for street light access. They also may be used as cages attached to 'giraffe' type fork-lifts or tractors. In this case they need anti-tip lock downs and stabilisers.

Because of their mobility and the relatively small base, the stability of the vehicle is paramount to avoid tipping the vehicle and its passengers from a height. MEWPs are controlled from the platform itself or from the ground via radio communication. Ground conditions and obstacles need to be checked before moving along. A trained banksman may also be required at ground level. MEWPs require stabilisers especially if they are cantilevering away from the base, which need retraction and reconfiguration of the lift before moving to another position. A methodology for the use of MEWPs is important so that there is a known plan of use for the reach of the MEWP.

WAHR (2005) regulation	Requirements for employer	Risk assessment
Regulation 2 Definitions		
Regulation 3 Application	Do the regulations apply to you?	Assess responsibilities Train others to comply where they do apply
Regulation 4 Organisation and planning	Ensures that work at height equipment and rescue are properly planned, appropriately supervised and carried out in a manner which so far as is reasonably practicable is safe	Supervisor and user to work on particular plan so both are happy with workplace. Resources must be enabled. Consider weather, rescue, supervision, access/egress, loading. Consider schedule 1/2/3 on general requirements
Regulation 5 Competence	Do not allow a worker, supervisor or any other to engage in work at height or use associated equipment unless competent or, if being trained, is supervised by competent person	A competent person is one with training and preferably previous relevant experience. Consider training, supervision and experience. A scaffolder requires registration
Regulation 6 Risk assessment to reduce or avoid risk	The hierarchy is to avoid work at height or if not possible to prevent or reduce risks of falls and finally mitigate the consequences of falls	Designer and contractor. There is always a risk in alternative methods, so comparisons of alternative methods should be made. Consider design and construction method
Regulation 7 Selection of work equipment (includes platforms such as scaffold)	Give priority to collective measures over personal protection. You must look to reduce the distance and consequences of a fall and plan any need for a timely rescue	Principal contractor to consider all users or restrict use to competent users. Consider choice in duration and frequency, rescue, loading and installation risks
Regulation 8 Requirements for *particular* work equipment cover the use of ladders, scaffolds, elevating platforms, airbags and nets, barriers and fall arrest systems	Comply with schedules. There are schedules for a range of equipment. Schedule 3 pt 1 work platforms Schedule 3 pt 2 scaffold Schedule 4 for airbags Schedule 5 work positioning, rope access, fall arrest, work restraint Schedule 6 ladders	Employer or principal contractor if collective Training or induction to gain competence in use by particular users Site rules to apply
Regulation 9 Fragile surfaces	Restrict work on fragile surfaces, e.g. roofing sheets, if it is feasible to work safely and under appropriate ergonomic conditions without doing so	Principal contractor and users. If required sufficient, access boards, support, barriers/handrails, protection and fall arrest if it is not possible to do so

Table 1 Work at Height Regulations, schedules and risk assessment

The ground covered can be tested and regularly inspected to be used safely. Under LOLER 1998 all equipment should be examined regularly by a competent person for wear and tear and any problems.

Box 3 Causations of MEWP accidents

According to the HSE accidents have especially occurred where:

1. MEWPs have toppled over because they are unbalanced or have lost stability on uneven ground.
2. Poor maintenance on vehicle mounted booms where a pin has slipped out between the boom and the cage allowing cage to rotate.
3. Forklifts are used with non-integrated platforms which have 'hiccupped' forward on restart and cages have slid off the forks, dropping their occupants to the ground. Occupants should ideally use hook-on safety harnesses as a backup to tipping accidents.

Mobile towers

These towers are often erected by non-scaffolders. Erectors however must have proper training and inspect towers prior to and during use. Towers feature patented frames which slot together in different combinations to allow varying height provision. A platform should allow internal ladder access through a trap door (or half platform landings), have a set of wheels with brakes and a set of external bracing members (outriggers) to extend the effective footprint of support and minimise toppling. They are often constructed of lightweight materials and so components are more easily subjected to damage and need to be inspected more regularly if you are using them.

Towers and their outriggers need to rest on firm *level* ground each time they are moved and inspected for stability and brakes properly applied. Only small amounts of material should be loaded onto the tower. The tower should be protected from others walking or driving into it. Ladders are normally part

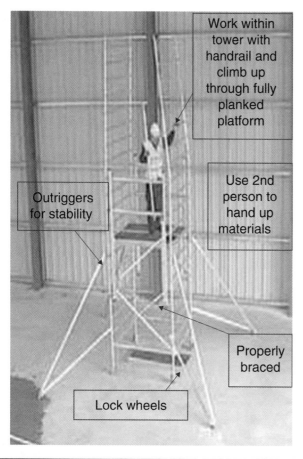

Figure 1 Mobile tower erection. Training is required to erect a mobile tower
Image courtesy of PASMA. Reproduced from HSE 2007, © PASMA

of the tower structure. Tall towers or those with activities that create strong lateral forces (e.g. heavyweight drilling) need anchoring against a structure.

The HSE Construction Information Sheet CIS10 gives extra supportive information for erecting, dismantling and inspecting towers (HSE, 2007) see **Figure 1**.

The use of ladders

To date one-third of all falls from height have involved the use of ladders, consequently they are recognised as a vulnerable part of the safety chain. The poor use of ladders has caused many major injuries due to them being:

- insecurely fixed so that they slip over;
- poorly used because of a temptation to stretch and unbalance;
- poorly maintained so that key bits break in use.

In scaffolding access, timber ladders are often now replaced by alloy staircases which rot less and also provide a hands-free ascent with a proper handrail. Scaffolding is preferred for accessing roofing work while mobile towers rather than step ladders are inherently more stable for work at ceiling height internally. The HSE has an employer's guide INDG402 (HSE,

2005b). It suggests that you should use only an industrial class 1 ladder. Contrary to popular belief they are not banned by WAHR and are often best used for simple short jobs at low heights. Induction in their use is important.

Safe use of ladders means you need to remember a few things (see **Figure 2**):

- Set the ladder at a maximum angle of 1 horizontal for every 4 in height.
- Tie the top of the ladder and fix its base whenever possible.
- Always use the ladder face on to the task.
- Use the ladder with three points of contact, i.e. two feet and at least one hand on the ladder – It you have a paint pot or an additional tool, hook it safely onto the ladder or wedge safely onto the platform of the steps. In any case don't overreach the ladder sideways so that your belt buckle extends outside the style of the ladder.
- If the ladder is unsecured then the whole ladder is in danger of slipping backwards or swaying sideways. You should not rest it on a back slope of more than 16° or a prop it level on more than a side slope of 6°. The feet may slip if the floor is smooth or slippery, or the feet have got dirty or wet.
- Ladders can get damaged or worn out so a pre-inspection is necessary of:
    - the feet to ensure a grip and that they have not slipped out
    - the rungs for structural strength to take weight (loose, splitting, worn, secure) or whether any are missing
    - the styles for straightness and cracks.
- You should ensure users are competent for the task.

All damaged ladders should be disabled or securely kept from use.

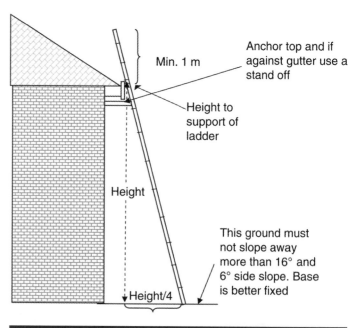

Figure 2 Ladder use requirements

ICE manual of health and safety in construction © 2010 Institution of Civil Engineers

Stepladders

Many contractors ban stepladder use and specify mobile stepped platforms on wheels internally with brakes and a handrail to the steps and the platform (see **Figure 3**). These are chosen for size or flexible for different heights and may be pushed around. For greater and varying height work they take the form of mini cherry pickers, which pass through doors

If you do use stepladders there are two HSE guides for their safe use. INDG402 (HSE, 2005b) for employers and a pocket size INDG405 (HSE, 2005c) for users. The main dangers are toppling the ladder or slipping off it.

Fall protection equipment

There are three types of equipment which are:

- work constraint to stop falling (Schedule 5 part 1 and 5)
- fall arrest which limits falling (Schedule 5 part 4)
- work positioning systems, e.g. cradles or rope access (Schedule 5 part 2/3).

The Work at Height Safety Association (WAHSA, 2006a) identifies ten considerations (**Table 2**) for the use of personal fall protection equipment which will normally be in the shape of a harness. There is however a range of equipment for different purposes.

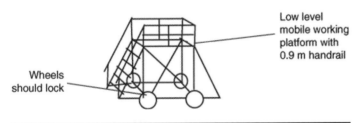

Wheels should lock

Low level mobile working platform with 0.9 m handrail

Figure 3 A wheeled platform is safer for use internally

It will be clear, in the use of rope access and fall arrest equipment, that a great deal of trust is put in the equipment working as expected and so a systematic use of equipment is required with a competent person to inspect equipment, a qualified person to use the equipment properly and a formal recording of equipment source, age and full inspections made available with the equipment.

The everyday use of equipment needs a pre-inspection of quality before each use and needs to be fitted safely. Two people working together provide a safe methodology to ensure the points in **Table 2**. Any doubt about equipment means that it should be withdrawn from use and destroyed or returned to the manufacturer.

Full inspection needs to be at least every 12 months according to BS EN 365: 2004 (BSI, 2004) but for continuous use in arduous conditions such as building sites WASHA recommends at least every three months (WAHSA, 2006b) where the abundance of sharp edges to abrade the rope and the opportunity for the rope to get nicked or mistreated is much greater. Information on suspended access cradles can be gathered from HSE Sheet MISC611 (HSE, 2003).

Scaffolding

Erection

Scaffolding is a skilled trade and those charged with erecting scaffold for others must have minimum training and possess a competency certificate for the level of work. In Britain, scaffolders are members of the National Access and Scaffolding Confederation (NASC) or similar associations which ensure certain minimum standards. Scaffolding should not be attempted by unskilled labour or supervisors. Scaffolders are classified as registered (under the Construction Industry Scaffolders Registration Scheme, CISRS) or as advanced scaffolders and there is dedicated training for specific structure types. In addition check that they have the competence and experience for the type

1. Have I got the right equipment and does it fit, i.e. is it fit for the purpose of saving you in the event of a fall?
2. Is it in good condition and properly fitted, i.e. is it fit for use?
3. Does it have a traceable history so I know it has been cared for properly and uses the right materials?
4. Is it compatible with other materials so that it functions properly, e.g. size of rope and fittings?
5. Are the anchorage points strong enough for the weight and length of the fall and are they regularly maintained and tested? Scaffold may need adapting.
6. Does the harness fit so that it is the right size and is comfortable for the person using it?
7. Is it secure and fastened properly? It should not release until required.
8. Does the equipment exceed its recommended lifespan?
9. Is there sufficient clearance to allow the equipment to deploy properly before hitting something else?
10. Is the product suitable for the particular situation?

Table 2 Ten considerations for the use of personal fall arrest equipment (WAHSA, 2006a)

ICE manual of health and safety in construction © 2010 Institution of Civil Engineers   www.icemanuals.com   171

and complexity or ensure they are supervised by a competent person. Trainees need to be supervised by a qualified scaffolder.

Design and structural safety

One of the most spectacular accidents to scaffold is the total collapse of a multi-storey scaffold. A method statement covering the design of scaffold and its erection and dismantling sequence is also needed. Tall scaffolds or scaffolds on unstable ground should be tied to the structure or braced. A weak point which fails can easily lead to a rapid progressive collapse of the rest of the scaffold as the load is transferred until good scaffold is overloaded. If weather sheeting is attached to scaffolds then this puts a 'sail effect' on the scaffold and the lateral load caused is an important design consideration. The NASC Guidance SG4 and the technical guidance TG20.08 (NASC, 2008) are key documents in basic scaffold design. Patented scaffolds need to follow manufacturer's erection guides.

Box 4 Case Study 2

Rose (2006) reports that in April 2006 a 15-storey scaffold collapsed in high winds at Witan Gate site in Milton Keynes . On this occasion one worker was killed and two received multiple injuries. The HSE investigated the scaffold design, the tying of the scaffold to the structure and the inspection regime. The scaffolding company, the principal contractor and the carpentry company who employed two of the injured were prosecuted because of shortcomings.

There are three main types of basic scaffold, as follows.

Putlog scaffold uses lateral flattened cross supports to build into the joints of existing masonry. This cuts out some of the internal standards. Some of the dead load is carried by the wall.

Independent scaffold has a set of internal vertical poles (standards) and all loading of the scaffold passes down the standards to the base plates so need to be plumb. This type of scaffold requires to be braced (diagonals) and tied to the vertical structure at occasional intervals to stop it falling away from the building. Ledgers (horizontal members) should be level.

Proprietary scaffold which has special components and joints such as Kwikstage and Cuplock are freestanding. These are generally easier to fix, but require similar design considerations for tying in, sequence and bracing as independent scaffold.

HSE requires that complex requirements must be designed by a competent person. Examples of more common designed scaffold are:

- shores and façade retention
- cantilevered scaffold and ladder beams to span openings
- staircases, fire escapes and supporting goods/passenger hoists
- access birdcages for large space access at high level, e.g. roofs

- falsework systems supporting in situ concrete
- protective fans
- any scaffold subject to high loading, vibration, high risk areas, long term duration or in high risk areas.

If in any doubt about a scaffold design, ask a competent designer.

Use of scaffold

A lot of people depend on the stability and safety of the scaffold and it is important for it to be erected and altered by trained operatives and for users to be deterred from altering scaffold. Those using scaffold have a duty to check, report and to not use scaffold if there are obvious problems. This control makes the daily inspection more credible and any tampering more obvious to others.

Typical problems are the removal of handrails and brick guards together with associated trip hazards, the poor tying on of ladders, missing or poorly supported boards (e.g. cantilevered) and the leaving of unsafe loads on the scaffold – these can blow off or cause scaffold collapse. It is also popular to use tied on plastic sheeting to shield workers from rain and wind. Sheeting needs to be checked.

Inspection

Scaffold should be inspected daily by a competent person and a report made that day and kept on the site. An inspection report should indicate any defects and corrective action even if these are rectified immediately to assist with investigating recurring problems. Incomplete or defective scaffolds should have relevant warning signs displayed at access points.

To inspect scaffold you should be trained and there are certified scaffold inspection classes. 'A non scaffolder who has attended a suitable inspection training course and has the necessary background experience e.g. a site manager can be counted as competent to inspect a basic scaffold' (HSE, 2008a). Inspections will pick up the simple requirements of TG20 (NASC, 2008) for basic scaffold, but tailor-made designs need more competence. The areas to look for in the inspection of basic scaffold are as follows:

- secure footings
- longitudinal and cross bracing in place and plumb and level verticals, horizontals and ledgers
- most scaffold should be tied in to a permanent structure
- guardrails and toeboards in place (and brickguards where loaded)
- free from slips and trips
- safe, non cantilevered and adequate width scaffold platform
- ladder access compliant
- putlogs should be inspected for safe connection and bearing on the wall. They still need tying in for higher rise scaffold.

Other inspections should be made after inclement weather and work postponed during unsafe conditions. The HSE scaffold checklist referenced above (HSE, 2008a) is very useful to point out common faults.

Working on roofs

This is a particularly vulnerable element as by definition it usually is the highest point of a building. A roofer is a specialist in laying a watertight skin to a building and has gained some experience in their trade – tiling, asphalting, felt or single ply, metal claddings and other work. More than a quarter of injuries caused by falls from height are to do with work on roofs. The HSE (2008b) has produced a book. The most common forms of roof accident are falling through fragile roofs.

General safety features

Normally you will have to provide nets or cushions for new roofwork to arrest the fall of anyone falling off or through a roof and as such must provide complete coverage and be in accordance with industry standards. Guidance on this is given by the Advisory Roofing Committee (ACR, 2008). Nets need to be fixed so that no holes larger than 100 mm^2 appear on impact. They need to be stretched tight between walls and fixed securely to the structure. Overlaps for smaller nets should be 2 m or nets stitched together. They are fixed by trained riggers. They are not designed to catch debris although debris nets can be laid on top. Cushions may be airbags or bean bags and are configured to cover the floor below completely. They do not stop falls, but protect from injury with falls up to 2 m if restrained at the edge. They may also be stacked to reduce fall heights. They are often used on the first floor of houses.

Edge protection is normally provided in conjunction with netting/airbags and traditionally consists of a 950 mm scaffold handrail an intermediate rail and a 150 mm toeboard on a scaffold fixed to the structure. The toeboard may stop material rolling over the edge or a person sliding down a sloping roof under the handrail There are however many proprietary products available which can be fixed to the edge of the roof to do the same thing, but material will need to be delivered safely by a telescopic handler to a safe platform (see **Figure 4**). Roof ladders may be used for short duration work on sloping roofs. Avoid working off fragile roofs where possible, but take precautions.

Nothing should be thrown from a roof and material should reach skips via a direct shoot or containers lowered to the ground. This being so, exclusion areas should also be created in any danger areas or scaffold fans formed for major and prolonged work. Debris nets may provide some protection against unintentional falling objects. All openings in the roof must have edge protection or be covered.

There are three types of roof to consider:

Flat roofs. You must provide a 900 mm handrail and toeboards similar to a scaffold if no scaffold is provided to the perimeter of the roof. If access only is required then a freestanding barrier 2 m from the edge is sufficient.

Sloping roofs. You will need an external scaffold or tower for access, with the use of roof ladders or attached crawling boards to access. Roof ladders hook onto the opposite slope of the roof with an attached ridge iron. They should not depend upon the ridge tile alone. Alternatively, on new or replaced tiled roofs it is possible to use the timber battens in place for tile fixing as a ladder up the roof. These need to be fixed safely initially working from the bottom up.

Fragile roofs. These are a particular problem in roof maintenance and consist of non load bearing or deteriorating roof sheets spanning between purlins on industrial style roofs. A common material is corrugated asbestos cement or galvanised sheeting which has rusted. Other fragile materials on these roofs are roof lights and other glass roofs, e.g. canopies and atria. You will need staging boards with handrails fixed both sides to span across at least three secure rafters or purlins. Platforms will be required for working.

If the roof covering is incomplete or fragile then safety nets need to be rigged under the roof by a trained specialist or cushions used. An alternative is to use a MEWP to access the roof from the edge or from under or alongside the leading edge of a new roof. This is both safe and recommended where access is available for the MEWP because it does not have to protect against slips or open edges. It is not dependent on the unknown strength of existing structures and fragile materials including thin roof sheets, wood wool slabs and worn out timbers. See **Figure 4**. In all cases the supply of materials should be carefully planned and loading towers set up if not using a crane.

Integrated nature of risk controls

Roofwork has a range of hazards as well as slip protection, depending upon the fixing systems. The table below is a risk assessment showing several significant risks and possible controls in laying a new flat roof on a tall building on top of a concrete base. Other things you might consider for each risk are who is at risk and at what level in order to assess controls. If the public are at risk in the 2nd risk then you would also consider fencing off no go areas at the bottom of the building. The assessment is an outline and is provided for illustration only and you would carry out assessment using your own knowledge, those of others that work with you and company procedures.

This is an indication of how a relatively simple job needs to be planned meticulously. A further amount of work needs to done on a method statement for laying the roof. Hazards by themselves are not risks they only point to significant risks which need controls to reduce that risk. It is important to allocate responsibility to those best able to manage it with an ultimate responsibility on the coordinating contractor. The responsibility for the method study is the specialist contractor, but for access and hoisting they need to liaise with the principal contractor (PC) as these may be shared facilities if other work is going on. The construction details for the roof should be ascertained early on, relevant courses for updating and innovation devised and inspection regimes devised especially for work on fragile roofs.

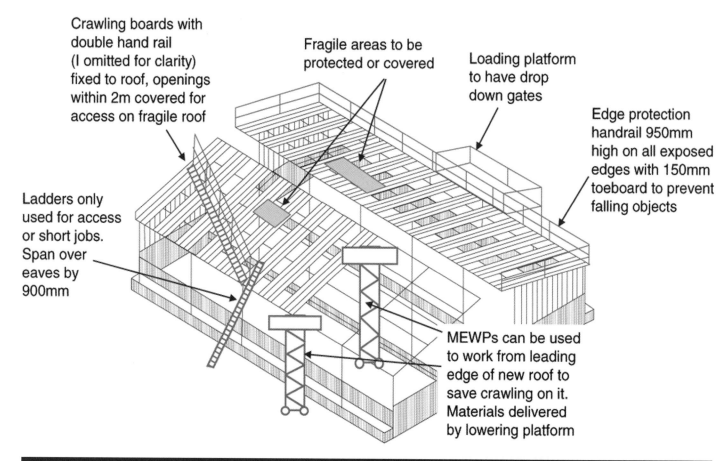

Crawling boards with double hand rail (I omitted for clarity) fixed to roof, openings within 2m covered for access on fragile roof

Fragile areas to be protected or covered

Loading platform to have drop down gates

Edge protection handrail 950mm high on all exposed edges with 150mm toeboard to prevent falling objects

Ladders only used for access or short jobs. Span over eaves by 900mm

MEWPs can be used to work from leading edge of new roof to save crawling on it. Materials delivered by lowering platform

Figure 4 Roofing an industrial building
Adapted from HSENI, 2005 © Crown Copyright. Reproduced with permission from Her Majesty's Stationery Office

Rope work

Generally this section deals with work positioning systems which use a two rope system. It is used for jobs such as window cleaning, external maintenance of tall buildings, permanent rock anchorage and structural surveys and tree felling. Because of the connection with lifting/loading some of the requirements come under the Lifting Operations and Lifting Equipment Regulations (LOLER 1998) where requirements include the lifting or lowering of people and materials. There is also a requirement to consider the working practices relevant to working at height (WAHR 2005) and the selection of suitable equipment (Provision and Use of Equipment Regulations 1998).

However, the technical nature of rope access means it is highly regulated by a system of training and certification run by the Industrial Rope Access Trade Association (IRATA) which sets certain minimum standards of operation and identifies three levels of skill defining allowable tasks. Safety issues to think of the regular inspection of equipment, the choice of the right ropes and equipment and the safe set-up and use of a two rope system. BS EN standards consider equipment specifications. The HSE has also provided some guidelines for different types of rope use in window cleaning (HSE, 2009c).

Safety depends upon planning proper anchoring and drops in the right places so that work is accessed without fraying ropes or snagging the main or fall arrest lines. There is also a need to plan rescue in the case of fall arrest being required. The industrial rope work system is a two rope system with a working line and a safety line (see **Figure 5**). The first rope attaches to a worker who controls his or her position vertically with a braking mechanism from one anchor point but may also lift their position. The second rope, from a separate anchorage, is a fall arrest system. A third braked rope may be used for tools or materials especially if they weigh over 10 kg. This will also be from a separate anchorage. Two technicians normally work together as an additional safety feature.

There are three types of manoeuvre which are used:

- Straightforward where the rope follows a straight path from anchorage to the ground, which involves relatively simple techniques for lowering, lifting and rescue.

- With deviation which involves pulling the rope away from vertical during descent with a slight increase in technique skill.

- More complex procedures to manoeuvre or climb over obstacles to vertical descent or suspended horizontal traverse (aid climbing).

Hazard	Significant risks	Mitigating control	Residual risk and responsibility
Height	Worker falling or sliding off the edge of, or through the roof	Edge barrier fixed to structure or scaffold around structure. Cover openings or edge barrier to fragile areas <2 m away	PC to check regularly and repair and record defects. Restrict access
	Materials or tools falling off edge or through roof	Fix handrail and toeboards. Stack heavy materials off roof or spread load	PC checks nets or cushions. Foreman induction
Removing waste	Danger of materials waste disposal and dust	Install covered waste chute. Wet down as necessary. Warning notices.	PC ensures skip changeovers and dust protection
Burning	Windy weather makes using hot mastic dangerous	Do not work in winds above 10 knots or put mastic pot behind shelter	Foreman to check weather regularly
	Burning from mastic	Training to reduce spills. Use of protective gloves, goggles and masks. Attend boiler to stop overheating	Specialist to give induction (PC) and toolbox talks
	Danger from flashback on torch	Torch guards and wear fireproof gloves and goggles	Regular maintenance by plant supplier
Removing asbestos sheets	Danger from dust and long term disability	Use respiratory device, remove whole asbestos sheets and undo fixings, pack in sealed bags, take to toxic waste site	PC checks work and excludes access during work. Contractor controls dust
Back injury	Lifting felt rolls or tiles. Hoisting mastic tank and materials	Training and lifting aid to bring up on lift	Training (PC) and toolbox talks (specialist)
Lifting gear failure	Hoist drops load	Regular inspections in place	Use of LOLER procedures to check lifting gear (PC and Foreman)

Table 3 Risk assessment for asphalt roofwork

The rope technician may also connect him/herself direct to framed structures with lanyards as shown in **Figure 6** where other suitable provision is made.

Inspection procedures

Inspection procedures will be based on similar procedures to those mentioned under fall protection and they will consist of:

- Pre-start induction and assessment of competence and qualification for level of work.

- Pre-checks which take place before the equipment is used each time and this should also ensure that equipment is fitting properly and correctly attached.

- Regular three monthly inspections of equipment recommended by WAHSA (2006a).

The pre-checks are particularly important where trainees are involved, but always involve a team approach of mutual checking each other, where a complete checklist approach is required each time. Nothing can be assumed to have remained the same from last time as equipment may have been moved or tampered with.

LOLER has very specific backstop requirements and thorough examination and daily pre-start checks are both referenced. All inspections should be recorded and kept on site and made available for subsequent in spections.

Box 5 Case Study 3: The Brunswick Centre, London

Problem: after installing feature neon lighting to eight water towers at the Brunswick Centre, future access for maintenance of the neon lighting was needed. Costs of erecting scaffold were deemed to be very expensive.

Solution: Rope Task technicians installed over 350 safety anchor points conforming to EN 795. These anchor points will allow future access by specialist teams of rope access technicians to carry out maintenance and replacement of the neon tubes.

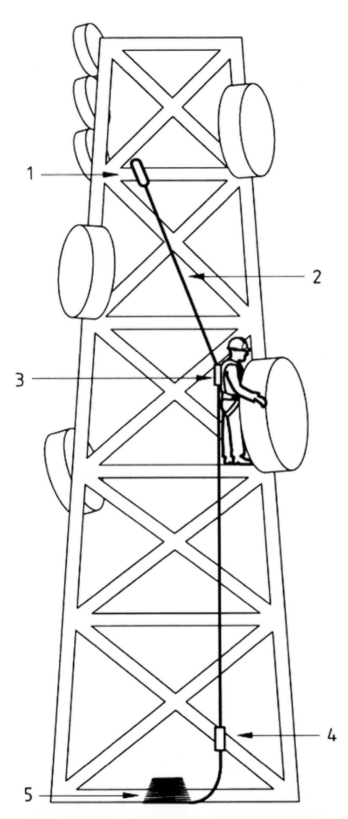

Figure 5 Example of working with in suspension
© **Rope** Task Limited, reproduced with permission
Key 1 = primary anchor point, 2 = temporary anchor line, 3 = guided type fall arrestor, 4 = tensioning weight, 5 = unused anchor line

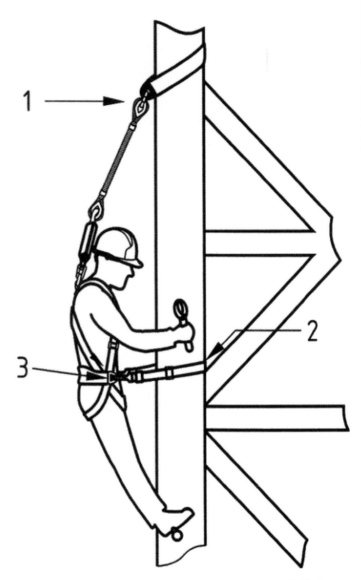

Figure 6 Lanyard support on framed structures
© **Rope** Task Limited. Reproduced with permission from the Rope Task Training Manual
Key: 1 = safety backup device 2 = work position lanyard, 3 = attachment to waist of full harness

Working procedures and practices

IRATA requires that there are at least two people in the team with one of them at supervisor level 3 (WAHSA, 2006b, p. 16) which means they have sufficient competence and experience to plan and risk assess the work and to have planned a method statement and rescue procedures. Companies need to train all personnel to the standards of BS 8454 Code for delivery of training and education for work at height and rescue.

Rope access is a positioning system to reach a place of working at height and as such each workplace needs to be risk assessed for the normal dangers and for specific danger to the rope and suspension equipment. These dangers include protruding obstacles, heat, abrasion, independent anchor points, lowering and handling of suspended materials, evacuation in

ICE manual of health and safety in construction © 2010 Institution of Civil Engineers

case of fire and positioning of power cables and gas lines for welding torches. Work rest periods and rescue procedures need also to be planned depending on the work type, the complexity of access, the severity of the working conditions and the effect of the weather.

Your work method is adapted when you are in suspension and supportive teams are required to supply materials at an appropriate rate. Communication is likely to be by radio. Tools used may need to be adapted such as welding torches and PPE because of the more precarious position and unique characteristics of working suspended on a rope! Exposure can also be more severe and generally less protected so rest periods should be regular and shifts appropriately shorter in poorer conditions.

You should issue a permit to work so that other people can be made aware of any dangers to themselves overhead and site adjustments made such as barriers. Sentries (no need to be formally trained) can stand guard on anchorage points and be in communication with workers. In addition anchor areas and working edges need to have barriers.

You need rescue procedures in the case of accident. If a fail safe system comes into effect it may leave a worker suspended and in many cases they can self recover. However, if they have no way to move themselves they will need communication and a rescue team specialist enough to reach them quickly. This will need to be coordinated with your site emergency procedures. The period of suspension after a sudden arrest should be limited because of the potentially fatal effects of suspension fatigue. The HSE has given advice for emergency first aid procedures (HSE, 2008c).

The use of tools on rope access is a particular skill and specific training and safeguard should be in place. For instance cordless tools are recommended and tools must be able to be safe in wet conditions. There is also a danger of dropping tools so these are attached to lines or to the worker, but not tangled with the working lines. Materials will be provided in suspended bags on their own ropes, but by their nature may also be dropped, so you need to make an exclusion zone at ground level due to the unexpected hazard to those below. Large tools over 8 kg in weight, such as grit blasters, need to be suspended on their own rope and fitted with a 'dead man's handle' to cut off power in the event of a mistake, accident or emergency.

On completion of the work there should be a proper inspection to ensure that nothing can drop off the work face. The nature of the work needs patience and systematic planning with teamwork backup so that forgotten tools or materials can be supplied efficiently and work is not rushed so as to endanger the rope worker or others. IRATA suggests that particular personal characteristics are necessary for the safe carrying out of such painstaking work. This may also mean vetting personnel who arrive at work in a stressed or unfit condition.

No scaffold policy

Scaffold is often considered expensive and has been the subject of some accidents because of misuse. Construction planning should get permanent platforms/ floors and stairs into place early

in the process to cut down on time spent on temporary platforms. Some of the prefabrication systems allow scaffold to be eliminated by working from the inside, but some such as timber frame require complete scaffold in place prior to the arrival of units, because superstructures go up in two to three days and there is need for a platform to apply a cladding such as brickwork.

Complete prefinished wall or floor panels may be delivered to site, craned and bolted together with very short exposure to height. Simple work constraint systems and pre-fixed edge protection are used. Some of these panels also include the cladding which is sealed together and precludes the use of external scaffold for the cladding. These have been used for multiple storey student accommodation where there is quite varied layout and heights of up to seven storeys. Windows are fixed afterwards from the inside and internal finishing work is done safely within 'four walls'.

> **Box 6** Case study 4
>
> The second Severn Crossing was built by cantilevering precast box units out from the 23 central piers, themselves precast, which had been constructed at intervals across the estuary width. Bridge deck units were post-tensioned and kept in balance for each pair of units stretching from the pier either side until they joined the cantilever from the adjoining piers. Temporary post-tensioning was carried out from within the box girder structure before the crane disengaged tying the last section to the previous ones. The balanced cantilevers provided a secure protected working platform able to support working equipment and at least one unbalanced unit. Permanent post-tensioning tied the span together

On large steel framed buildings it is possible to avoid the use of scaffold by providing a level ground access for large scissor type MEWPs to fix the cladding. Flooring and other pre-cladding work is carried out using edge protection. There are hazards involved with the use of MEWPs, but these are balanced against the danger of erecting and dismantling the scaffold, the competence in the use of cladding fitters to use the MEWPs safely and the saving in time and money.

Roofing systems such as Kalsip® can also be operated from a single tower access because they are fed out from a continuous roll from the apex or eaves of the roof and fixed from MEWPs and by running the former up the length of the standing seam.

Summary of main points

- Working at height hazards occur at all levels and so a wide range of jobs should be risk assessed for compliance with the safe procedures set out in the Work at Height Regulations.

- It is good to develop a culture of awareness and improvement by looking out for each other, because more than half the fatalities and many major injuries are caused by falls from height. Low falls produce more serious injuries and absence from work because of less awareness of the dangers.

- It is essential to provide training for all those using access equipment and certificate competence.

- Workers involved in the erection, use or inspection of scaffolding must be competent and where necessary appropriate training provided. Encourage reporting of unsafe equipment and reckless or unsafe usage, by all those using it.

- Roofwork is particularly dangerous and produces the largest number of accidents and needs a greater awareness of good practice to avoid dangers.

- You need to be particularly aware of the dangers of working on existing fragile roofs and have nets or fall constraints in place.

- Construct permanent platforms and staircases as early as possible and integrate lifting plans to access materials.

- Successful rope work has to comply with the lifting regulations (LOLER 1999) and also WAHR 2005 and must have a rescue strategy in place as well as coordination with other work carrying on underneath it.

Acknowledgement

The author would like to thank Haydn Gamble of Rope Task Limited, West Bromwich, UK, for the information he provided on rope work and for supplying Figures 5 and 6, and case study 3.

References

Advisory Committee for Roofwork (ACR). *ACR[CP]003:2008 Rev1 – Recommended Practice for Use of Safety Nets for Roof Work*, 2008. London: ACR. [Also known as 'the Blue Book']. Available online at: http://www.roofworkadvice.info/html/publications.html

Griffiths, S. Swansea Site Manager Fined £10,000 After Worker Hurt in Fall. *Building* 2009, 27 January.

Health and Safety Executive (HSE). *Safety in Window Cleaning Using Suspended and Powered Access Equipment*, HSE information leaflet MISC611, 2003, London: HSE. Available online at: http://www.hse.gov.uk/pubns/misc611.pdf

Health and Safety Executive (HSE). *The Work at Height Regulations 1995 (Presentation)*, 2005a, London: HSE. Available online at: http://www.hse.gov.uk/falls/downloads/1and2.pdf

Health and Safety Executive (HSE). *Safe Use of Ladders and Step-ladders: An Employer's Guide*, HSE leaflet INDG402, 2005b, London: HSE. Available online at: http://www.hse.gov.uk/pubns/indg402.pdf

Health and Safety Executive (HSE). *Top Tips for Ladder and Step Ladder Safety*, HSE pocket card INDG405, 2005c, London: HSE. Available online at: http://www.hse.gov.uk/pubns/indg405.pdf

Health and Safety Executive (HSE). *Tower Scaffolds – Construction Sheet Information No. 10* (CIS10) Revision 4, 2007, London: HSE. Available online at: http://www.hse.gov.uk/pubns/cis10.pdf

Health and Safety Executive (HSE). *Scaffold Checklist*, 2008a, London: HSE. Available online at: http://www.hse.gov.uk/construction/scaffoldinginfo.htm

Health and Safety Executive (HSE). *Health and Safety in Roofwork: Health and Safety Guidance HSG33*, 3rd edition, 2008b, London: HSE.

Health and Safety Executive (HSE). *First Aid Management for Harness Suspension When Working at Height*, 2008c, Available online at: http://www.hse.gov.uk/falls/harness.htm

Health and Safety Executive (HSE). *Falls and Trips in Construction – Guidance*, 2009a. Available online at: http://www.hse.gov.uk/construction/tripsandfalls/index.htm

Health and Safety Executive (HSE). *Results of Construction Division Intensive Inspection Initiative – March 2009*, 2009b, London: HSE. Available online at: http://www.hse.gov.uk/construction/tripsandfalls/results2009.htm

Health and Safety Executive (HSE). *Safety in Window Cleaning Using Safe Rope Access,* HSE Information Sheet MISC612, 2009c, London: HSE. Available online at: http://www.hse.gov.uk/pubns/misc612.pdf

Health and Safety Executive Northern Ireland. *Guidance Document – Roof Work*, 2005, Belfast: HSENI. Available online at: http://www.hseni.gov.uk/roof_work.pdf

National Association of Scaffolding Contractors. *SG4:05 Preventing Falls in Scaffolding and Formwork*, 2008.

National Association of Scaffolding Contractors. *TG20:08 – A Guide to Good Practice for Scaffolding with Tubes and Fittings*, 2008. London: NASC.

Rose, J. HSE On Site for Investigation into Tragic Scaffold Collapse. *Building* 2006, Issue 16. Available online at: .http://www.hse.gov.uk/construction/scaffolding.htm

Work at Height Safety Association (WAHSA). *TGN01 – Considerations for the Use of Personal Fall Protection Equipment*. Technical Guidance Note 1, 2006a, Whitchurch: WASHA.

Work at Height Safety Association (WAHSA). *TGN03 – Guidance on Inspecting Personal Fall Protection Equipment*. Technical Guidance Note 3, 2006b, Whitchurch: WASHA.

Referenced legislation

BS EN 365: 2004. *Personal Protective Equipment Against Falls From a Height. General Requirements for Instructions for Use, Maintenance, Periodic Examination, Repair, Marking and Packaging*, London: British Standards Institution.

BS 8454: 2006. *Code of Practice for the Delivery of Training and Education for Work at Height and Rescue*, London: British Standards Institution.

Construction (Design and Management) Regulations 2007, London: The Stationery Office.

Lifting operations and Lifting Regulations. Statutory Instrument 1998.London: The Stationery Office.

Provision and Use of Work Equipment Regulations 1998 Reprinted February 2004. Statutory instruments 1998 2306, 1998, London: The Stationery Office.

Work at Height Regulations. Statutory Instrument 2005, No. 735, London:The Stationery Office.

Further reading

BS EN 362: 2004. *Personal Protective Equipment Against Falls From a Height. Connectors*, London: British Standards Institution.

BS EN 12277: 2007. *Mountaineering Equipment. Harnesses. Safety Requirements and Test Methods*, London: British Standards Institution.

BS EN 1263–1: 2002. *Safety Nets. Safety Requirements, Test Method*, London: British Standards Institution.

BS EN 1263–2: 2002. *Safety Nets. Safety Requirements for the Positioning Limits*, London: British Standards Institution.

BS EN 12841: 2006. *Personal Fall Protection Equipment, Rope Access Systems, Rope Adjustment Devices*, London: British Standards Institution.

Health and Safety at Work Inspectorate. *Working Safely with Scaffold*, 2002. Available online at: http://www.gov.im/lib/docs/dlge/enviro/scaffolding.pdf

Health and Safety Commission. *Safe Uuse of Lifting Equipment. Lifting Operations and Lifting Equipment Regulations 1998. Approved Code of Practice and Guidance* (L113), 1998, London: HSE Books. [LOLER applies to the inspection and use of rope access and MEWPs.]

Health and Safety Commission. *Safe Use of Work Equipment – Provision and Use of Work Equipment Regulations 1999. Approved Code of Practice and Guidance*, (L22), 2008, London: HSE Books.

Health and Safety Executive (HSE). *The Lifting Operations and Lifting Equipment Regulations 1998 ('LOLER') How They Apply to Rope-based Access Systems for Work at Height June 13v2*, 2007, London: HSE. Available online at: http://www.hse.gov.uk/falls/downloads/ropeaccess.pdf

Health and Safety Executive (HSE). *The Working at Height Regulations (Amended), A Brief Guide* (INDG401REV1), 2007, London: HSE Books. Available online at: http://www.hse.gov.uk/pubns/indg401.pdf

Health and Safety Executive (HSE). *The Selection and Management of Mobile Elevating Work Platforms* (HSE information sheet CIS58), 2008, London: HSE Books. Available online at: http://www.hse.gov.uk/pubns/cis58.pdf

IRATA. *Guidelines on the Use of Rope Access Methods for Industrial Purposes*, 2nd edition, 2000. Borden, Hampshire: IRATA.

IRATA. *International Guidelines*, 2nd edition rev 1 (01/100), 2000. Borden, Hampshire: IRATA.

Websites

British Standards Institution (BSI) http://www.bsigroup.com

Eurocodes Expert – making Eurocodes easier
http://www.eurocodes.co.uk

Falls from Height, HSE http://www.hse.gov.uk/falls

Health and Safety Executive (HSE) http://www.hse.gov.uk

Health and Safety Executive Northern Ireland (HSENI)
http://www.hseni.gov.uk

Industrial Rope Access Trade Association (IRATA)
http://www.irata.org

Institution of Civil Engineers (ICE), Health and safety
http://www.ice.org.uk/knowledge/specialist_health.asp

National Access and Scaffolding Confederation (NASC)
http://www.nasc.org.uk

Chapter 15

Excavations and piling

Michael Battman Gardiner and Theobald LLP, Manchester, UK

doi: 10:10.1680/mohs.40564.0183

Most types of construction project will require some excavation works. There are many safety hazards thrown up by excavation work; these are both directly as a result of the works (state of the excavation sides, sufficient support for the loads imposed, groundwater ingress, surcharges from adjacent loads) and indirectly from the working area (access, presence of contaminants, confined spaces, falls from height, buried structures), and all have to be considered when undertaking an excavation.

Hazards are inherent in piling works; the safety hazards posed by these works such as noise and vibration, working near plant/machinery, working at height, manual handling, mechanical lifting and concrete handling, are discussed.

The Designer plays a role in reducing hazards by the design they choose. Safety hazards need to be assessed and planning and sequencing of works undertaken to allow the various control measures to be implemented and to prevent hazardous work becoming dangerous work.

CONTENTS

Introduction

Box 1   HSE press release

The Health and Safety Executive (HSE) is reminding construction workers of the dangers they face when working in excavations following recent fatalities caused by trench collapses.

There have been three fatal incidents since April where workers have been killed due to trenches collapsing on top of them. These could have been avoided if the appropriate safety measures had been taken.

HSE Specialist Inspector Nigel Thorpe said:

"Trench collapses are entirely avoidable. Without suitable support, any face of an excavation will collapse; it's just a matter of when. The steeper and deeper the face, the wetter the soil, the sooner the collapse.

Trenchless technologies are available which avoid many of the hazards of excavation, but if a trench is required modern proprietary systems allow the ground support to be installed without the need to enter the excavation."

HSE, Press release: E118:04-19 August 2004. HSE urges greater awareness of trench collapse dangers

It is not the first time that such advice has been given: groundworks are dangerous operations and if not properly managed and controlled accidents will happen.

On most construction sites excavations and piling works are the first significant works to be carried out; they can also be some of the most dangerous works that will occur on a project. It is easy to be deceived by the appearance of the sides of an excavation; the condition of the excavation can suddenly change and collapse can happen without warning – it is not an exact science.

Virtually every type of construction project involves some excavation works; even refurbishment projects tend to involve upgraded utility supplies that have to be delivered to the building.

Excavations come in many shapes and sizes: from relatively shallow and narrow trenches for cables to massive basement foundations and cofferdam constructions. Excavations are one of the most hazardous activities undertaken by civil engineers; it is an activity that has in recent years claimed an average of more than six lives and many serious injuries each year.

Because the works are normally at the front-end of any construction project it is highly likely that this element of work will be included in the Construction Phase Health and Safety Plan that is to be forwarded by the Principal Contractor to the Construction Design and Management (CDM) Coordinator (CDM-C). This is an opportunity for the CDM-C to make an impact by ensuring that the front-end works are adequately planned and resourced.

The 2007 revision of the Construction (Design and Management) Regulations (CDM 2007) encompassed the Construction (Health, Safety and Welfare) Regulations 1996 and outlines specific duties relating to excavations, cofferdams and caissons, and reports of inspections.

Hazards

The most common form of hazard associated with excavations is collapse of the sides, which often happens without warning (see **Figure 1**). A cubic metre of earth, depending on its composition, will weigh in the region of 2 tonnes. Excavation sides will deteriorate over time and they should be regularly inspected.

Excavations can and will collapse if:

a) **The sides of the excavation are not self-supporting**

Excavation sides will vary in their ability to resist collapse dependent on factors such as the type of material, the amount of groundwater present and the loads imposed on the excavation by plant, adjacent building or the spoil from the excavation.

Rock will in most cases support itself; clay and other cohesive materials are often look safe, but are liable to sudden and catastrophic collapse – granular materials will need support from the outset.

b) The supports provided are insufficient to cope with the loads imposed

Poor or no design of the temporary support system can lead to collapse; incorrectly constructed support systems also can and will fail.

c) Surcharges from spoil, adjacent foundations/buildings, stored materials, plant or temporary works imposed loads overload the ground adjacent to an excavation

Additional weight added to the sides of the trench increases the forces that contribute to the collapse of the excavation. It is easily done: dumpers, excavators and excavated material are the biggest culprits; but proximity to adjacent buildings can lead to collapses. In some cases, especially with an old building, it may be necessary to monitor the adjacent structures for movement while the excavation continues.

d) Groundwater ingress reduces the strength of the ground and can lead to unexpected inundation of excavations

Groundwater destabilises the existing soil structure, making it more likely to slip and collapse. This can be especially dangerous after sudden rain showers. If a water main is damaged in an excavation it can seriously affect the structure of the ground and render a previously safe excavation dangerous. If there has been movement of water through the ground over a period of years it is not unusual to find voids in the ground; these have to be managed to ensure stability.

An additional danger of water in excavations is that standing water hides the true extent of the depth of an excavation. Water should routinely be pumped out of excavations.

e) Excavation supports are removed prematurely, to facilitate backfilling or compaction

The removal of trench and excavation supports should be carried out in a planned and considered manner. It should only be carried out in conjunction with sequenced backfilling and compaction; due consideration should be given to surrounding structures. Any sign of movement may affect them.

Additional hazards include:

f) The presence of contaminants, which may be harmful to health, whose levels cannot always be assessed by sight or smell

This is especially relevant in areas known to contain contaminated ground. Redevelopment sites are often prone to contamination and enabling works may need to be undertaken to remove and treat such material. You should check whether this operation requires a waste management licence. Hygiene procedures should be of

very high standard, with personal protective equipment (PPE) and welfare facilities available to enable workers to clean their hands prior to handling food, cigarettes, etc. which are then put in the mouth (also see Chapter 11 *Controlling exposure to chemical hazards* and Chapter 12 *Controlling exposure to biological hazards*).

g) Gases migrating into excavations and creating explosive or poisonous atmospheres, which fall under the category of confined spaces

An often forgotten hazard of excavations is that the atmospheric conditions for workers inside them can deteriorate and confined space precautions should always be taken. Care should be taken with the location of petrol or diesel powered plant such as pumps, compressors and generators, to ensure that the exhaust fumes are not going into the excavation.

Regular air monitoring should be carried out and all staff should be trained in recognising confined space conditions and how to deal with them. Awareness of what a 'confined space' is should be part of any ground workers' competence (also see Chapter 16 *Confined spaces* and Chapter 20 *Fire and explosion hazards*).

h) The presence of underground or buried utility services

One could write a book on this topic alone; suffice to say that buried utility services cause innumerable additional problems when excavations are being carried out.

The problems given by each type of utility are varied, ranging from fire and explosions to electrocution, health hazards as well as severe monetary penalties. Most buried services are fairly shallow; apart from drains you can normally expect to encounter all other services in the first 1.5 m from the surface.

It is essential that all available information is obtained from the utility companies about the whereabouts of their pipes and cables; remember that apart from the usual water, gas, electric, telecom and drainage there may be private chemical and gas pipelines and supplies. Additionally, once on site, managers and operatives should use their experience to look for signs of buried services; surface boxes, reinstatement in road surfaces, gas risers, electrical sub-stations, marker tape and backfilled material, which are all clues to the possible presence of a service.

Personnel on site should have access to emergency phone numbers of all the utility companies and should not hesitate to call them if their property is damaged in any way.

Utilities also hinder trench support systems and modifications may have to be made to ensure the pipes/cables can continue to pass through the trench undamaged.

i) The presence nearby of other excavations or other voids

The stability of any excavation relies heavily on the stability of the ground surrounding it. If there are voids and other excavations in the vicinity these stable conditions are compromised and collapse can more readily occur. Extra

ICE manual of health and safety in construction © 2010 Institution of Civil Engineers

Figure 1 Säo Paulo metro, Pinheiros station site, tunnel collapse. Seven people died when the walls of the shaft collapsed creating a crater on 12 January 2007. The investigation into the cause of the collapse is ongoing [January 2010] (reproduced from *New Civil Engineer*, 2009)

vigilance should be taken in areas where old mine workings are known to exist.

j) Access and egress

Once you have excavated and made your excavation safe, it is equally important that workers can get in and out in a safe manor. Properly secured ladders and, where necessary, landing stages should be constructed and maintained. Remember that, as excavations progress, the access arrangements will need to be changed to match the new conditions.

k) Falls into excavations – work at height

Falls from height are normally considered to relate to working on buildings or structures, but the drop can be just as far if you fall into an excavation. Consequently it is essential that the edges of all excavations are well guarded to prevent people and equipment falling in. It is particu-

larly important that the protection around access points is well guarded and designed. Also see Chapter 14 *Working at height and roofwork* and Chapter 13 *Controlling exposure to physical hazards.*

l) Interaction and conflict with the public

This is particularly relevant to utility works in the streets where often the excavations are quickly dug and quickly backfilled, the site being very much transient in nature. On busy highways this can lead to major interaction with the public that needs constant management and vigilance for the ever-changing conditions. The public need to be well informed of alternative routes to take and signage to Chapter 8 of the Department for Transport's Manual (DfT, 2009b) or the *Safety at Street Works and Road Works: A Code of Practice* (DfT, 2009a) is an essential part of safe working in areas where the public can access. See Chapter 19 *Transportation and vehicle movement* for more information.

If children are likely to visit the area, additional precautions may need to be taken, plating excavations over night or even backfilling them at weekends.

The public can be inquisitive, so viewing panels should be considered to enable the public to safely view the excavation.

m) Noise and vibration

A by-product of most construction techniques is exposure to excessive levels of noise and vibration, which can lead to health hazards that affect people long after they encountered the hazard. Piling operations are particularly hazardous in this respect as they are both noisy and can cause direct and indirect vibration. Where this is likely it may be advisable to carry out pre-construction surveys of buildings close to the excavation as well as noise and vibration surveys.

The vibration of jack-hammers is particularly hazardous and can lead to circulation problems, in particular 'vibration white finger'; modern, well maintained equipment will help reduce the hazard. Also see Chapter 13 *Controlling exposure to physical hazards*.

n) Proximity of vehicles and plant

Almost without fail excavations will include the need, as a minimum, for excavators and wagons or dump trucks to remove the waste; other plant, generators, compressors, compaction equipment, etc. will often be involved. Additionally, many excavations are not happening in isolation, but are part of a larger construction site with much more plant and equipment.

The presence of plant of any sort is an additional hazard that needs to be recognised, controlled and managed (see Chapter 19 *Transportation and vehicle movement*).

Control measures – options

a. Trenchless technology

Not always a practical solution but can massively save on excavation for pipes and cables; it does not totally eliminate excavation, as launch and reception pits will still normally have to be dug. Types of trenchless technology include ground moles (see **Figure 2**), pipe bursting, headings and tunnelling; all these techniques have their uses in the right environment. But there is always the risk that the mole could hit existing buried services. You would be advised to discuss this option with local utility companies.

b. Battering

Battering is a solution that is only feasible in open spaces and relies on the ability to excavate the ground to an angle greater than the angle of repose of the material being excavated. It will lead to greatly increased volumes of excavated material, which can cause its own safety issues.

c. Ground support

Ground support is the most common form of controlling excavations. There are many ways in which it can be achieved, varying from simple ad-hoc timber supports to patented systems.

Common patented systems include Trench Sheets, Sheet Piles, Tank and Manhole Braces; Struts and Wailer systems, Trench and Drag Boxes.

Choice is governed by a combination of criteria including depth, ground conditions and the number of utilities that cross the excavation; it is often the case that ground support types in trenches will change with the changing location.

d. Inspection

Excavations change and deteriorate over time, it is therefore important (and a legal requirement) that regular inspections are carried out of all excavations.

Inspections should take place: at the beginning of every shift; after any significant modification to the support system; after any significant fall of material or any other event that may affect ground stability. Additionally a written report must be made after each seven day period or if any defects are noted. These inspections and reports must be made by competent persons who understand excavations and the processes that can lead to their collapse.

Work should be stopped if any defects are noted until the problem is resolved.

Planning

As with all construction work, time spent planning the works before construction starts is time well spent.

Avoidance of excavation

Why is the excavation necessary? Can the cable go overhead instead of underground? Is there an existing duct that the cable can be threaded down? Can the two trenches be combined into one? These are the types of questions that should be asked before commencing any excavations; the answer will more often than not be 'no', but on the few occasions that it is 'yes', lives could be saved.

Collation of information

It is essential that all possible information which will enable the excavation to be controlled and better planned is collected, analysed and made available to the contractor. Remember there may be an existing health and safety file for the site. Check with your client. The types of information that can ensure a safer operation include:

- trial hole logs
- utility drawings
- site investigation reports – including contamination information
- foundation details of existing buildings
- munitions surveys
- abandoned mine workings surveys.

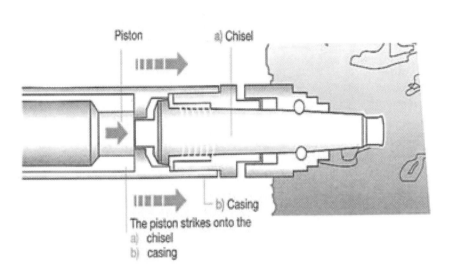

Figure 2 Example of trenchless technology – schematic of a working mole and photo of a mole in situ. Principle of a working mole: the piston strikes the chisel head assembly, which is propelled forward as the piston accelerates and impacts against the rear of the chisel. The piston then strikes the casing pulling it into the previously created borehole. This is repeated creating a hammer motion forcing the mole through the ground
© **Essential** Water Services. Reproduced with permission, http://www.essentialwaterservices.co.uk

Sequencing of works

Inevitably, excavation works require plant and machinery; they need room for materials to be moved in and out of the construction area, consequently they can affect other construction activities and have a knock-on effect on the programme.

If it can be implemented it may be beneficial to carry out excavation and utility works ahead of the main works as enabling works. This can bring about other problems as the design of the footprint and foundations will normally have to be 'fixed' to ensure that later costly changes do not have to be made.

Risk assessments and method statements

The Management of Health and Safety at Work Regulations state, '*Every* employer shall make a suitable and sufficient risk assessment of the risks to health and safety of his employees … and … persons not in his employment …'. Risk assessments for excavation works should, as always, be carried out by a person trained and competent to do so. The assessment shall consider all the issues listed earlier, together with any other situation such as other contractors in the area that may affect the safety of the workers and the public. Based on the risk assessment a method statement or safe working method is created which should address all of the key issues that arose and how they will be managed or controlled.

Risk assessments and method statements should not be too verbose or complicated; their sole purpose is to communicate the risks and hazards of the operation to those carrying out the works and those in the proximity of the works. They should not include obvious hazards that competent employees will know well, they should highlight unexpected and unusual hazards that may not be obvious.

As work progresses risk assessments should be reviewed and updated as necessary.

Competence of staff and operatives

Groundworks have too often in the past been treated as unskilled work that can be done with unskilled operatives. Fortunately things have changed and the industry now recognises that excavation work needs trained and competent workers to safely carry out the works. Managers and supervisors also need to be competent.

The skills needed are wide and varied, including confined space awareness, a basic knowledge of soil mechanics, a knowledge of support systems, understanding signage, etc.

Competence can be proven in various ways: Construction Skills Certification Scheme (CSCS) cards, new roads and street works qualifications, confined space training are just some of the essential qualifications needed by at least one gang member.

Designers – what can they do? (See also the section on planning)

Historically designers designed and contractors built, including sorting out safety on site. The CDM Regulations, which first saw the light of day in 1995, give designers specific duties to consider risks and hazards and to eliminate or reduce them where possible. The decisions designers make can have profound effects on the construction process and as such all of the issues that a contractor will face need to have been considered by the designer. Any opportunity to modify the design to help eliminate or reduce the requirements for excavating is worth pursuing (see Chapter 3 *Responsibilities of key duty holders in construction design and management*).

Piling

General

The name piling covers a variety of load supporting systems; they may be temporary or permanent; timber, concrete or steel; bored or driven. The purpose of a pile foundation is to transmit a foundation load to a solid ground or to resist vertical, lateral and uplift loads.

Legal

Excavation and piling can fall under many health and safety regulations, but listed below are the ones that are most likely to be relevant to excavation works. The likes of the Provision and Use of Work Equipment, Manual Handling and other cross-operational regulations have been omitted for brevity.

Health and Safety at Work etc Act 1974

Construction (Design and Management) Regs 2007

Management of Health and Safety at Work Regs 1999

Control of Substances Hazardous to Health Regs 2002

Confined Space Regulations 1997

History

Piling is not new; since the days of early civilization, timber piles have been driven in to the ground by hand or holes were dug and filled with sand and stones to strengthen the ground. The technology of the industrial revolution led to steam and later diesel driven piling machines; nowadays the drive to utilise areas with poor ground conditions has given rise to many advanced techniques of pile installation.

Box 2 Types of piles by materials

Steel piles

Steel piles are suitable for handling and driving in long lengths. Their relatively small cross-sectional area combined with their high strength makes penetration easier in firm soil. They can be easily cut off or joined by welding. If the pile is driven into a soil with low pH value, then there is a risk of corrosion, but risk of corrosion is not as great as one might think. Although tar coating or cathodic protection can be employed in permanent works.

Concrete piles

Pre-cast concrete piles or pre-fabricated concrete piles: usually of square, triangle, circle or octagonal section, they are produced. They are pre-cast so that they can be easily connected together in order to reach to the required length. This will not decrease the design load capacity. Reinforcement is necessary within the pile to help withstand both handling and driving stresses. Pre-stressed concrete piles are also used and are becoming more popular than the ordinary pre-cast, as less reinforcement is required.

Driven and cast in place concrete piles

Two of the main types used in the UK are:

- West's shell pile: pre-cast, reinforced concrete tubes, about 1 m long, are threaded on to a steel mandrel and driven into the ground after a concrete shoe has been placed at the front of the shells. Once the shells have been driven to specified depth the mandrel is withdrawn and reinforced concrete inserted in the core. Diameters vary from 325 to 600 mm.
- Franki pile: a steel tube is erected vertically over the place where the pile is to be driven, and about a metre depth of gravel is placed at the end of the tube. A drop hammer, 1500 to 4000 kg mass, compacts the aggregate into a solid plug which then penetrates the soil and takes the steel tube down with it. When the required depth has been achieved, the tube is raised slightly and the aggregate broken out. Dry concrete is now added and hammered until a bulb is formed. Reinforcement is placed in position and more dry concrete is placed and rammed until the pile top comes up to ground level.

Timber piles

Used from earliest recorded time and still used for permanent works in regions where timber is plentiful. Timber is most suitable for long cohesion piling and piling beneath embankments. The timber should be in a good condition and should not have been attacked by insects. For timber piles of length less than 14 m, the diameter of the tip should be greater than 150 mm. If the length is greater than 18 m, a tip with a diameter of 125 mm is acceptable. It is essential that the timber is driven in the right direction and should not be driven into firm ground, as this can easily damage the pile. Keeping the timber below the groundwater level will protect the timber against decay and putrefaction. To protect and strengthen the tip of the pile, timber piles can be provided with toe cover. Pressure creosoting is the usual method of protecting timber piles.

Composite piles

A combination of different materials in the same pile. As indicated earlier, part of a timber pile that is installed above groundwater could be vulnerable to insect attack and decay. To avoid this, concrete or steel pile is used above the groundwater level, while wood pile is installed under the groundwater level.

Types of piles by systems

End bearing piles

End bearing piles transfer their load on to a firm stratum located at depth below the base of the structure and they derive almost all of their carrying capacity from the penetration resistance of the soil at the end of the pile. The pile behaves as an ordinary column and is designed as such.

Cohesion piles

Cohesion piles rely on skin friction to transmit most of their load to the soil. If these piles are driven close to each other in groups it greatly reduces the porosity and compressibility of the soil within and around the groups. During the process of driving the pile into the ground, the soil becomes moulded and, as a result, loses some of its strength. This leads to the pile not achieving its ultimate design strength right away, as it is not able to transfer the exact amount of load. The soil will normally regain much of its strength three to six months after it has been driven.

Friction piles

Friction piles also transfer their load to the ground through skin friction. The process of driving such piles does not compact the soil to the extent that cohesion piles do. These types of pile foundations are often known as floating pile foundations.

Combining friction and cohesion piles

Various piling systems rely on a combination of both friction and cohesive piles; often this is achieved by forming a 'bulb' of concrete in a softer stratum immediately above a firmer layer to give an enlarged base.

Piling – safety issues

Piling is skilled work that should only be carried out by specialist firms with the correct equipment and a trained and competent workforce. Some of the health and safety issues cannot be totally eliminated, and can only be controlled and managed to reduce the effect; in particular noise and/or vibration is inherent in all systems.

Noise and vibration

These two hazards are treated together as they invariably arise together; there are systems that cut down on noise at the expense of an increase in vibration and vice versa. Any system that is using percussion will inherently be noisy and cause vibration; all operatives will be expected to wear suitable ear protection. As with excavations property conditions surveys, vibration and noise monitoring is often a necessary aspect of the management of piling operations. Areas of the site close by where others may be working should also be deemed 'ear protection areas' and management should rigorously enforce the wearing of protection.

Planned maintenance of the plant and equipment can help to reduce the noise and vibration; similarly the use of modern plant will help.

In-situ concrete piles, once installed, often need to be broken down to the correct height; this is because they should always be cast too high so that after trimming a sound connection to any foundation can be made. There are various ways of safely breaking down the excess concrete, as discussed below.

The use of hand-held jack-hammers or breakers should be avoided because of the unacceptable health issues caused. Instead there are a range of hydraulic pile breakers available which although not suitable for all pile sizes will prevent the need for

jack-hammers. Hydro-demolition techniques, using high pressure water to remove the concrete but not the steel can be used, but they do bring their own hazards and are rarely cost effective.

Additionally, passive techniques rely on the installation of some form of de-bonding system at the break-off level; two examples are the Elliot System and the Coredek System.

Finally, there are active techniques which rely on systems activated after pouring to cause cracking at the correct level after the concrete has cured. The systems include inflatable tubes and chemical injection techniques.

Working near plant and machinery

Working in the vicinity of any plant is hazardous: piling plant is noisy, lifting will be going on and there will be trailing hoses and pipes to consider and avoid. Operatives, as stated above, should be wearing ear protection, this can reduce the ability to verbally communicate instructions and warnings, consequently all people involved need to remain vigilant and aware of the changing dangers around them. The use of recognised signalling systems is recommended.

Also be aware of other site personnel wandering into the area and other plant working close by which may encroach on the works. Segregation of the area, even by relatively simple methods such as barrier tape, may assist in preventing unauthorised access.

Working at height

Because of the construction of rigs and the need to connect and disconnect piles and equipment, there is usually an element of work at heights involved with piling works. The use of harnesses and fall restraints should be considered in such cases (also see Chapter 14 *Working at height and roofwork*).

Manual handling

Although piling rigs will usually be working in conjunction with a crane, there are many manual-handling operations that will be needed to operate efficiently. Often the processes will be repetitive, carried out many times during the shift.

Operatives should be aware of good lifting practices and also have knowledge of the weight of pieces of equipment they are handling. Training in lifting techniques should be given to all operatives (also see Chapter 13 *Controlling exposure to physical hazards*).

Mechanical lifting

Many piling rigs are not only lifting equipment, they may well be converted cranes adapted for use in piling operations; other piling equipment such as winches and ropes is deemed to be lifting equipment.

All this equipment comes under the Lifting Operations and Lifting Equipment Regulations (LOLER) which means that there must be an appointed person, an authorised person and a competent person nominated for all piling operations. Additionally a slinger/signaller will also be required to guide the lifting process. Most importantly a lifting plan will be required (see Chapter 19 *Transportation and vehicle movement*).

Underground services

As with excavations, the hazard of damaging underground services is always present and the same precautions should be taken as for excavations. An additional hazard is that it is possible to damage services and not realise at the time; this can lead to excessive repair bills or, in the case of sewers, seepage and contamination going untraced.

COSHH issues

If concrete is being used and gets onto exposed skin it can cause numerous severe dermatitis issues, additionally the plant and machinery will require many oils and greases that can easily come into contact with operatives. The wearing of gloves and good hygiene procedures are essential to avoid health issues. Concrete, oil, grease or anything else if it finds its way onto your skin should be cleaned off immediately.

Exposure to grease and concrete can be due to pipes and hoses failing, so may not be planned or predictable; regular visual checks of pipes and hoses should be carried out (see Chapter 11 *Controlling exposure to chemical hazards*).

Other issues

Spoil will often be ejected or fall from some types of piling rigs that rely on drilling techniques to remove spoil prior to piles being inserted. As well as standard PPE, eye protection will normally be required. Additionally, many projects are on 'brownfield' sites and the arisings from bored and augured piles can be toxic or contaminated in some way; consequently good hygiene procedures are required.

When piling occurs near or over water, safety boats and/or other rescue procedures may have to be provided (see Chapter 21 *Working on, in, over or near water*).

Summary of main points

Excavation and piling can be some of the most hazardous works carried out on a construction site. Excavations need to be inspection regularly for stability. Additional loads must be monitored and considered; remember these loads are not always constant (e.g. materials, plant or temporary works). There are many indirect hazards as a result of excavation (e.g. confined space or utility services) and these have to be assessed. Time spent planning is time well spent; the sequencing of works plays an important part in reducing the effect of the direct and indirect hazards. In piling, many of the safety hazards are inherent (e.g. noise and vibration), and need to be controlled and managed to reduce their effects.

For excavation and piling works, segregation of the work area is important to help prevent unauthorised access. If you are not involved in the process STAY CLEAR.

References

Department for Transport. *Safety at Street Works and Road Works: A Code of Practice*, 2009a, London: The Stationery Office.

Department for Transport. *Traffic Signs Manual Chapter 8 – Part 1: Design. Traffic Safety Measures and Signs for Road Works and Temporary Situations*, 2nd edition, 2009b, London: The Stationery Office.

Department for Transport. *Traffic Signs Manual Chapter 8 – Part 2: Operations. Traffic Safety Measures and Signs for Road Works and Temporary Situations,* 2nd edition, 2009c, London: The Stationery Office.

Health and Safety Executive. *HSE Press Release: E118:04 – 19 August 2004. HSE Urges Greater Awareness of Trench Collapse Dangers*, 2004, London: HSE. Available online at: http://www.hse.gov.uk/PRESS/2004/e04118.htm

New Civil Engineer. Catalogue of Failures Led to São Paulo collapse. *New Civil Engineer* 2009, 25 June. Available online at http://www.nce.co.uk/news/transport/catalogue-of-failures-led-to-so-paulo-collapse/5204029.article

Referenced legislation

Confined Spaces Regulations 1997. Statutory instruments 1997 1713, London: The Stationery Office.

Construction (Design and Management) Regulations 2007 Reprinted March 2007. Statutory instruments 320 2007, London: The Stationery Office.

Control of Substances Hazardous to Health Regulations 2002 Reprinted April 2004 and March 2007. Statutory Instruments 2677 2002, London: The Stationery Office.

Health and Safety at Work, Act 1974 Elizabeth II. Chapter 37, London: HMSO.

Lifting Operations and Lifting Equipment Regulations 1998. Statutory instruments 2307 1998, London: The Stationery Office.

Management of Health and Safety at Work Regulations 1999 Reprinted February 2005. Statutory instruments 1999 3242, London: The Stationery Office.

Manual Handling Operations Regulations 1992. Statutory Instruments 1992 2793, London: HMSO.

Provision and Use of Work Equipment Regulations 1992. Statutory Instruments 1992 2932, London: HMSO.

Further reading

Federation of Piling Specialists. 2000 Code of Industry Best Practice – Lifting Operations and Lifting Equipment Regulations 1998 (last revised 2007), Beckenham: Federation of Piling Specialists. Available online at: http://www.fps.org.uk/fps/guidance/LolerGuidance.pdf

Websites

Construction Skills Certification Scheme (CSCS) http://www.cscs.uk.com

Federation of Piling Specialists (FPS) http://www.fps.org.uk

Health and Safety Executive (HSE) http://www. hse.gov.uk

Institution of Civil Engineers (ICE), Health and safety http://www.ice.org.uk/knowledge/specialist_health.asp

North American Society for Trenchless Technology (NASTT) http://www.nastt.org

United Kingdom Society for Trenchless Technology (UKSTT) http://www.ukstt.org.uk/

Chapter 16

Confined spaces

Philip McAleenan Expert Ease International, Downpatrick, Northern Ireland, UK

doi: 10:10.1680/mohs.40564.0191

Confined spaces can be extremely hazardous environments with the potential for multiple fatalities. Hazards arise as a result of the nature of confined spaces which either because of their design, function, location and/or what they contain have the capacity to contain and accumulate air contaminants (including explosive gases), prevent the free circulation and change of air, trap and insulate against heat dissipation and, in particular circumstances, present a flooding hazard to workers inside. When any of these conditions arise, the whole of the confined space environment is compromised, as is the safety of those working in or near the confined space.

Confined spaces are defined as containing or having the potential to contain one or more specified risks that by definition are particular to, and only exist in, confined spaces. As we explore these hazards and how to eliminate and control the work activities we will deal only with these risks, however the reader should bear in mind that many other hazards will also exist in respect of any particular confined space that they encounter. These hazards must also be considered as part of the assessment and appropriate controls put in place, as failure to consider them may precipitate the confined space incident that your controls were designed to prevent.

Introduction

Confined spaces fatalities and injuries reportable under the Reporting of Injuries, Diseases and Dangerous Occurrences Regulations (RIDDOR) 1995 are a relatively infrequent occurrence in recent years. Of the 229 fatal injuries on the Health and Safety Executive (HSE) list for 2008–2009, one was as a result of an incident in a confined space (HSE, 2010). However the University of Glasgow (2005) has estimated that 15 people are killed annually in confined space accidents in the UK, and HSE figures over the period since the introduction of the Confined Spaces Regulations (Great Britain, 1997) indicate that multiple fatalities each year are not uncommon. In the 1990s, being the period when the current confined spaces legislation was being considered and drafted, fatalities and injuries in confined spaces were more frequent, albeit still at lower rates than other types of accidents in all industries.

The concern for safe working in confined spaces stems from the peculiarities of confined spaces and the consequences of incidents. Confined spaces accidents have the potential at all times to injure many individuals rather than one and this is because the incidents in general stem from the environmental conditions rather than the simple failure of equipment or behavioural lapse, i.e. the injury is more often not directly caused by the equipment failure but by the effect of the failure on the environment (e.g. the atmospheric conditions are compromised by a leak from a gas line which in turn affects all who are in the environment).

Second, the fatality rate for confined spaces incidents is much greater than for other types of incident. Research indicates that two out of three confined spaces incidents lead to a fatality (Surada et al., 1994) compared with a 1:50 rate for incidents arising as a result of falls from a height (HSC, 2009); again this stems from the 'peculiar' nature of confined spaces incidents and their consequence on the environment, e.g. the release of toxic gases/flammable gases or flooding in constricted spaces.

Because of this there was an awareness by employers, legislators and operators that additional care must be taken when working in confined spaces, and the legislative history shows that in both industry and construction obligations were placed upon employers to effectively control the atmospheric conditions in confined spaces (cf. Factories Act, Docks Act and the Construction Regulations (Reg. 22))

Legislation

Post the Health and Safety at Work Act 1974 (HSW, 1974) and the Health and Safety at Work (Northern Ireland) Order 1978 (HSW Order 1978) when the general duty to provide safe working environments was prescribed, the problem lay in determining what constituted a confined space. In the first instance the three acts/regulations mentioned above applied to different industry sectors permitting the argument to be made that their requirements were exclusive. Second, the legislation in total did not provide a conclusive definition of what actually constituted a confined space. Confined spaces were described by a list of examples rather than given a universal definition which was unambiguous and applicable across industry sectors. Confined spaces entry supervisors and managers looked at the situations presented to them and compared it with sizes and configurations, suitability for human activities in normal circumstances, as well as the presence of gases and materials that would harm workers, in order to determine whether it would be a confined space or not. It meant that in some circumstances extremely large spaces were discounted as they were not viewed as being 'confined' (as used in the sense of being constricted), e.g. a dry dock. But likewise very small spaces were disregarded as workers were not capable of wholly entering the space, e.g. shallow trenches, and other spaces were discounted because people normally worked in them even though an exceptional work activity would have been taking place that changed the normal to abnormal circumstances, e.g. fumes from painting.

The 1997 (United Kingdom (UK)) and 1999 (Northern Ireland (NI)) Confined Spaces Regulations changed all that and removed the confusion. Working on the basis of what actually caused the injury or fatality in a confined space incident that was unique, a definition was arrived at based on unique hazards rather than configuration or human presence (although the 2001 regulation in the Republic of Ireland retained an element in the definition that referred to the configuration of the space).

The current definition refers to confined spaces as being a place where one or more of five *specified risks* are present or are likely to arise in the course of the work operation. The definition for a confined space and a specified risk is the same in both the Confined Spaces Regulations 1997 (Great Britain, 1997) and the Confined Spaces Regulations (Northern Ireland) 1999 (Great Britain, 1999):

> confined space' means any place, including any chamber, tank, vat, silo, pit, trench, pipe, sewer, flue, well or other similar space in which, by virtue of its enclosed nature, there arises a reasonably foreseeable specified risk:
>
> Reg. 1 (2) 'specified risk' means a risk of –
>
> (a) serious injury to any person at work arising from a fire or explosion;
>
> (b) without prejudice to paragraph (a) -
>   (i) the loss of consciousness of any person at work arising from an increase in body temperature;
>   (ii) the loss of consciousness or asphyxiation of any person at work arising from gas, fume, vapour or the lack of oxygen;
>
> (c) the drowning of any person at work arising from an increase in the level of a liquid; or
>
> (d) the asphyxiation of any person at work arising from a free flowing solid or the inability to reach a respirable environment due to entrapment by a free flowing solid.

Notwithstanding the requirement to control other hazards such as noise, slippery surfaces, biological organisms, heights, etc. which are dealt with in other chapters, this chapter will focus on these five specified risks: how to assess the potential for harm, the development of controls, the roles and duties of entry workers, supervisors and managers and the requirements for appropriate equipment.

The following examples described do not constitute a definitive list of circumstances and hazards in relation to confined spaces. Entry workers must assess each situation prior to and throughout the period of work in a confined space.

Serious injury to any person at work arising from a fire or explosion

In 1984, 44 people, including eight employees and 36 visiting dignitaries, entered an underground valve house of a water transfer scheme in Abbeystead, Lancaster. As part of the presentation, water was to be pumped over a regulating weir into the river. Shortly after pumping commenced there was an intense flash followed by an explosion. The explosion was caused by an accumulation of methane and air which was pushed into the valve room when pumping commenced. Sixteen people died and 28 others were injured in the explosion. The gas had seeped into the porous pipeline from the soil. Methane will form an explosive mix with air at 5–15% (BBC, 2008).

The loss of consciousness of any person at work arising from an increase in body temperature

Core body temperature is 98.2°F (± 1.3°F), 36.8°C (± 0.7°C). When the body temperature rises rapidly and uncontrollably, hyperthermia results. At 104°F, 40°C, this is life threatening and requires immediate medical intervention. This situation can occur when the ambient environmental temperature is high (e.g. working inside partially cooled boilers, or enclosed spaces in hot weather conditions); the work being undertaken is onerous and the work clothing insulates the body.

The loss of consciousness or asphyxiation of any person at work arising from gas, fume, vapour or the lack of oxygen

In 1983, a graduate engineer on work experience at the new Carsington Reservoir in Derbyshire, entered an inspection chamber to test for seepage. He collapsed. Three work colleagues separately attempted to rescue him and as each entered the chamber, they too collapsed. All four young men died.

A build-up of fume/vapour from equipment such as combustion engines, or from welding/cutting operations in an enclosed space can accumulate such that the breathable air is replaced or the level of toxic fume rises above a safe threshold. By way of example, 2 contractors working in a shipyard were overcome by carbon monoxide fumes when using a petrol powered road cutter in a tented enclosure.

The shipbuilding sector has been made aware of 5 fatalities in the 18 months to May 2003 involving petrol powered equipment being used in confined spaces. Three of these deaths were due to petrol-powered generators and one of the incidents involved a double fatality. (HSE, 2003)

It is unsafe to work in atmospheres with greater than a time-weighted average (TWA) of 50 ppm carbon monoxide. In less than 1 hour: 4000 ppm will kill.

In a substantially enclosed space with little or no ventilation, or poor circulation of air, the O_2 in the air may be used up by the workers over a period of time and replaced by exhaled CO_2 to the extent that it is no longer safe to remain in the area. O_2 is at 20.9% in the atmosphere and, although it will not normally become dangerous until it is substantially below that, when it falls to 19.5% workers must evacuate the space and not re-enter until it has been sufficiently ventilated and that ventilation maintained. This can happen in deep trenches, wells and manholes.

Contaminated soils may react to the atmosphere and sunlight on exposure and release gases into excavations where they may remain posing a hazard for any worker who enters the excavation.

The drowning of any person at work arising from an increase in the level of a liquid

Any environment where liquids can either flow through or collect, e.g. pipelines connected to the water or sewage systems, trenches and excavations capable of flooding due to rainfall or seepage through the soil.

The asphyxiation of any person at work arising from a free-flowing solid or the inability to reach a respirable environment due to entrapment by a free-flowing solid

Free-flowing solids are granular materials that separate easily when pressure is applied, or restraining barriers are removed such that they can engulf an operative wholly or partially thus preventing breathing through the replacement of air or through compression on the chest.

Classification of confined spaces

Confined spaces may be classified on the basis of the presence of one or more of the specified risks, or the likelihood that one or more will arise during the work operation.

Where one or more of the specified risks exists, the confined spaces will be classified as immediately dangerous to life or health (IDLH) and entry should be prohibited until the risk(s) is/are eliminated. The only exceptions to this prohibition are where it is necessary to enter the space to eliminate the risk itself or to affect a rescue. In these circumstances only those fully trained in the use of all the appropriate equipment, including respiratory protective equipment (RPE), vapour or chemical

suits, and any other equipment necessary to protect the entry worker. Rescue workers must be fully trained in current rescue methods. (Generally this means the public rescue services or a specialist rescue company or organisation. Do not assume that the public rescue services have the capability to affect confined spaces rescues. This requires specialist training, which they may not have. If you propose to use them in your rescue plan, confirm the details with them first.)

Where the assessment of the space shows that the specified risks have been eliminated but there remains a possibility (however remote) that one or more of those risks may arise during the work operation, either because of the work that is taking place, because of disturbance of sludge, ingress of gases or liquids (seepage, rain, etc.) or some other unknown factor, entry into the space with require appropriate monitoring and emergency equipment such as gas detectors, escape RPE, and higher level communication, and the provision of non-sparking tools and work clothing.

Where the specified risks do not exist or have been eliminated and there is no probability that they will arise during the work operation, the space can be de-classified as a confined space and the work proceed with such other safeguards appropriate to the situation.

Duties in respect of confined spaces

One of the first considerations when faced with a confined space is whether the work requires entry into the confined space. If it can be carried out externally then that option must be followed, unless it is not reasonably practicable to do so.

If it proves necessary to enter the confined spaces, then it must be done in accordance with a safe system of work that in relation to the specified risks renders the work safe and without risks to health. Among other things the safe system of work will include information on the precautions to be taken prior to, during and at the end of the entry operation, who should be authorised to enter the space, what external safeguards are to be in place, how communications are to be maintained, what isolations, if any, are required and what the emergency procedures are.

Prior to entry priority must be given to determining what hazards exist and eliminating these before deciding on the precautions necessary for entry. It is advisable to incorporate confined spaces working into the permit-to-work system in use on the project (see Chapter 6 *Establishing operational control processes* for further detail on permits to work).

Emergencies arise, whether as a result of any specified risk or other cause, and will continue to arise in confined spaces. The operation of safe working and permit systems isn't a guarantee against emergencies arising, therefore it is a requirement that every confined space entry procedure be accompanied by emergency arrangements. The arrangements will depend upon the nature of the confined space, its location, the type of emergency that may occur, the numbers of people likely to be affected, directly or indirectly, and the availability of emergency services. The emergency arrangements must also be designed to safeguard the health and safety of anyone required to put them into operation.

In addition to arrangements that are specific to the confined spaces being worked in, all arrangements must stress the need for workers to:

1. Immediately evacuate the space on hearing the alarm or becoming aware of any unexpected condition or situation.
2. Remain outside and not to re-enter the confined space unless and until authorised to do so.
3. AVOID any attempt to rescue a colleague unless they have:
   a. been trained
   b. been provided with all the appropriate equipment to affect that rescue, and
   c. been authorised to do so.

Design issues

For designers of structures or plant, the requirement for workers to avoid entry where it is not necessary has particular importance. As Chapter 9 *Assessing safety issues in construction* describes, the safety of the construction worker, the end users and maintenance workers, and eventually those charged with the demolition of the structure at the end of its life, is enhanced when the designer gives consideration to the hazards that they will face and, in as far as it can be achieved, designs out hazards and designs in safety elements. Consult with those who will be required to work in the confined spaces for their views on what would be required.

With regard to confined spaces, where they are a necessary element or an unavoidable consequence of the design, the following considerations may assist:

- Look to solutions that will permit inspections to take place externally, e.g. through the placement of inspection windows at various points along the confined space, or design the interior to allow for remote controlled CCTV inspection. Include facilities that will allow as much as possible for cleaning and maintenance to take place remotely; this may include insertion points for hose and water sprayers. Consider also the inclusion of insertion points for gas detection equipment.

- Allow for the free flow of air through and around all points within the confined space. This may require two or more ventilation points. This will allow for the continuous supply of breathable air, the dispersal of contaminants such as fume or gases, and the maintenance of comfortable levels of heat.

- Where the structure is to be sited below ground level, or the confined spaces within the structure are below ground level, consider the use of non-porous construction materials that will prevent both seepage of liquids and the ingress of gases.

- As entry into confined spaces will at some point be required, consider the means of access, the distance to the nearest exit point, the number of exit points for the size and configuration of the space, the ease with which workers can move through and about the space, bearing in mind that they will generally be carrying equipment (work, personal protective equipment (PPE) and RPE).

Horizontal entry is preferred to vertical, steps are safer than ladders and for bridge inspections for example a door at either end of a long walkway is preferable to sending a worker unnecessarily back along the same route.

- If the nature of the confined space is such that on occasions there will be substantial time spent by workers inside it, consider the communications element and whether it may be appropriate to install hard-wired communications lines, particularly for large complex spaces.

- Consider how workers may affect both self and assisted rescue. Self-rescue will require ease and speed of exit, whereas assisted rescue may require two or more rescue workers to enter with equipment, administer on-site first aid and cardio-pulmonary resuscitation (CPR), as well as possibly stretchering injured workers out.

These are examples of the type of design considerations and ultimately it is in consultation with the contractor and the end user that the specific requirements will be detailed and sufficient knowledge gained to allow for an appropriate design solution.

Pre-entry assessment

Prior to commencing work in or near a confined space, every effort should be made to determine what the existing conditions are and an informed judgement made of the likely atmospheric and environmental conditions within the space. The current or previous contents of a confined space will give an indication of some of the potential specified risks. Silos and tanks, for example, may have flammable or toxic content. Signs of rusting on the rims of metal tanks may indicate oxygen depletion, although the absence of rust should not be taken to mean that all is well.

Old pipelines or sumps are likely to contain sludge which when agitated will release gases such as hydrogen sulphide (often fatal at 1000 ppm, work should not take place at levels above 20 ppm). In newly constructed tanks water and organic matter accumulating at the bottom will rapidly decompose and present the same hazards as sludge in older pipelines and sewers. Decomposing organic matter will also deplete the O_2 content of the atmosphere creating an irrespirable atmosphere.

Soil analysis will provide information on contaminants that have the potential to react to the atmosphere and create toxic conditions when a new excavation is made. It is useful to check historical records as dump sites from, for example, the Victorian era may not be immediately apparent on a visual inspection. Even ancient industrial sites from as far back as the Roman period and long since lost may mean that greenfield sites could harbour deep soil contaminations with high levels of toxicity.

Check with the site owners and obtain as much information as possible about the confined space. Utilities records and independent surveys will provide information on the location of water, sewers, gas, electrical or other lines crossing or contained within the confined space.

Armed with this information, the next step is to carry out tests on the atmosphere within the space. Many gas detectors come complete with sampling rods that can easily be inserted into the space without the need for a man-entry. Draeger tubes, a glass vial filled with a chemical reagent that reacts with specific chemicals or families of chemicals, may also be used to test for atmospheric, soil or water contamination.

Testing the internal atmosphere must be thorough and take into consideration the physical properties of the gases that may be present. Atmosphere monitors generally sample air within a 2–3 ft (1m) radius of the intake. It will be necessary then to ensure that the air is sampled every few feet by lowering the detector on a lanyard and letting it rest for a minute at each point to ensure enough air passes over the detector to give an accurate reading. On the horizontal an operator may walk through using a 1 m aluminium tube attached to the detector pump, again resting for a minute every few feet. Ensure that depressions on the floor or voids above head height are tested, as gases may be either lighter or heavier than air and remain in the voids/depressions awaiting the unwary worker. During such testing the operator must wear his escape RPE and don it immediately the alarm activates and exit the confined space.

The presence of an atmospheric contaminant or the depletion of O_2 will require appropriate remediation before entry is permitted.

O_2 depletion may be resolved by opening ventilation points such that there is a sufficient flow of uncontaminated air through the confined spaces. Ensure that open ventilation points are suitably safeguarded against ingress of contaminants and from access by unauthorised persons. Be aware that on many sites the source of fresh air may be contaminated by the presence of exhaust fumes from vehicles working or passing through the site. Changes in wind direction during the day may mean that previously safe locations for the air intake point are now in the path of contaminants that will be pumped into the confined space.

If it is not possible to have two or more ventilation points because of the configuration of the space, forced air ventilation is appropriate ensuring that the air is pumped beyond the working area and is forced back through the working area to the entrance. This will provide a breathable atmosphere for workers and take away low levels of contaminants generated by or during the work operation.

Failing these two options, it will be necessary for entry workers to be provided with a personal source of breathable air either through some form of self-contained breathing apparatus (SCBA) (see **Figure 1**) or a line-fed breathing apparatus (fresh-air hosed (FAHBA) or compressed airline (CABA)).

Where the air contaminants are toxic, these must be removed following an appropriate method. Some may be diluted in the ventilation process and dispersed safely into the atmosphere; others, however, pose a serious hazard even at extremely low concentrations, and in these circumstances

Figure 1 Example of breathing apparatus
RPE equipment image supplied courtesy of Draeger UK Ltd

must be removed by a specialist firm. Take care not to ventilate contaminated atmospheres into someone else's breathing space, e.g. from the confined space into another closed room or chamber.

Where the contaminants are flammable or explosive in nature, the atmosphere must not be ventilated by the introduction of fresh air as the oxygen explosive gas mixture will at some point be within the flammability range, thus increasing substantially the likelihood of an explosion occurring. The first step is to purge the confined space using an inert gas such as nitrogen. This will dilute the gas below the lower flammable limit and thus, in the absence of O_2, ensuring that an explosion cannot take place.

The second step is to ensure that the purged atmosphere is ventilated with fresh air as per above. On 19 March 1981 during launch preparations at NASA, Columbia's aft engine compartment was under a nitrogen purge to prevent the build-up of oxygen and hydrogen gases from the propulsion system. Six technicians entered the aft engine compartment, five lost consciousness due to the lack of oxygen in the compartment. Two died, the other four workmen were treated and released (Indopedia, 2009).

Entry controls – safe system of work, permits and authorised persons

The safe method for entry into, working in and exit from a confined space must specify all the control measures that must be followed by all persons working in or near the confined space. Preparation should take as long as is necessary to ensure that all the hazards are identified and considered, all the controls are identified and in place, and that all of the work team are aware of the requirements for the task.

Work inside the confined space will be limited to what has been agreed and recorded on the permit, including any particular methods that are to be followed. Any deviations are to be undertaken only following assessment and on the instruction of the permit issuer/entry supervisor.

The work team for any and every confined space operation must comprise persons who are experienced and trained in the specific aspects of confined spaces work that they are required to undertake. The key players in any team are the employer, the permit issuer, the entry supervisor or permit receiver, the authorised attendant or standby operative and the authorised entrant. (Note: a trained operative is not automatically authorised to enter into any confined space, or participate on any confined space team. He must be specifically authorised on the permit to work on any particular job.)

The employer is responsible for nominating and training confined space workers.

All workers must know and understand the hazards associated with confined spaces working, be able to assess and put in place the controls necessary for safe working, be competent in the use of all equipment that will be required, and understand the causes of and be able to react immediately to emergencies. All workers must be competent and fit for the work. Fitness or medical conditions that will affect their ability to work in confined spaces should be notified to the company immediately. Claustrophobia, lung and heart diseases may render a worker permanently unfit for confined space entry, whereas short term illnesses such as flu may require temporary withdrawal from confined space entry.

The permit issuer must provide a clear definition of the work to be done, positively identify the equipment or area boundary to be worked on or in, and determine the effects of the work or other activities so that the necessary actions may be undertaken and the residual risks understood prior to issuing the permit (see **Figure 2** for an example permit).

The entry supervisor is responsible for ensuring that the pre-entry conditions detailed on the permit are in place, that the atmosphere and other necessary tests have been conducted, verifying that the rescue services are available and the means of summoning them are operable, removing unauthorised persons from the vicinity of the confined space, and determining, whenever the responsibility for a confined space operation is transferred and at intervals dictated by the hazards and operations performed, that the operations remain consistent with the terms of the entry permit, and that acceptable entry conditions are maintained. He is responsible for terminating the entry and cancelling the permit at the end of the operation or at any time during the operation should a deviation from the specified conditions arise.

Authorised attendants are in place to ensure the safety of those who are inside the confined space. They must be aware of the behavioural effects of hazard exposure so that he may recognise from the observed behaviour of those within, whether all is well or an evacuation is necessary. He must maintain visual or auditory contact with those within in order to monitor their status or to alert them of the need to evacuate. He must maintain an accurate count of who is inside the confined space and ensure that there is a means to accurately identify them.

He is responsible for monitoring the activities inside and outside the confined spaces to ensure that no hazards affect the safety of the workers inside, including responsibility for removing all unauthorised persons from the vicinity of the confined space. He must summon the rescue or other emergency services when the situation requires it and is authorised to perform non-entry rescues as specified on the emergency procedures.

He must remain in situ at all times until relieved by another authorised attendant, and must not perform any other activities that will interfere with his sole duty to the workers inside the confined spaces.

Authorised entrants must ensure that they have all the correct equipment, are able to use it, and do so. They must maintain constant communication with the attendant to enable him to carry out his monitoring role, remain alert to

Permit number ....

Permit to work in a confined space

The following confined spaces controls are required to make the work operation safe: (strike out those that do not apply)

Confined Spaces Permit:					
Type of hazards:	Flammable atmosphere	Oxygen deficiency	Toxic atmosphere	Free flowing solids	Flowing liquids
Is the confined space entry required	Yes	No	If 'No' entry is prohibited		
Has the space been vented and tested	Yes		Is ventilation satisfactory?	Yes	No	If 'No' the following artificial ventilation is in place
Results of testing:					
Continuous atmosphere monitoring is required	Yes	No	If 'Yes' state how:		
Have all necessary isolations been made?	Yes	No	N/A	Have all sludges been removed?	Yes	No	N/A
Health professional has been informed	Yes		Safe means of access and egress has been provided	Yes	
Rescue plan is in place and attached	Yes				
The following are permitted to enter	Name		Name		
External attendant	Name				
All the above precautions have been put in place	Signed:		Date:	Time:	

Figure 2 Sample permit
NB: This is an abstract from a full permit to work. To use this it will be necessary to incorporate it into existing permit to work documentation.

any warning sign or symptom that would indicate exposure to a dangerous situation or prohibited condition, and exit the confined space using the approved procedure as quickly as possible whenever an emergency or potential emergency arises.

Where it is practicable to do so, the normal entry and exit point should be one and the same; this will facilitate the accuracy of checking workers in and out of the space. It is not unheard of for workers to be closed inside a confined space in the mistaken belief that they had already left. In an emergency the nearest safe exit point is to be used with workers returning to a pre-ordained assembly point. The method of entry will also be decided upon and must be followed, e.g. a ladder entry with workers attached to a fall arrest system may be appropriate for descent into deeper shafts.

In a lone worker situation it may be a requirement that immediately prior to entry the worker contacts his base and notifies them that he is entering the confined space, and then contacts them again immediately on exit. The requirement may extend to having fixed intervals between contacts with the base. Failure to make these contacts will result in the initiation of the emergency procedures.

Work inside the confined space will be limited to what has been agreed and recorded on the permit, including any particular methods that are to be followed. Any deviations are to be undertaken only following assessment and on the instruction of the permit issuer/entry supervisor.

Some situations may require the workers to remain attached to a lifeline, although this becomes impracticable where two or more workers are in the confined space, or where there are obstacles such as fixed plant or equipment that may cause an entanglement, thus rendering the lifeline useless.

Entry controls: energy sources and atmosphere monitoring

Any energy system that interfaces with the confined space and is a potential source of one of the specified risks must be identified, de-energised, isolated and locked off against accidental or unauthorised re-energisation. This includes any fluid or liquid line, electrical energy and pressure lines, with the exception of those integral to the work being done. Throughout the work operation and while the confined space is under the control of the work team, the keys to the isolation locks must remain under the control of the team and the locks and isolations removed in accordance with lock-out principles and only when it is clearly established that all have vacated the confined space.

Where there is a potential for an IDLH atmosphere arising, all persons inside the confined space must have both a personal gas detector and an escape RPE set. As detectors sample the air within the breathing space of an individual, sharing a single detector by placing it in a central location is unsafe. The detector cannot determine the source of the gas and by the time it detects a gas it may already have risen to fatal levels within the breathing area of one or more workers closer to the source than the detector.

Where there is the possibility for flammable gases to occur, a continuous flow of air through the confined space may be sufficient to keep the level of flammable gases below the lower explosive limit, and appropriately calibrated gas detectors will provide ongoing confirmation of the gas levels, with the alarm activating whenever the level approaches an unsafe condition. From time to time, however, higher concentrations of gas may be released, e.g. whenever sludges are disturbed, presenting an immediate and absolute unsafe condition. When there is this potential, careful consideration must be given to the elimination of all sources of ignition.

Static electricity is generated when two materials are in contact with each other: the charge from one moving across to the other thus causing an imbalance. When the materials move apart one is positively and the other is negatively charged. If the material is able to conduct the charge away it will dissipate and recombine with little electrical effect. However, if the materials move apart quicker than the charge is dissipated the charge will build up and eventually create a noticeable electrical effect, normally a discharge creating spark.

While all materials in contact will generate static electricity, some are better able to conduct electricity away. Clothing comprised of synthetic materials does not conduct static build-up as effectively as cotton clothing, and thus is more likely to generate sparks when worn and surfaces between arms or legs rub against each other. Check with the supplier of work clothing as to which materials are resistant to static sparking and provide these for confined space workers.

Because water content has a large impact upon a fluid's dielectric constant, i.e. its ability to concentrate electrostatic flux, hose washing of tanks that previously held flammable fluids increases the risk of explosion. A reduced flow rate reduces the likelihood of a static build-up.

Metal tools used inside a confined space may also create sparks; in this case percussion sparks as minute hot pieces of the tool or object to which it has been applied fly off. Non-sparking tools are available comprised of alloys of beryllium-copper or aluminium-bronze.

Electrical appliances and equipment all generate sparks when switched on or off, or where there are motor brushes, connecters, etc. inside them. Non-essential electrical items such as phones, personal digital assistants (PDAs) and electric watches should not be used in confined spaces. Essential electrical equipment, such as torches, communication radios or the gas detectors themselves, should all be intrinsically safe for use in explosive environments. Intrinsically safe equipment has voltages or thermal energy below the level necessary for ignition and the energy supply, wires etc. are encased within an isolation barrier.

(Note: if batteries fail inside the confined spaces, they must not be changed until they have been taken outside and at a safe distance from the confined space. All seals must be correctly restored before the item is taken back into the confined space.)

It goes without saying that cigarettes, matches, lighters and any other ignition source must not be taken into, or used in the vicinity of, a confined space. Where burning operations are necessary the items to be worked on should be taken outside the confined space to a designated hot work area and welding, etc. carried out before re-installing the finished object.

Entry controls: communication

Efficient and speedy communication between those working inside the confined space and between them and those on the outside, as well as for raising help from the emergency services is essential to ensure the quickest possible initiation of the emergency procedures should anything go wrong or appear to do so. The attendant, becoming aware of an unsafe condition, whether one of the specified risks or some other hazard, must alert the workers inside immediately who must respond without delay, donning their RPE as appropriate and vacating the confined space immediately. Workers inside the confined space need to be able to communicate with each other clearly and unambiguously to ensure the effective conduct of the task and to warn others of any unsafe conditions that they become aware of. In other circumstances, where an individual collapses because of the rise of an IDLH atmosphere there may be only minutes before death or severe irreversible brain damage (as with O_2 depletion) occurs.

Communication is also helpful to workers who experience isolation or who are beginning to panic, especially those who may be wearing RPE to carry out the work. The attendant can talk through a problem with a worker or help maintain his calm through reassuring conversation.

There are many forms of communication systems from speech, through tugs on a rope, telephone and radio (which should be intrinsically safe where there is a risk of flammable or explosive atmospheres). Whatever method is used it must be appropriate for the confined space, taking into consideration the work being conducted, for the workers and any PPE and respiratory protection that they are using (e.g. ear defenders or RPE hoods). Note: tugging on a rope is ineffective unless the length of the rope is short, is free from entanglements and does not bend around corners. As a principal means of meeting the communication requirements it has long been superseded by reliable technological means of communication.

When considering the work environment, take into account its location, depth, whether it is underground, near any plant or equipment that will interfere with communication. What is inside the space that could impede visual or speech communication? Is the confined space large with bends that prevent line of sight contact, or too long for the worker to have effective rope-tug communication or for the voice to carry? Although the attendant can see the workers inside, they may not see him or any visual warning because of their attention to the work that they are doing.

The noise inside the confined space, as well as noise in the vicinity of the attendant, will impact on the effectiveness of communication, often drowning it out or making it garbled. Machinery and plant, the work being carried out inside the space, the noise from ventilation systems provided for safe entry all contribute to a loss of effective communication.

When using radio or telephone communications, check to see if the confined space is constructed of a material or situated in an area subject to electromagnetic interference, radio frequency interference or a Faraday shielding effect. Check your equipment in the space at the outset to ensure that it is fully functional all the time (Ibbetson, 2004).

Auditory signals are reduced when wearing ear defenders, the use of which may be unavoidable in noisy environments. Speech is obstructed when wearing face masks; even with those that have incorporated voice diaphragms the sound is reduced and partially garbled. The noise of inhalation/exhalation on some masks will also contribute to reduced speech and auditory communication. An overall effect may be the creation of a 'Chinese Whispers effect' – the effect that occurs when something is misheard and through word of mouth, a variation on the original message or topic of discussion is communicated, as commonly happens in the childhood game of Chinese Whispers.

The requirements for effective communication will take these factors into account and the choice will be dictated by the effectiveness of the chosen method to provide real-time communications that are clear, intelligible and unambiguous. The range of methods includes:

- Direct speech is easiest but limited to situations where the distance is short, low noise conditions and hearing and voice are unobstructed by PPE. Where the distances are long, as in a tunnel, attendants may be placed at appropriate intervals along the route to relay communications; this, however, creates a time delay, and may be more costly than alternative suitable equipment.

- Visual contact, although limited in terms of being principally one way, open to misinterpretation if there is a lot of activity going on, or the space is dark or obscured by contents. Use of hand signals, sign boards, flags, etc. may be used to support verbal communication.

- Portable radios and mobile phones have the advantage of not requiring connecting cables, can accommodate a number of users and can be adapted to work with PPE. However, there are disadvantages in that the signal may be lost due to the configuration, location and material of the confined space, it may be interrupted by others using the same channel, present an explosion risk if they are not intrinsically safe or the seal has become damaged through the normal use of the device. These devices are generally not 'hands-free' and even those that claim to be still require the worker to stop what he is doing to activate it via a push-to-talk switch. Voice activated switches can be activated by higher levels of noise in the space and, unbeknown to the user, lock the device in to a transmission mode, rendering it ineffective.

There is a range of devices available for communication when wearing a face mask, including those that are integrated into the mask and throat microphones. Again there are positive and negative aspects for these types of devices and the choice to use them will depend upon them having been tested for effectiveness in the particular confined space being worked in.

Shielded cable or hard wire portable systems have an advantage over radios and mobile phones in that they are protected against interference, dead-spots and fading. They can also be incorporated into RPE and chemical suits and are fully hands-free. The disadvantages are the limit on the number of users, the distance over which it is effective and, of course, the fact that it requires the user to be attached by the cable.

At the end of the day, communication is essential to any safe system of working procedure in confined spaces and must be given equal prominence with any other safety control. This requires a full assessment of the communication requirements and the effectiveness of any particular method to meet those requirements. Any aspect that will reduce the effectiveness of communication must be addressed and remedied before entry into the space commences.

Entry controls: respiratory protection

The normal atmosphere that we breathe is a mixture of gases with the prominent gases being nitrogen (N) and inert gas present at approximately 78% and oxygen (O_2) at 20.9%. When we inhale, the body abstracts from the volume of air the percentage of O_2 necessary for effective metabolism. This is not the whole volume of the O_2 but approximately one-fifth of it or 4% of the total volume of the air inhaled. On exhalation the volume of O_2 we have abstracted is replaced with the end product of respiration, namely carbon dioxide (CO_2).

Other gases in the atmosphere play no part in metabolism and are therefore exhaled in full. **Table 1** gives approximate concentrations (by volume) of these gases in inhaled and exhaled air.

Box 1 Oxygen volume and partial pressure

For the body to function it needs oxygen in a volume and at a partial pressure compatible with the amount of work demanded. The partial pressure (pp) is essential to the gas exchange process that occurs within the body and is vital to metabolism. It is determined by the percentage volume of O_2 in the air, decreasing as it enters the trachea, lungs and eventually to the blood and cells of the body. In normal air the ppO_2 is 160 mmHg reducing to 150 mmHg in the trachea (as it mixes with additional gases such as water vapour and CO_2), 105 mmHg in the alveoli and 100 mmHg in arterial blood and 40 mmHg in venous blood and <40 mmHg in the cells. As gases diffuse from higher pressure to lower, too low a starting partial pressure will result in a point being reached in the respiratory process when O_2 cannot diffuse across cell walls, thus preventing metabolism.

Breathing is not stimulated by low levels of oxygen in the body but by increased levels of CO_2 which make the blood acidic and thus elicits the need to draw a breath. Although in a given volume of air there may appear to be sufficient oxygen present to inhale, it is the CO_2 levels in the body combined with the pp of oxygen that determines whether autonomic breathing occurs and, if so, the degree to which the body experiences hypoxia.

Gas	Inhaled	Exhaled
Nitrogen	78.08%	78.08%
Oxygen	20.95%	16.95%
Carbon dioxide	0.038%	4.038%
Other trace gases	0.932%	0.932%
Total	100%	100%

Table 1 Approximate concentrations (by volume) of atmospheric gases in inhaled and exhaled air

Although when breathing in all persons follow the same process, the rate of breathing, and thus the quantities of O_2 required, will depend upon the amount of work being undertaken and is affected by a range of factors such as health, age and fitness. It is important to understand this point in order to be able to make an accurate assessment of the time available to different individuals who are using SCBA, particularly escape sets that have no warning alarm to indicate how much air is left.

Table 2 illustrates the approximate air (not O_2) consumption of an average healthy male at different levels of activity. The darker area of the table represents a hypothetical average rate of work used to calculate an average amount of time available to users of RPE of given cylinder capacity (see section on RPE below).

As the work level increases the body requires more O_2 to metabolise energy and facilitate cellular respiration. Increased metabolism generates more CO_2 which in turn triggers increased breathing, drawing greater volumes of air into the lungs and thus providing the body with the O_2 it needs.

Increased fitness improves the body's efficiency in O_2 consumption thus reducing the volume required for similar amounts of work. Conversely, decreased fitness reduces efficiency and increases the volume needed for effective metabolism. Likewise other factors that will adversely impact upon the efficiency of the body to utilise oxygen include illness (which will extend through the recuperation period), lung capacity and functionality (smoking, asthma and other lung conditions should be taken into consideration), age, stress (being inside the confined space increases stress level and the sounding of the alarm will increase the adrenalin, a 'fight or flight' hormone that increases the supply of O_2 to the brain and muscles) and of course elevated work levels in constricted and uncomfortable spaces.

Activity	Approximate air consumption of an average healthy male		Sensation
	Gallons/minute	Litres/minute	
At rest in bed	1.8	8	No conscious sense of breathing
Standing at rest	2.3	10.5	No conscious sense of breathing
Walking at 2 mph (3.2 kph) Strolling	4	18.5	May be conscious of breathing from time to time
Walking at 3 mph (4.8 kph) Normal walking pace	5.5	25	May be aware of breathing, especially after a prolonged walking
Walking at 4 mph (6.4 kph) Brisk walk	8.5	38	Will experience laboured breathing after prolonged walking (20–30 minutes)
Walking at 5 mph (8 kph) Fast paced walk	13.5	61	Will experience laboured breathing after a few minutes and discomfort after 10–15 minutes

Table 2 The approximate air (not O_2) consumption of an average healthy male at different levels of activity

A simple test to ascertain an individual's personal capacity provided by a tank of air is to elevate his or her consumption through a short period of work, for example ten minutes of step aerobics carrying (but not breathing through) the RPE. When the rate of breathing has increased sufficiently, the mask should be donned, the air flow on and the full cylinder breathed either to its complete draining or through a fixed volume, e.g. from 200 bar pressure to 100 bar pressure. This period should be timed. This will give an approximate time available for that individual to exit a confined space before a full cylinder is drained. (This test should be carried out annually and at a time when the user is in an average state of health.)

Although an individual may have for example ten minutes of breathing time in a two litre cylinder, it is important to realise that the nearest exit point should be approximately one minute away from the point of work. This allows a substantial margin of error, as in an emergency confusion may reign, visibility may be reduced and obstacles may be encountered. Always aim to leave the space with the cylinder more than half full.

Entry controls: breathing apparatus

There are a number of choices of breathing apparatus available and the type selected will depend upon the circumstances of the confined space and the requirements of the task. Breathing apparatus is an independent source of breathable air and is used where there is present or likely to be present, even momentarily, an IDLH atmosphere, i.e. where the O_2 levels fall below 18%, where toxic gases are present at dangerous levels or both.

Fresh-air hosed (FAHBA) and compressed airline (CABA) breathing apparatus systems have airlines stretching from the wearer to a source of air outside of the confined space. These have limitations, particularly on construction sites, in that they considerably reduce manoeuvrability.

Self-contained breathing apparatus, (SCBA), comprise a face mask, airline and controllable source of air carried on the back or over the shoulder. The wearer is therefore independent of anyone else and able to manoeuvre with a high degree of flexibility. They come in two types: full working sets which contain one or two 6 and 9 litre cylinders, and escape sets with a single 2 or 3 litre cylinder. The former are used when entry is required into a known IDLH or an atmosphere that has been untested and where the conditions are unknown. The latter is used to enter breathable environments where there exists a potential for an IDLH atmosphere to arise (and are used purely for escape purposes).

It is important to select RPE that is suitable for the task at hand and the conditions that entry workers can expect to find. At the core of the choice is that the mask will fit the wearer comfortably and provide him with a good seal against the ingress of any hazardous atmosphere. There are a number of approaches to determining whether masks will fit snugly, the most effective being to have them specifically fit-tested to the wearer. Fit-testing is normally carried out by specialist companies who will travel to an employer's sites and conduct the testing in situ, or by a competent person employed by the company who can continually ensure that workers are issued with the correct masks.

An alternative is to use a hood which fits completely over the head and seals at the neck. This has advantages in that it can be pulled over the head and requires no re-adjusting of straps to ensure a good fit, an important consideration for emergency escape sets.

All equipment requires regular inspections and periodic testing to ensure that it remains in a condition which is safe to use. The interval between tests will be given by the manufacturer, although equipment that is in constant and heavy use should be submitted for testing at earlier intervals, or when they have suffered some damage. For pressurised air cylinders there is a test period at five-yearly intervals when the cylinder will be subjected to hydrostatic and other tests. The date of the last test will be stamped onto steel cylinders, and attached to carbon composite cylinders, usually in the format of month/year. If it is more than five years from this date and there is no other date discernible, the cylinder must not be used but sent off for testing. Any cylinder that fails this test will be destroyed. Steel cylinders can last indefinitely as long as they pass this five year test, whereas carbon composite cylinders have a service life of 15 years which may be extended only by the written authorisation of an approved inspection body and the manufacturer.

Carbon composite cylinders are much lighter than steel cylinders and contribute to ease of movement, lessening of heat stress and are more comfortable to carry.

Cylinders used for RPE sets normally have a working pressure of 200 bar (with a test pressure of 350 bar). As a rule of thumb, air under pressure occupies a volume in inverse proportion to the pressure, e.g. double the pressure and air will occupy half the volume, quadruple the pressure and air will occupy a quarter the volume. Thus a 2 litre cylinder may contain 400 litres of air under full working pressure of 200 bar. And by reference to the average man's average rate of work requiring 38–40 litres of air per minute, a 2 litre cylinder, when full may provide ten minutes supply of air, the minimum amount of time required to affect a safe exit in an emergency.

A 9 litre cylinder will contain 1800 litres of air under maximum working pressure, or about 45 minutes average breathing time. A low pressure whistle calibrated to 50 bar pressure will sound when there is about 11 minutes left, indicating that work should cease and the worker exit to obtain a replacement cylinder.

Pre-use visual checks should be carried out by the user at the start of each work period when the equipment is to be used. This is best carried out at the depot/stores, as alternative sets can be issued should any faults be discovered. The primary rule is that should any fault be discovered on inspection, the equipment must not be used until a replacement part is obtained, or the equipment repaired by a competent person.

It is useful to develop a sequence for carrying out the inspection as this becomes familiar and ensures that nothing is missed, alternatively a checklist may be used. All checks should be thorough and sufficient time taken to ensure this.

For escape and full working SCBA, the tests are the same with a couple of additional tests for the working set, the first being the back-plate used to carry the cylinder (for the escape set the check will be on the carrying bag) and the second being the high and low pressure tests to check the integrity of the air hose and the low pressure whistle.

Check the back-plate to ensure that there are no cracks or breaks on the plate itself or on the buckles used to secure the cylinder or the whole kit to the wearer. Look for any signs of corrosion, and on the fabric straps ensure that there is no fraying of the material or unstitching.

Similarly, check the carrying bag for wear and tear on the straps, tears or corrosion on the bag, cracks on or broken buckles. Ensure that the Velcro and holding loops for the cylinder are present and in good condition. Some escape sets have a valve lock attached to a D-ring in the bag. If this type of set is being used, ensure that the D-ring is securely attached to the bag and in good order.

For all cylinders the checks in **Table 3** are to be followed.

At the end of the working day, ensure that the equipment is washed clean and free from grit and contaminants before storing away. This is good hygiene and good manners for those users who come next to use it. Warm soapy water will suffice and gentle handling, especially around delicate diaphragms, will ensure that no damage occurs. Do not use harsh detergents or chemicals to clean the equipment. In situ, alcohol medical-wipes may be used to keep it clean.

If there is time, air drying is recommended, otherwise pat the equipment dry with soft cloth or tissue. Avoid rubbing as any residual grit may cause damage to parts essential to the face seal, or the diaphragms.

As with any other item of equipment, RPE comes with a manufacturer's handbook which will provide substantially more and equipment specific detail than this chapter has space for. Read these handbooks carefully for use, maintenance and (for competent persons) repairs. If they have been lost, contact the manufacturer with make and model details, and they will send a copy out.

Donning and using RPE

Prior to entry into the IDLH confined spaces the full working set must be donned. Using a partner to assist with lifting the equipment onto your back adjust the positioning and straps ensuring that it is comfortable and well balanced before closing the buckles. Open the demand valve at the cylinder before putting on the mask (otherwise there will be no flow of air). Place the mask onto your face and take a breath to activate the flow of air. Holding it in place, pull the straps over your head and adjust by pulling back on the end of the strap. There are five straps, above and below the ears and one over the crown of the head. Test to ensure you have a face seal. This can be done simply by closing and holding closed the valve button at the mask; this

will stop and prevent the flow of air. With a good face seal the mask will pull into your face as a vacuum is created by your inhalation of the last of the air inside the mask. If this happens, release the button and air will flow immediately.

If with the button closed you can still obtain an air flow, this is likely to be coming into the mask by way of a break in the face seal. Release the button, adjust the mask and straps and repeat the test until you successfully get a face seal. With a little practice this will come easily. All things being equal in terms of other requirements, you are now ready to enter the confined space.

In respect of the use of escape RPE, this must be worn in conjunction with a personal gas detector. The detector is worn to the front of the body, facing out and unobstructed.

The escape set, if in a carrying bag, is worn over the head and either to the front and below the gas detector, or under the opposite arm with the gauge side to the front where it can be read. Other escape sets can be carried attached to a belt.

Ensure that the valve is open, or that the appropriate release mechanisms are attached to the D-ring inside the bag before entering the confined space, that the mask is attached to the hose, the head straps folded back and the bag packed for ease of access.

On hearing the alarm inside the confined space, cease your work activity and, without moving from the work point, remove and don the mask. Escape sets have two head straps and these are to be pulled back (not sideways) to tighten. Check your face seal by bending and squeezing the air hose tightly; this will stop and prevent the flow of air. With a good face seal the mask will pull into your face as a vacuum is created by your inhalation of the last of the air inside the mask (or hood).

If you can still obtain an air flow, this is likely to be coming into the mask by way of a break in the face seal. Release the hose, adjust the mask and straps, and repeat the test until you successfully get a face seal. With a little practice this will come easily. Only when you have a face seal is it safe to make your exit from the confined space and this should be done without delay.

Emergency procedures

Every confined space entry procedure must have attached a procedure for responding to emergencies that may occur. A generic emergency procedure will not suffice as each confined space will present its own peculiar requirements based on hazards, configuration, location, type of emergency that arises and any other number of factors that differentiate one confined space situation from another.

There may be features in common and these may inform a general template for an emergency procedure, but these must be made specific to each confined space by adding information to the procedure that will assist the operatives, the response team and the rescue services survive the emergency without loss of life.

There are a number of principles that are to be observed in any procedure. The first is that when an emergency arises all workers inside the confined space must cease what they are

Item	Check	Comment
Cylinder	Check that it contains breathable air	It will state 'Air 'or 'Breathing Air' on the cylinder. All cylinders for breathing air must have black and white quadrants painted around the top of the cylinder. These indicate that the contents are medically safe. Air cylinders without these quadrants are not safe to use as breathing apparatus
	Ensure that it is within the five-year test-period	Look for the most recent date stamp (there may be several as older ones may not have been erased)
	Check the water capacity	This is normally written as WC 2l indicating a Water Capacity of 2 litres (or whichever number is appropriate). Escape sets are normally 2 or 3 litres; working sets 6 or 9 litres
	Inspect for signs of damage	Wear and tear such as scrapes on the paintwork may be ignored, but indentations, cuts etc. will render the cylinder unsafe to use. Look for signs of corrosion and rusting, especially around the valve
Valve and hose assembly	Check the high pressure demand valve and hose assembly	Depress the diaphragm to ensure that it is not sticking. You will hear a hiss as air escapes, this will cease when you release the diaphragm
Pressure gauge	Inspect for signs of damage.	Ensure that glass is not cracked or any other damage to the instrument. Check the reading; it must read at or above 80% full (160 bar) in order for there to be sufficient to use within the confined spaces. The working set must be above 80% full at the start of the work operation and entry is permissible on multiple occasions during that shift until the volume of air reaches the low pressure level (25% full or 50 bar) when the whistle will activate indicating the need to change the cylinder
Air hose	Inspect for signs of damage	The air hose may be rigid with a small diameter (on working sets), or soft, flexible and corrugated on escape sets. Look out for cuts, signs of corrosion, powdering or hardening of the material (which results from exposure to UV light or storage in hot temperatures)
	Check the O-ring	Where the hose attaches to the face mask or hood, detach and check that the O-ring is intact and flexible. Re-connect and ensure that it can turn freely
Face mask	Inspect for signs of damage or wear	Thoroughly check the soft parts of the face mask to ensure that there is no corrosion, hardening of the materials or cuts, scratches or grit that will break a face seal. Look inside and outside. Ensure that the ori-nasal mask (inner mask) is soft and pliable and correctly fitted. Check the diaphragms inside the ori-nasal mask to ensure that they are intact and pliable
	Check the visor	The visor may be laminated. Check for cracks, other than minor scratches, both inside and out. Check the gasket for corrosion, and the screws to ensure that they are tight
	Check the head straps	The head straps may be rubber, elasticated fabric, or a combination of both and with or without netting. Check for stress fractures, especially at the connection points, fraying or unstitching of the fabric, or hardening and dosing of the rubber. Ensure that any buckles etc. are intact. Some metal ones may have bent thus preventing adjustments to the head piece
Full working SCBA	High pressure test	For the full working set, it is not possible to see pin-sized holes in the air hose. To determine if there are any, use the high-pressure test. This involves closing the valve at the mask end and opening the valve at the cylinder end to fill the hose. Close this valve. A certain expansion of the hose will occur and this will show as a slight decrease in pressure on the gauge. Continue to observe the gauge for one minute. If the pressure drops by more than 5 bar in that time, there is a leak in the hose and it cannot be used. Replace it and test the new hose in the same way
	Low pressure test	The low pressure test follows naturally from this point. This test will determine if the whistle will activate at or around the 50 bar mark. Ensure that the whistle is free from obstructions then slowly release the air from the hose by opening the valve at the mask end and observe the gauge
Escape SCBA	Valve activation	On escape sets with a valve locking device ensure that it is detached from the D-ring in the carrying case before taking it out to examine. Once the examination is complete, replace the cylinder securely in the case and re-attach the D-ring. In use, by pulling open the case flap the locking device will be extracted and air will flow immediately
	Valve activation (alternative design)	A variation of this is a release cord wound around the valve which turns open the valve when the flap is opened. Some systems require manual opening of the valve. Ensure that it is opened after the inspection and ready with a supply of air once the user dons the face mask

Table 3 Tests and checks

doing, don their RPE and vacate the confined space immediately without stopping to assist anyone else. There is limited air in an escape set, sufficient only to get the user out. Any delay to assist a colleague increases the likelihood that such an act of charity will lead to fatality. If a worker has collapsed due to an irrespirable or toxic atmosphere, there is little that can be done for them by an untrained and ill equipped colleague with little air in their breathing set. On the other hand if they have collapsed for some other reason, it is quite likely that the worst of the injury has already occurred and their condition may not deteriorate substantially more before an authorised rescue team arrives. Vacate the space and stay outside.

A second principle is that any rescue (other than self-rescue) is to be carried out by an authorised team who are appropriately equipped to conduct the rescue. Some years ago at a well in Ballymoney, NI, two would-be rescuers who were normally employed in the emergency services carried out an unauthorised rescue attempt without equipment and both died. Authorised rescue workers include the confined spaces attendant who is authorised to assist people out of the confined space without himself making an entry, generally by operating the winch to extract people, an on-site rescue team (who are fully trained and equipped for emergencies), although such teams are rare outside of large scale fixed sites such as airports and refineries, and finally the public emergency services, fire, ambulance, etc.

Third, all confined space workers must be trained in self-rescue techniques. This involves being able to recognise and respond to emergency and unplanned events, use of RPE and any other escape equipment, and being able to develop and understand the specific procedures for each confined space they are working in.

Although you may have assessed all the hazards and put in place the necessary controls to safeguard against them, the emergency arrangements begin with the question, 'What could go wrong?' The answer to this question will guide the development of the remainder of the procedure. Information about the nature of chemicals, atmosphere conditions, flow of liquids, work activities etc. will give some indication of what emergency equipment may need to be on hand, or provide the emergency service with some vital information.

Will any potential emergency be contained within the space, or will it impact upon the vicinity outside and if so how extensive would the impact be?

Who will take immediate charge and will they have the authority to order an evacuation of the rest of the site? Note that once the emergency services arrive on site they will take control of the site and emergency and will not hand that control back before they are satisfied that the emergency is over. They may retain control for a longer period if there is an investigation, and indeed other bodies may take control at different times, e.g. the police, especially if there is a fatality, and the HSE or Health and Safety Executive Northern Ireland (HSENI).

What is the location of the confined space, (a) on the site and (b) geographically? Someone at the entrance to the site will need to be available to guide the emergency services to the exact point of the emergency. In remote rural areas it would be useful to have Ordnance Survey grid references of the location written onto the procedure as this is the most accurate guide for the emergency services; in an urban area, write in the exact address of the site.

Once an emergency has begun, it will be necessary to establish who is to initiate the procedure, who is to call the emergency services, who is to notify others on-site, to order

an evacuation, etc. These individuals must be named on the procedure, be informed of their responsibilities and capable of carrying them out. Speed of response and clear unambiguous awareness of who is responsible for various activities is vital to saving lives and delimiting the emergency. Example: In 1984 a fire broke out at an LPG terminal in Mexico. Workers tried to deal with the emergency themselves and it was about five–ten minutes after the initiation of the emergency before someone pressed the emergency shutdown. Some 500 people died.

Write down the key elements of the emergency procedure and append them to the safe working procedure and ensure that all people on the operation are fully aware of its contents and understand their responsibilities. Take time to go through it and review it daily.

Summary of main points

When approaching work in confined spaces it is a requirement that consideration is given to being able to carry out the work operation externally. If that is not possible and entry is required a full assessment of the confined space must be conducted prior to entry, with a safe entry and exit procedure agreed upon and conveyed to all members of the work team. The procedure must cover:

- the pre-entry preparations, including:
    - testing and venting of the confined space, and
    - inspection of the equipment
- the methods of entry and exit
- the work to be carried out inside
- the means of communication
- the PPE and RPE requirements
- the role of each person on the team, and
- the emergency procedures.

Reading this chapter is an important introduction to the central issues concerned with safe working in confined spaces, however it is not to be regarded as a substitution for suitable and sufficient training in confined spaces entry and the use of respiratory protection and other equipment. It is a legal requirement that all confined spaces workers and supervisors are trained to the appropriate level with sufficient refresher training to ensure on going competence.

References

BBC. *1984: Villagers Die in Water Plant Blast*, 2008. Available online at: http://news.bbc.co.uk/onthisday/hi/dates/stories/may/23/newsid_2969000/2969125.stm

Heath and Safety Commission. *Statistics of Workplace Fatalities and Injuries, Falls from Height*, 2009, London: HSC.

Health and Safety Executive (HSE). *Shipbuilding and Ship-repairing Health and Safety Consultative Committee paper SSHSCC/47/A: Work in confined spaces*, 2003, London: HSE. Available online at: http://www.hse.gov.uk/aboutus/meetings/committees/ships/ 090403/ 47confinedspaces.pdf

Health and Safety Executive (HSE). *Number of Fatal Injuries, as Notified to HSE and Local Authorities*, 2010. Available online at: http://www.hse.gov.uk/statistics/fatalquarterly.htm

Ibbetson, A. *Communication During Confined Space Entry*, 2004, Vancouver: Con-space Communications Ltd.

Indopedia. *List of Space Disasters*, 2009. Available online at: http:// www.indopedia.org/Space_disaster.html#Ground_crew_fatalities

Surada, A. J., Casatillo, D. N., Helmkamp, J. C. and Pettit, T. A. Epidemiology of Confined Space Related Fatalities, in *Workers Deaths in Confined Spaces*, 1994, 11–25, Atlanta, GA: NIOSH. Available online at: http://www.cdc.gov/nioshj/94-103.html

University of Glasgow. *Working Safely in Confined Spaces*. University of Glasgow, 2005. Available online at: http://www.gla.ac.uk/ seps/workplaceandbuildings/confinedspaces.html

The Confined Spaces Regulations 1997, London: The Stationery Office.

Confined Spaces Regulations (Northern Ireland) 1999, London: The Stationery Office.

Further reading

BP Safety Group. *Confined Space Entry,* 5th edition, 2005, London: Institution of Chemical Engineers.

British Safety Industry Federation. *Respiratory Protective Equipment Fit Test Providers' Accreditation Scheme*, 2001, London: BSIF.1

Health and Safety at Work Act 1974. London: Her Majesty's Stationery Office.

Health and Safety at Work (Northern Ireland) Order 1978. London: Her Majesty's Stationery Office.

Health and Safety Executive (HSE). *Specification for Large Seamless Steel Transportable Gas Containers,* 1995, London: HSE.

Health and Safety Executive (HSE). *Safety Work in Confined Spaces, ACoP*, 1997, London: HSE.

Health and Safety Executive NI. *Safety Work in Confined Spaces in Northern Ireland, ACoP*, 1999, Belfast: HSENI.

Health and Safety Executive (HSE). *Specification for Fully Wrapped Carbon Composite Containers*, 2002, London: HSE.

Health and Safety Executive (HSE). *Fit Testing of Respiratory Protective Equipment Facepieces*, OC282/28, 2003, London: HSE.

Health and Safety Executive (HSE). INDG258 *Safe Work in Confined Spaces,* 2006: London: HSE.

Health and Safety Executive (HSE). *A Guide to the Reporting of Injuries, Diseases and Dangerous Occurrences Regulations 1995*. 3rd edition, 2008, London: Health and Safety Executive. Available online at: http://books.hse.gov.uk/hse/public/saleproduct. jsf?catalogueCode=9780717662906

McAleenan, C. and McAleenan, P. *Confined Spaces Expert*. 1999, Downpatrick: Expert Ease International.

Sargent, C. *Confined Space Rescue,* 2000, Tulsa, OK: Pennwell Books.

Veasey, A. D., McCormick, L. C., Hilyer, B. M., Oldfield, W. O., Hansen, S. and Krayer, T. H. *Confined Space Entry and Emergency Response,* 2006, Chichester: WileyBlackwell.

Websites

Health and Safety Executive (HSE) http://www.hse.gov.uk/

Health and Safety Executive, RIDDOR pages http://www.hse. gov.uk/riddor/

Health and Safety Executive Northern Ireland (HSENI) http://www. hseni.gov.uk/

Institution of Civil Engineers (ICE), Health and safety www.ice.org.uk/knowledge/specialist_health.asp

ice | manuals

doi: 10:10.1680/mohs.40564.0207

Chapter 17

Falsework

John Carpenter Consultant, Manchester, UK

This chapter discusses the special characteristics of falsework that give rise to a safety risk, and the responsibilities of Designers (of both Permanent Works and Temporary Works) to manage the process.

Falsework is often complex and usually related to work at height. Designers and users are governed by Regulation and guided by British Standards and industry guidance. This chapter outlines the key aspects of these various documents but does not cover every aspect. It is essential to consult with the actual references if engaged in this area of work.

History has demonstrated that falsework demands careful attention if failure is to be averted. Although the industry has advanced significantly since the 1960s and 1970s, which saw major collapses (leading to the introduction of specific guidance and a British Standard on falsework), the industry cannot afford to be complacent. It is essential that all those involved in falsework are cognisant of the risks and the necessary measures to ensure safety. In particular Designers should understand how these temporary structures behave and their requirements for stability.

Introduction

Temporary Works is the generic name given to:

i) **Falsework**, i.e. temporary supporting structures used to support 'formwork' (see item ii below) and/or the permanent works until the latter is self-supporting and stable. Falsework may be applicable to all construction materials.

ii) **Formwork**, i.e. the temporary containment to in situ concrete, and its immediate supporting members, pending the concrete gaining sufficient strength to sustain its own weight and/or act as a structural element.

This chapter concentrates on falsework. It is not a technical design guide. Instead it outlines the special characteristics that differentiate falsework from Permanent Works, the responsibilities of Designers and highlights key aspects of the management of falsework, such that safety risk is mitigated. Potential ill health effects arising from the design or use of falsework are no different from other forms of construction.

Falsework is usually provided on a temporary basis and removed from site after use. However, occasionally some falsework components may become part of the Permanent Works on completion.

Simple falsework may consist of scaffolding or proprietary falsework systems supporting formwork of ply sheets on timber joists. Any number of variants may come into play as the structure increases in size or complexity: bespoke systems, climbing systems, travelling systems or modular systems. These may be constructed of timber, steel or aluminium. In Hong Kong bamboo scaffolding is commonplace and an accepted structural solution.

Falsework is an element of the project for which the Permanent Works Designer may have some responsibilities. The precise split and/or extent of involvement with the Temporary Works (falsework) Designer will depend upon the nature of the project itself.

Falsework characteristics

Falsework has a number of characteristics not normally present in the Permanent Works, viz:

■ It is usually required for short durations.

■ It is often subjected to significant loading which may be less well controlled than for the Permanent Works.

■ The ratio of dead load/live load is markedly different from most permanent structures, making live load impact and lateral loading in particular more critical.

■ Accuracy of assembly may be less than for the Permanent Works and hence the design, specification and supervision need to reflect this.

In addition:

■ Most falsework system design relies on the Permanent Works for its lateral stability.

■ Some Permanent Works designs require careful thought and input by the Permanent Works designer in order that the necessary falsework may be designed safely and economically by others.

■ Many falsework systems are proprietary products, often originating from outside the UK. It is important that the clarity and quality of safety related data, e.g. safe loads, is suitable and sufficient.

■ Falsework is usually designed by a party within the contractor supply chain and this party is unlikely to have any contractual connection with the Permanent Works designer.

For these reasons, the management of falsework is particularly important. The consequences of any collapse are always likely to be very severe in human terms and in economic loss. This chapter discusses some of the issues associated with the management of safety risk arising from falsework structures (see **Figure 1**).

Background issues

The design of Temporary Works is covered by BS 5975: 2008 Code of Practice for Temporary Works Procedures and the Permissible Stress Design of Falsework (BSI, 2008) and BS EN 12812: 2004 Falsework: Performance Requirements and General Design (BSI, 2004) which outline a limit state design approach.

BS 5975 was originally issued in 1982 which was the first time formal guidance had been made available; it followed on from some ground-breaking work undertaken by the Institution of Structural Engineers and the Concrete Society (The Concrete Society, 1971), followed by the Advisory Committee on Falsework which published what became known as the Bragg Report, named after its chairman: the Interim Report (Bragg, 1974) and the Final Report (Bragg, 1975).

The Bragg Committee was established following a number of serious falsework collapses which had occurred in the preceding years. Several of these incidents are summarised in the final Bragg Report. One example is the Loddon viaduct, where in October 1972 falsework supporting part of the A329 dual carriageway collapsed under the load of concrete being poured, killing three and injuring ten (Bridle and Porter, 2002).

Both these reports are still relevant as reminders of what can occur if risk is not managed appropriately. These collapses highlighted several significant shortcomings in the way falsework was dealt with, in particular:

■ A lack of clarity in the division of responsibility.

■ A failure to reflect the 'high risk' nature of falsework in the design and site controls.

■ The essential need to provide robust means of ensuring lateral stability, and specifically so at forkhead level.

■ The need to discard damaged or distorted items.

■ The need for stiffeners at all load-bearing points in steel grillages.

The Bragg Committee made some pertinent comments which still hold good today. These include:

Falsework requires the same skill and attention to detail as the design of permanent structures of like complexity, and indeed falsework should always be regarded as a structure in its own right, the stability of which at all stages of construction is paramount for safety. (paragraph 8 Interim report)

The Report highlighted some key concerns:

i) *Competency*: having persons of the appropriate competence to procure, design, erect and utilise the falsework.
ii) *Design procedures*: the need for clarity of brief and adequate checking.
iii) *Design responsibilities*: having clarity as to which party is responsible for which aspect of the project, at all construction stages.
iv) *Communication and coordination*: the essential need for taking appropriate measures to ensure information is transferred and the design coordinated.
v) *Inspection and supervision*: as for all structures, falsework should be subject to adequate independent checks and supervision before, and during use.
vi) *Lateral stability*: a crucial aspect of most falsework structures and a key example of where coordination between Permanent and Temporary Works is needed.

These items were identified from the study of falsework failures and practice at the time. Many of these issues have been incorporated into BS 5975, and all remain pertinent to today's projects. Despite the time lapse since these earlier documents were produced it is a useful exercise to read them so as to be reminded of the core principles of good risk management in this field of activity, acquired from the bitter experience of actual collapses. The Bragg Report describes a number of these incidents.

Notwithstanding the introduction of a code of practice, concerns remain regarding current practice. In 2001, research for the Health and Safety Executive (HSE, 2001) stated that concern arose from 'a lack of understanding of the fundamentals of stability of falsework ... at all levels' and went on to comment on 'a lack of adequate checking and a worrying lack of design expertise'. In 2002, the Standing Committee on Structural Safety (SCOSS) issued a topic paper entitled *Falsework: Full Circle?* (SCOSS, 2002) as it was considered that many of the lessons learned over the previous 30 years or so had been forgotten. This paper stated that 'it is only a matter of time before a serious event occurs'. Since the SCOSS paper was written, serious events, such as the collapse of falsework in Milton Keynes in 2006, have occurred.

Box 1   Milton Keynes scaffold collapse 2006

A multi-storey office block was being refurbished and was provided with a scaffold access structure around the perimeter of the existing building. This collapsed one afternoon killing one worker and seriously injuring two more.

The Health and Safety Executive (HSE) investigation revealed significant overloading of the scaffold and a significant deficiency in ties to the building façade. HSE issued an alert which stated:

The warning aims to alert those working on similar projects to the importance of their arrangements to provide and maintain stable scaffolds. HSE recommends that those arrangements

Figure 1 Formwork at biomass station at Stevens Croft, near Lockerbie, Dumfries and Galloway. Harsco Infrastructure (formerly SGB) was contracted by E.ON UK plc to develop a complex shuttering, falsework system and back propping system to facilitate the construction of an obstruction free 1.2 m high slot 48 m long which, when constructed, will enable fuel to be fed onto conveyors by way of short travelling Archimedes screws at the rate of 235 cubic metres per hour during peak load.
image courtesy of RED Reproduced with permission, © Harsco Infrastructure, http://www.sgbformwork.com; www.harsco.com. Further detail available online: http://www.sgbformwork.com/sgb/newsroom/show-article?article_id=20andviewpage=4andPHPSESSID=c12eb3ec913717974ebcd8866f7b66f2

are reviewed regularly and that reviews take account of factors which include, but are not limited to:

- scaffolding design implementation;
- arrangements for securing scaffolding to structures;
- intended and actual loadings on scaffolds, including the impact of wind;
- the risk of direct impact by construction plant or vehicles;
- the frequency and thoroughness of scaffold inspection arrangements;
- systems in place for the handover of new or adapted scaffolds;
- the training and competence of scaffold erectors;
- the adequacy of the scaffold foundations; and
- the prevention of unauthorised modifications.

In particular, the 2002 SCOSS paper highlights:

- Few 'main contractors' now have their own temporary works departments; the norm is for the temporary works responsibility to fall to a subcontractor or to a specialist contractor/supplier. This can result in a lengthy supply chain. In the 1970s almost all main contractors would design their falsework in-house.

- The concept of the management contractor, a lead role that dispenses with the main contractor, was not known in the 1970s.

- The nature of falsework has changed. Whereas in the 1970s most falsework construction was made using 'tube and coupler', the proprietary system now dominates the market. The design skills, and the knowledge of the abilities of the systems, lie with the specialist organisations.

- The gradual but inexorable loss of traditional skills within the industry has meant that the site foreman, with a lifetime's experience of 'what works', has been largely lost.

- Procurement routes are usually chosen to maximise commercial benefit and have little regard to information flow considerations. The difficulties experienced with long supply chains are further exacerbated when erection responsibility is split from design responsibility, and design or supply briefs do not include for site inspection.

- The industry exists in a very much harsher commercial climate than 20 years ago.

The paper went on to quote from the research commissioned by the Health and Safety Executive (HSE, 2001) which indicated shortcomings in the industry. These were:

- A lack of understanding of the fundamentals of stability of falsework and the basic principles involved. This shortfall occurs at all levels.

- Wind load is rarely considered, and, if included, it tends not to be to BS 6399.

- A lack of clarity in terms of design brief and coverage of key aspects such as ground conditions.

- The lateral restraint assumptions made by designers is often ignored/misunderstood by those on site.

- There is a lack of adequate checking and a worrying lack of design expertise.

- Erection accuracy leaves much to be desired.

These are all fundamental to the safety of falsework and give emphasis to the need for competency, adequate information and supervision in particular.

BS 5975 has recently been updated and revised (2008). It deals with management and procedural issues, as well as technical matters, in some considerable detail. The Foreword gives a useful summary of its history. At the time of its first publication, drawing heavily on the Bragg Report and the Joint Committee of Institution of Structural Engineers and Concrete Society, it was the first known attempt in the world to set down good practice for falsework. The current code:

- emphasises that the '… success of falsework is closely tied up with its management';

- makes reference to the associated duties under the Construction (Design and Management) Regulations (CDM) and other health safety regulations;

- makes reference to the Eurocodes (for example in respect of wind loading).

As noted in the opening sentence to this sub-section, the BS code is complemented by BS EN 12812. This deals with the limit state design of falsework, and places falsework into one of three categories for this purpose:

- Class A applies dimensional limits to restrict this class to smaller applications. No specific design rules are given as it is envisaged that proprietary equipment will be used in conjunction with experience and rule of thumb.

- Class B1 based entirely on the Eurocodes. The standard of design and documentation is assumed to be no less than that for the permanent works.

- Class B2 based on a lower (simplified) level of design. An additional factor (1.15) is applied by way of compensation.

This code limits itself to technical matters and does not include the advice on management aspects provided in BS 5975.

Responsibilities

It is essential that all those engaged in a construction project understand their responsibilities for any falsework that may be required. This requirement needs to be understood within the framework of the differing requirements of contract and statute. Contract may introduce a number of stipulations but these cannot override or discount statutory obligations. The CDM Guidance for Designers (Construction Skills, 2007) gives a comprehensive commentary on statutory responsibilities and their relationship to contract.

Figure 2 Road bridge, near La Massana, Andorra, collapsed during a concrete pour. On 7 November 2009, during a concrete pour, a 20 m section of the bridge under construction collapsed, killing five and injuring six people. Investigation into the cause of the collapse is ongoing [January 2010] (reproduced from *New Civil Engineer*, 12 November 2009)

In simple terms, whereas the contact will often allocate responsibility for falsework to the Contractor, statute requires all structural Designers (no matter who employs them) to play a part dependent upon the nature and complexity of the project and hence the scale of associated hazards and risks. For simple projects no involvement will be necessary from the Designers of the Permanent Works; for complex projects there may need to be some considerable involvement.

Permanent Works design

Permanent Works Designers (PWD) may be employed by the client, consultants, contractors or suppliers. No matter who employs them they have a responsibility to discharge their duty under regulation 11 of CDM, i.e. to eliminate hazards and to reduce risks, and to supply information. This requirement applies, inter alia, as much to the consequences of their design decisions upon the method of construction, and hence the use of falsework, as it does to the Permanent Works in use.

In practice, where the structure and falsework are straightforward, using standard industry solutions and techniques (e.g. generic scaffolding structures), no action will be required by the PWD. However, where this is not the case, the PWD may need to consider one or more of the following as noted in

Table 1. In all cases, the involvement of the PWD should be proportionate to the risk.

The Permanent Works design should be checked, and also overviewed by an engineer of experience, to ensure that 'buildability' issues have been adequately considered. This is a sensible risk management procedure which will not only help to produce an economic solution but will also mitigate safety risk.

Temporary Works (falsework) design

Falsework Designers will usually be employed by a contractor or supplier of falsework. However, their actions are governed by CDM regulation 11 (see Chapter 3 *Responsibilities of key duty holders in construction design and management*) which makes no distinction between temporary and permanent structures.

It is important that Falsework Designers:

■ Are competent in this specific field of design.

■ Have a specific design brief and all the necessary data for them to produce a competent design.

■ Understand how the Permanent Works functions, whilst being constructed and in its final state.

■ Understand any site constraints or limitations.

- Provide comprehensive information on their design, e.g. assumptions and limitations, to those who are going to erect and use it.

Although the design of individual standard proprietary components, e.g. props and table forms, are outwith the scope of CDM, the designer utilising these items has a duty to be satisfied that they are suited for the intended purpose. The suppliers also have statutory obligations to ensure they are suitable for their stated purpose and that sufficient information is provided to allow others to incorporate them safely into the overall falsework design. Notwithstanding, particular care is required when components are sourced from outside the UK.

Specific issues to consider are shown in **Table 1**.

Robustness

Robustness may be defined as 'the ability of a falsework structure to withstand unintended events (such as impact or human error) without being damaged to an extent disproportionate to the original cause'. This is an important concept.

Robustness is different from 'strength', and unlike strength is not covered so explicitly in codes. In practical terms it means applying good engineering judgement to the design. Such judgement only comes with experience. Some illustrations of this are shown in **Table 3** which uses the SCOSS concept of the '3Ps', i.e. **P**eople, **P**rocess and **P**roduct to illustrate the wide range of influences.

Managing the process

Because of the high risk nature of falsework erection and use, it is recommended that the contract is specific on related management matters.

Main construction contract

There should be a requirement for:

- falsework to comply with BS 5975: 2008 (and other relevant codes and industry guides);

- a Temporary Works Coordinator (TWC) to be appointed. This person should meet specified competency levels. For larger projects it may be beneficial to include similar requirements for a Temporary Works Supervisor.

- The TWC to liaise with the CDM Coordinator.

It will also assist if the contract requires details of all falsework designers to be passed to the CDM Coordinator, for Notifiable projects, or directly to the Permanent Works Designer, if not notifiable (i.e. less than 30 days on site or less than 500 person days input on site).

Depending upon the complexity of the project, there may need to be a stipulation that the falsework design is reviewed by the Permanent Works designer.

Sub-contracts

Sub-contracts between the 'Main' contractor and those who will have responsibility for designing and erecting the

Aspect to be considered	Action
Access and/or lack of space	If a standard industry solution (e.g. scaffolding, mobile towers) will not work through lack of space or access, the PWD will need to ascertain how this may safely (and economically) be overcome. This may necessitate discussions with contractors or temporary works suppliers
	The PWD should consider how falsework is to be delivered to its required location, or moved from one location to another. In particular, tunnel and table forms can be problematic if this aspect is not thought through
Temporary instability	If the Permanent Works are likely to be unstable at any point during the construction phase, this should be pointed out to the contractor on the drawings. The PWD should consider whether a proportionate change or addition to the Permanent Works could lessen the risk
Temporary supports/propping	Where it is known that temporary supports/propping will be required the PWD should normally: - identify this fact - consider where they can be detailed as part of the Permanent Works or otherwise accommodated - consider the consequences of the propping loads on the Permanent Works
Sequence	If a particular sequence of construction is required in order to ensure stability, this should be pointed out to the contractor on the drawings
Loadings	Any unusual or exceptional loadings, occurring during construction, should be identified
Use of the Permanent Works for support	Where it is likely that the Permanent Works will be used to provide support (and specifically lateral restraint) for the falsework, the PWD should design the Permanent Works to accommodate this, or otherwise identify the safe loads on the drawings

Table 1 Aspects of falsework for consideration by the Permanent Works Designer

ICE manual of health and safety in construction © 2010 Institution of Civil Engineers

Aspect	Comment
Use of components	Ensure manufacturers' data are understood and unambiguous. This is particularly important with imported equipment
	Is the proprietary component being used outside its stated function, e.g. in respect of height, load, assembly?
	Ensure that used components are checked for damage or misalignment
Interfaces	Risk thrives at interfaces. These may be procedural, e.g. between contracts, or physical e.g. between the Temporary Works and the Permanent Works. At these points it is essential that there is clarity of action and responsibility
Load paths	It is essential that the load paths (for vertical and horizontal actions) are made clear and accommodated
Sequencing	Where a specific sequence of construction is required or results from constraints, this should be made clear by the PWD (if originating from the Permanent Works design), or by the falsework designer if originating from the Temporary Works design
Lateral restraint	It is crucial that the means of lateral restraint is noted on the drawing and that the ability of the Permanent Works to accommodate the forces (real or notional) is also determined
Ground conditions	The ability of the ground to accommodate the likely falsework loads, and the prevention of scour or loss of support, should be ascertained during the design phase
Impact	Falsework should be protected against foreseeable impacts from vehicles or other actions

Table 2 Design aspects of falsework

The contribution to robustness from	Comment
People	
Competency	The design of falsework requires a team of designers (one or more persons in one or more organisations) with the appropriate level of overall competence. Competence infers experience in this field
Resource	The availability of adequate resource in order to complete the design, to the necessary standard of skill and care, within the project timescale
Process	
Design and checking	Ensuring design is to appropriate standards and given an adequate check and review
Procurement	Ensuring falsework is procured from organisations which can demonstrate competency, and in return are provided with a clear brief, sufficient information and adequate time to plan their work
Inspection on site	Ensuring falsework is inspected on site to ensure that the design intent is achieved. Inspections are discussed later in this chapter
Coordination	Appointment of a TWC (and as required TWS)
Product	
Redundancy	Providing alternative load paths so that in the event of a failure of one member, the load may be distributed through others
	Where it is not possible to provide redundancy, e.g. large single spans, particular attention should be paid to the supports and security at bearing beam level
Sensitivity	Ensuring that small variations, e.g. in positioning or verticality of members, do not negate structural design assumptions. (This is why stiffeners should always be provided to steel members at points of support)
Disproportionate collapse	Ensuring a 'domino' effect will not result from a local failure. Bracing is effective in this respect
Component size	It is unwise to skimp on critical materials by using the smallest size the calculations will 'allow'. Any critical member should be robust in its own right

Table 3 Robustness of falsework

falsework should include, as a minimum, additional requirements relating to:

- provision of design brief and information
- timing
- design standards
- specific performance requirements, e.g. deflection limits
- checking and review
- authority and role of the TWC and TWS
- division of responsibility between sub-contracts.

Contractual clarity

As with all contract matters, clarity is essential in respect of the procurement, design, erection, use and dismantling of all falsework. Although examples are given in the preceding text, each project should be subject to its own critical consideration.

Temporary Works Coordinators (TWCs) and Supervisors (TWSs)

As noted above, the concept of the 'coordinator' was first mooted in the Bragg Report, and subsequently incorporated into BS 5975. The rationale for such an appointment is that the high risk nature of falsework, and serious consequences of failure, demand a designated, competent person with an overseeing role, to ensure these risks are appropriately managed throughout the supply chain. It is important that the appointees are divorced from the commercial and contractual pressures of a construction project. Their decisions on design, 'ready to load' and striking times should be based on technical considerations alone.

On larger sites, or on more complex projects, this individual may be supported by a Temporary Works Supervisor (TWS) who is there to support the TWC and assist in some of the tasks. The precise split of duties will be project dependent (and should be in writing), but the TWC should always have overall responsibility.

The role of the TWC is set out in BS 5975. In summary it is to ensure that:

- All those involved are aware of their responsibilities and that activities are coordinated.
- An adequate design brief is prepared and that the necessary information is available.
- That the design is competent, compliant and checked, and that the falsework is subsequently built and maintained in accordance with this design.
- That those on site are carefully apprised of the necessary information, limitations and guidance necessary for the safe erection, use and dismantling of the falsework.
- No loading occurs without the TWC's permission.

For projects where a CDM Coordinator is appointed there will need to be close liaison between that duty holder and the TWC as the roles involve a degree of overlap.

Although the appointment of a TWC or TWS is a decision for the contractor, it should also be considered by the Permanent Works Designers and, if considered necessary or desirable, written into the contract (as noted above). In either case it is recommended that the contract should also stipulate minimum acceptable competency levels.

Both TWC and TWS (if applicable) should generally be appointed by the Main (Principal) Contractor. The appointment should only be by a sub-contractor if that party has complete control over the structure.

Small projects

On smaller projects, falsework may still be required but the scale and simplicity often means that standard solutions are used, e.g. scaffolding, proprietary towers. There may not be any formal design and there will not be a TWC. However, 'risk' shows no particular respect to small projects, hence a professional approach by all those involved is essential. This infers that the selection, use and dismantling of such falsework should be overseen by someone with appropriate competence.

Competency of TWCs and TWSs

Good business management suggests, and the law demands, that those undertaking the roles of TWC and TWS should be competent. In practical terms this can be interpreted to mean:

- Knowledge of the relevant structural engineering design and construction processes.
- Knowledge of relevant safety legislation and industry 'good practice'.
- Substantial site experience of falsework of similar type and complexity.

It is not a role which should be allocated on the basis of project management seniority or other arbitrary means.

Inspecting falsework

The inspection of falsework structures is an important element of safety risk management. The key aims of any inspection are to check that:

- It has been constructed in accordance with the design and within acceptable tolerances (or, in the case of small, simple falsework structures, in accordance with good practice).
- It is being used in accordance with the design assumptions, e.g. loading.
- It has not been adversely affected in some way which would affect its safety, e.g. as a result of vehicular impact.

CDM 207 requires all temporary structures to be of 'such design and so installed and maintained as to withstand any foreseeable loads which may be imposed upon it ...' And also, 'no part of a structure shall be loaded so as to render it unsafe to any person' (regulation 28). This clearly requires the use of competent persons and a management process to ensure these requirements are met. The use of a TWC (and if required a TWS) will assist in this respect.

The Work at Height Regulations 2005 also place obligations on those responsible for falsework as the activities will inevitably be 'at height'. These relate, *inter alia*, to:

- planning, organising and supervision by competent persons
- inspection requirements
- strength and stability.

The Regulations require inspections after completion of erection, if it suffers from significant deterioration, and, without prejudice to the foregoing, for falsework over 2.0 m in height, every seven days. Inspection records have to be kept until completion of the construction work and thereafter for three months.

BS 5975 also gives comprehensive advice in respect of workmanship, accuracy and inspection of falsework. The research by HSE (Health and Safety Executive, 2001) illustrated poor practice in respect of accuracy of erection. Although this should not occur in the first instance, a thorough inspection will identify these and similar shortfalls.

Needless to say, inspections should be recorded in writing and project procedures should ensure that any point of action is dealt with before acceptance of the falsework for loading.

Health and safety hazards

The hazards of which the designers should be aware are as for Permanent Works design and indicated in the CDM Guidance for Designers supplemented by those aspects scheduled in **Tables 1** and **2**. One risk, which applies to Permanent Works design, but which has special characteristics in connection with falsework, is 'collapse of the structure'. This chapter has given emphasis to this and outlined the broad approach required in order to eliminate it, or to reduce the resulting risks, so far as is reasonably practicable.

The means of dealing with safety and ill health hazards should follow the ERIC process or any of the other design risk assessment techniques outlined in the chapters in Section 4 *Health hazards* and Section 5 *Safety hazards*.

Summary of main points

- Falsework has particular characteristics which differentiate it from those of the Permanent Works.
- Although the Contractor has the prime responsibility on site, the Permanent Works Designer may also have an important role.
- Permanent Works Designers need to consider the interaction of the Permanent Works with that of the likely falsework arrangements and construction sequence.
- Falsework Designers should be aware of the pitfalls in the design and use or falsework, outlined in this chapter, and of the reasons for previous collapses.

- The appointment of a competent Temporary Works Coordinator is an essential step for all substantive falsework structures and one which adds value in addition to ensuring the management of safety risk.
- Careful inspection of falsework, by competent persons, is essential, and may be a legal requirement.

References

Bragg, S. L. Interim Report of the Advisory Committee on Falsework, 1974, London: HMSO.

Bragg, S. L. Final Report of the Advisory Committee on Falsework, 1975, London: HMSO.

Bridle, R. and Porter, J. (eds). *The Motorway Achievement: Frontiers of Knowledge and Practice*, 2002, London: Thomas.

The Concrete Society. *Falsework – Report of the Joint Committee*, The Concrete Society and the Institution of Structural Engineers, Technical Report TRCS 4, London, July 1971.

Construction Skills and Health and Safety Executive. The Construction (Design and Management) Regulations 2007 Industry Guidance for Designers, London: CITB; King's Lynn: Construction Skills. Available online at: http://www.cskills.org/supportbusiness/healthsafety/cdmregs/guidance/Copy_5_of_index.aspx

Health and Safety Executive (HSE). *Investigation into certain aspects of falsework*. Contract Research Report 394/200, 2001, London: HSE Books. Available online at: http://www.hse.gov.uk/research/crr_pdf/2001/crr01394.pdf

New Civil Engineer. Andorra Bridge Collapse, *New Civil Engineer* 12 November 2009. Available online at: http://www.nce.co.uk/home/structures/andorra-bridge-collapse/5210764.article

Standing Committee on Structural Safety (SCOSS). *Falsework: Full Circle?*, 2002, London: SCOSS. Available online at http://www.scoss.org.uk/publications/rtf/SCT0201_falsework.doc

Construction (Design and Management) Regulations 2007, reprinted March 2007. Statutory instruments 320 2007, London: HMSO.

Work at Height Regulations 2005. Statutory instrument 2005 735, London: HMSO.

Referenced standards

BS 5975: 2008. Code of Practice for Temporary Works Procedures and the Permissible Stress Design of Falsework, London: BSI.

BS EN 12812: 2004. Falsework: Performance Requirements and General Design, London: BSI.

Further reading

See 'Other Publications' and 'Further Reading' in BS5975: 2008.

Carpenter, J. *Designing for Safer Cncrete Structures*, undated, London: The Concrete Centre.

Websites

British Standards Institution http://www.hse.gov.uk

Concrete Society http://www.concrete.org.uk

Health and Safety Executive http://www.hse.gov.uk

Institution of Civil Engineers (ICE), Health and safety http://www.ice.org.uk/knowledge/specialist_health.asp

Institution of Structural Engineers (Istruct E) http://www.istructe.org

Standing Committee on Structural Safety (SCOSS) http://www.scoss.org.uk

Chapter 18

Demolition, partial demolition, structural refurbishment and decommissioning

Brian Neale Consultant, Cheshire, UK

doi: 10:10.1680/mohs.40564.0217

CONTENTS

This chapter will look at demolition in the wider context of partial demolition as well as the more commonly perceived whole demolition of a facility where perhaps the reader will have preconceived images of dramatic and spectacular reduction of buildings and other structures to piles of rubble – albeit these may be quite large piles of rubble! The latter is just one example of demolition and one where those show-piece events often, by their very nature, attract significant attention and publicity. Many benefits can arise from this approach as well as potential disbenefits, perhaps.

An essential feature of this chapter is that it provides an overview of many pertinent topics and gives some references for the reader to develop further their knowledge of these as they wish and require. This chapter looks at the meaning of demolition in its wider context as well as the more particular interpretation for some legal compliance. It includes demolition and partial demolition overview and some basic considerations. Meanings of words can vary in industry sectors and these will be mentioned to aid effective communication and help clarity.

Box 1   Key learning points

- Overview of demolition, partial demolition and structural refurbishment.
- Planning, procuring and managing considerations including the procure/source/build/maintain/refurbishment/demolish cycle.
- Understanding the importance of effective knowledge of the site or facility.
- Risk management and health and safety considerations in context.
- Principles of demolition, partial demolition and structural refurbishment.
- Approaches to decommissioning and environmental considerations.
- Structural stability approaches and demolition methods and techniques.
- Materials handling and processing; completion of the project and training.

Introduction

The perceptions of what may be called *demolition* are many, where the term can mean many things to different people – often depending on particular contexts. Indeed the differing perceptions may be fuelled by what could be considered a 'common sense' thought process through to that which may sometimes be promulgated in guidance by official bodies. The reason for the latter is usually where there is a need to create descriptions or definitions that might be required for compliance with particular legal requirements – or perhaps for codification to help with effective management – including for health and safety purposes, of course.

In its simplest form, demolition can be seen as the removal of material or materials from all or some part of a built facility, which very often will be some type of building or maybe some other type of structure, e.g. a bridge – or other facilities such as an energy plant. Hence this is an invasive action on a facility where hazards will arise with the need for appropriate considerations to manage those hazards and ameliorate any associated risks. As an example, for overall health and safety purposes, management would be refined to look at any related health and safety hazards, but, importantly, these need to be considered in context with other risk issues, such as environmental considerations. Thus health and safety management should not be considered in isolation for demolition activities because a wider approach can give significant gains through, for example, the fundamental examination of 'why demolish' rather than exploring a refurbishment strategy or approach. The latter may be more sustainable and thus give further benefits to the Client and others.

Demolition can therefore be whole or partial, small or large, and a significant part of a facility or a minor part. There may also be whole demolition of part of a facility, which may or may not be a significant part of the original. Most of the examples given here, where there is partial demolition, will be as a prelude to refurbishment. Where the material removed is structural and potential instability can be created, it is usually considered that structural refurbishment is taking place. Hence a well considered approach is required for any particular set of circumstances. Additionally, where a particular part of a facility is to be worked upon, appropriate decommissioning should be considered and planned into the process on a timely basis.

Historically, statistics have shown that the health and safety record during, or as a result of, demolition activities has been one of the poorest of any industry sector. Guidance has been produced by standards organisations and legislators. Additionally, professional bodies and trade bodies also produce guidance – often within a wider context but where health and safety management is a central consideration. Thus topics, covered in this chapter to varying extents, include:

- introduction, including whether demolition, partial demolition, structural refurbishment and decommissioning overview and basic considerations;

- risk management, including demolition is the best option, plus whole life thinking and sustainability;

- planning, procuring and managing, including the need for effective knowledge of the facility;

- principles of demolition, partial demolition including exclusion zones, approaches to structural demolition and retaining structural stability;

- approaches to demolition methods and techniques plus materials handling and processing;

- completion of the project and training.

The text also includes some waste issues including mention of site waste management plans (SWMPs), the Institution of Civil Engineer's (ICE) *Demolition Protocol* (ICE, 2008) and auxiliary works (for temporary states, e.g. as per Structural Eurocodes). The significance of including environmental issues is for information and also because of the crossover of the disciplines and processes for assessment purposes and some legal compliances. This requires that the overall management and execution of the demolition and partial demolition for health and safety activities are not undertaken in isolation but that other factors need to be considered also. This chapter is written in a way that is intended to minimise or avoid topics that will be mentioned more particularly elsewhere in the book, such as working at height (see Chapter 14 *Working at height and roofwork*) and other general health and safety issues (see Section 4 *Health hazards* and Section 5 *Safety hazards*).

What can be considered to be demolition?

What can be called demolition? What do you, the reader, think it might mean? Do others that you converse with think the same – do they agree? Would you know if they agree?

So why is it important to consider this? There are a number of reasons, that include clarification of whether certain activities may, for example:

- attract particular legal compliance, such as in criminal legislation;

- contribute to structural instability;

- expose potential health risks.

A common sense approach might be to consider it as the act of taking something down or perhaps the lesser option of removal of something from a built facility, even though the term *demolition* may not appear appropriate or obvious for some reason.

A particular concern that has been identified when this happens is when searches are performed, perhaps on the internet, for information and 'demolition' is not included. This can lead to highly relevant documents such as BS 6187: 2000 (BSI, 2000) on demolition not showing, thereby denying those involved important information that may significantly help them with their project – a critical omission.

> **Box 2** Example – façade retention workshop
>
> At a workshop on refurbishment attended by a dozen or so building professionals from a wide spectrum of the industry, discussions centred on large projects where most of the building would be removed with just the façades remaining. Those attending included senior people who were designers (both engineers and architects), developers in clients role, specialist contractors, management contractors, auxiliary works contractors and designers, all with extensive experience.
>
> The processes to be undertaken were being discussed when they were asked about their awareness of demolition and the British Standard on demolition, BS 6187: 2000 (BSI, 2000). Just one person was aware of it and even that person had not referred to it. The reason given by those present was that they did not consider demolition was taking place even though just the façades were to remain, with the rest of the building removed.
>
> *Why was this?* The reason given was simple. They did not consider that the extensive removal of most of the building was called demolition as some of the structure was remaining. Many words were used to describe what was happening – such as, removal, remediation, remodelling and, of course, refurbishment.
>
> *The consequence* of not thinking in terms of 'demolition' was that when searching for appropriate references for guidance, that important word was not used and hence the Code of Practice (BSI, 2000) did not show. Thus much valuable information and guidance was not made available to them.
>
> *Lesson:* it is important to ensure that understandings of words are clear to all.

There are two aspects to consider initially:

- Definition or description: there are very few definitions of 'demolition', including no definition in the British Standard for demolition, BS 6187: 2000 (BSI, 2000) – which was a deliberate decision for reasons given below. Some publications (Addis and Schouten, 2004) include a description, however.

- The industry sector in which demolition activities reside is construction, and this may lead to inappropriate communication and possible omission of proper and effective considerations. Within the health and safety context and more particularly within the Construction (Design and Management) Regulations 2007 (CDM 2007) 'demolition' is clearly within the 'construction sector' and is seen as a 'construction activity' within the construction industry (HSE, 2007). Hence, this can sometimes lead to demolition activities not being thought of when 'construction' is considered in its wider context and needs to be addressed to ensure that it is – or is not – relevant to matters being considered.

Thus questions to ask oneself include:

- What do I understand it might mean?
  … and particularly when communicating with others …

- Do we *assume* that we all agree … because we don't question what the terms mean to those using them?

- Have I made sure that everyone is clear and understands?

As a guide, the Health and Safety Executive (HSE) has suggested for CDM 2007 compliance, purposes that for activities to be considered demolition, they should relate to a substantial part of a facility, although substantial was not described or defined. In the past, HSE has stated, in an attempt to help Duty Holders comply with CDM 1994, that demolition relates to the removal of 2 tonnes or more of material, although this was later withdrawn as being inappropriate. The difficulty in defining demolition, however, has been cited as one of the reasons why no specific requirements were included in CDM 2007. As a background, the view was that demolition activities attract requirements just as do other 'construction' activities for appropriately assessed health and safety management and do not need special categorisation.

From a hazards/risks point of view it could be seen that identifying the potential for structural instability and managing this together with any resulting residual structural instability are the essential criteria, bearing in mind any potential health hazards as well.

A description that has been offered for demolition is:

The process of intentional dismantling and reduction of a building, or part of a building, without necessarily preserving the integrity of its components or materials. (Addis and Schouten, 2004)

This can be seen as a wide-ranging description that can include non-building structures and brings in a sustainability aspect. It can also apply to a variety of project types, significance and size, of course.

It could thus be seen as going beyond the scope of application of CDM 2007:

For pragmatic purposes however, demolition can be seen as a process – and one which involves removal of some or all of a facility, the significance of which will have consequences that need to be assessed and controlled, that is, effectively managed within given criteria.

Sometimes, however, words are used that, in a broader sense, can mean or include a form of *demolition* and there is a need to be aware that this may not be immediately evident to those involved.

Thus, to summarise, the scope can be significant, perhaps:

- the whole of an existing facility … what size?

- the major part of a facility … how significant – could it be part of existing facility?

- smaller demolition … but how small – small structures/buildings?

- work on a structure … that may or may not give rise to health risks;

- work on a structure … that may or may not affect stability.

Hence, it can be seen that refurbishment and renovation activities may include demolition activities – such as structural refurbishment.

What is meant by structural demolition, partial demolition and structural refurbishment?

When looking at **Table 1** it becomes noticeable that demolition can occur in many situations apart from the wholesale (full) demolition of facilities. This is often known as partial demolition and can be seen to take many forms and with a variety of terminology.

Partial demolition can be significant in many ways within the context of health and safety risk management, particularly as:

Refurbishment	Deconstruction	Alterations
Renovation	Dismantling	Augmenting
Rebuilding	Disassembly	Conservation
Remodelling	Demounting	Enlargement
Reconstruction	Development	Extending
Redevelopment	Partial demolition	Modifications
Restoration	Upgrading	Structural alterations

Table 1 Examples of some terminology used in the construction industry where there may be demolition activities

- Any poorly managed removal of materials may result, to a lesser or greater extent, in unplanned:
  - safety hazards such as collapse of materials (and thus including structures or parts thereof)
  - exposure to health hazards.

- The extent and type of activities on existing structures tends to be more than for full demolition.

- People tend to work much closer to the work activity and within the facility and thus not as remotely as with full demolition projects.

Hence such partial demolition can vary in extent, particularly where there is structural refurbishment and this could include the removal of, for example:

- the entire building except the façades;

- all or part of a wall known to be load bearing;

- all or part of a wall *not* known to be load bearing *but is*;

- the parapets of a bridge and its supporting structure in readiness for widening;

- a whole wing of a building (or more) and which may be giving support to part of a remaining structure.

In all instances, the effects of removal need to be known as part of the management strategy.

What is meant by decommissioning?

Decommissioning is a term that tends to be associated with specific industries and perhaps not thought of as applicable where any demolition or partial demolition activities are considered, such as outside those industries. The usual, and simplified, meaning of decommissioning could be seen as the taking out of service a facility, or part thereof, and rendering it suitable for being worked upon – such as for demolition or refurbishment activities. That state would need to be documented and all relevant information made available in good time to those about to tender for and then undertake the works.

However, there is much more involved than that simple description which can be considered as applying to what might be thought of as land based environments such as from 'ordinary' premises through to petro-chemical facilities in varying degrees of complexity. There are more specialist applications for the nuclear industry.

Box 3   Example of partial decommissioning

This example looks at health risks from a small facility where there had been partial decommissioning in that safety hazards such services had been dealt with – but the health hazards had not been considered.

A three-storey row of terraced Victorian houses was to be demolished in a piecemeal way to aid recovery of materials for later reuse. Some of the houses were still residential and some had been converted into offices. One of those that was in use as an office had however had a previous use. The use had been a dental practice where some of the rooms had been converted for use as surgeries before later further conversion to office facilities. Residual health hazards from the activities, such as the use of mercury, had not been considered, nor had the limited application of asbestos been considered.

Why did this occur? Those involved in planning the work had not considered previous uses of the facilities adequately.

The consequences were that the potentially harmful materials were discovered during the demolition process when workers removing parts of the structure noticed the materials mentioned above. They acted responsibly and reported what they saw and work stopped, pending further investigation. Appropriate measures were then adopted to deal with those materials.

The lessons learned include ensuring that adequate investigations are undertaken into previous uses of facilities.

Hence other meanings of the word, for example in the nuclear industry, extend the application to the removal of materials – and thus into the meaning of demolition. In the offshore oil and gas industry, the expression 'abandonment' is often similar to the use of 'decommissioning' in the nuclear industry.

Whereas scale and extent of decommissioning are relevant, it could be applicable in most cases where there are any demolition activities and should be considered an integral part of planning for any job, although the significance from a risk point of view will vary considerably.

Understanding the 'industry' – the context of demolition

Demolition is undertaken by many organisations from, for example, those that concentrate solely on demolition projects, such as demolition contractors, through to those that might primarily have some other role and undertake small amounts of demolition (in its wider sense), such as those involved with building services. Demolition, in terms of numbers of sites, is often undertaken by, or under the auspices of, organisations whose principal activities are centred on constructing facilities – and not demolition. These will include the constructor and also perhaps design companies, for example.

Within the context of demolition, the term 'demolition industry' is sometimes used. It is helpful to understand the background to this term, the way in which it tends to be used and thus what is often meant by those using the term. This will provide a useful insight into demolition historically and the way it is helpful to extend the thinking beyond those more traditional approaches.

There are many bodies that support those undertaking demolition activities, many of which are membership organisations. These have a wide range of interests from demolition in general, to those with specific interests in elements of demolition processes, such as from concrete cutting and sawing, to those focusing on perhaps health hazards from specific types of fibres such as asbestos (see HSE Asbestos webpage for further information) where, for example, particular requirements apply. Their interests may also extend beyond demolition processes, however.

The term 'demolition industry' tends to be used by the major contractors associated with demolition who often undertake set-piece wholesale demolition of structures. They are a valuable part of the sector with extensive experience, often with a lineage back to World War II for war damage clearance, where many started life as family businesses – with some still in that position. Some within this grouping see themselves as a separate sector to the 'Construction' sector because of the different processes involved – and, of course, the idea that the activities are often the reverse of those of creating (or building) a structure or other facility. This thinking can, however, have adverse effects as mentioned above.

Many contractors in the demolition industry are members of a trade body. That body, the National Federation of Demolition Contractors (NFDC), promotes itself as the voice of the UK demolition industry. It has a number of objectives which include representing the industry, particularly its member companies of course. Not all demolition contractors are members however.

Another organisation – one that is based on membership of individuals, rather than companies or organisations, who are involved with aspects of demolition processes – is the Institute of Demolition Engineers (IDE). This exists to promote and foster the science of demolition engineering. The main objectives include:

- being a qualifying body for individuals;
- encouragement of safer methods of working.

There are many involved with demolition and ancillary processes, who are fundamentally important to the process, but who may not regard themselves as part of the 'demolition industry' and indeed may be seen by those in the 'demolition industry' as outside the industry. What is relevant, of course, are the activities involved and appropriate competencies and awareness of possible misunderstandings which in this context could lead to detrimental health and safety consequences.

What is required – and does exist to some extent – is a view that there are many involved with demolition processes such as clients, procurers, managers, specialists, contractors, etc., some of whom consider themselves to be part of the industry and others not! They may perhaps be in a sub-sector. The facts are that many people, trades and professions can be involved with demolition processes whether whole or partial demolition, including structural refurbishment and decommissioning and are mentioned in this chapter for health and safety to give a broader overview.

Evolution of guidance

As an historical background, guidance written for demolition activities with health and safety in mind has evolved since the first code of practice was published in 1971. Publishers of such guidance include the British Standards Institution (BSI) and the British Government departments that dealt (or deal) with occupational health and safety – the former Department of Employment (DE) in which the HM Factory Inspectorate resided and which was followed by the Health and Safety

Executive (HSE) and the Health and Safety Commission (HSC). (Note: HSC merged with HSE in 2008 to form a body known as the Health and Safety Executive where the former HSC could be seen as the newly formed Board of HSE.)

Historic chronology of some principal publications, include:

1971	CP 97: 1971 *Demolition BS Code of Practice* (BSI)
1973	DE booklet 6E – *Safety in Construction Work: Demolition* (DE)
1970s (mid)	Joint Advisory Committee (JAC) *Report on Demolition* (JAC set up in 1973) (HSC)
1982	BS 6187: 1982 *Code of Practice for Demolition* (BSI, 2000)
1984/5	HSE *Guidance Notes GS29 Parts 1–4* (HSE)
	o *Part 1: Preparation and Planning*
	o *Part 2: Legislation*
	o *Part 3: Techniques*
	o *Part 4: Health Hazards*
2000	BS 6187: 2000 *Code of Practice for Demolition* (BSI, 2000)
2005	BS EN 1991-1-6: 2005 *Eurocode1 Actions on Structures. General Actions. Actions During Execution* (which includes demolition and refurbishment in its scope) (BSI, 2005)
2008	NA to BS EN 1991-1-6: 2005 *National Annex* (BSI, 2008)
In preparation	(revision of BS 6187: 2000) *Code of Practice for Demolition and Partial Demolition for Structural Refurbishment* (provisional title)

Some of the above have since been withdrawn, but it shows the development of some relevant guidance. There are also many other publications, including from further publishers, with scopes of content that relate to demolition and structural refurbishment activities.

Risk management approach – is demolition the best option?

The risk hierarchy

As part of an holistic risk management approach, health and safety is just one (but patently an important one) of a number of risk factors that need to be thought through when considering projects. Project Clients may, for example, wish to have a green image and include sustainability principles in works they may be planning, perhaps over and above legal requirements (see **Figure 1**). Indeed there is increasing legislation that needs to be considered when developing projects. There can be direct implications for health and safety, however, so the significance of including environmental considerations is essential. As part of complying with the laws there are the related associated costs for

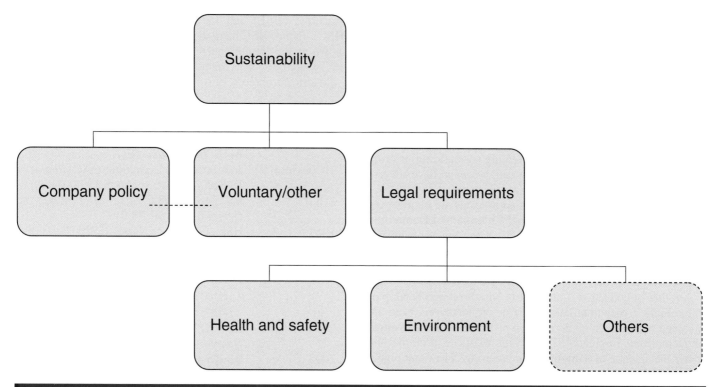

Figure 1 An example of the sustainability influence in context of health and safety

the disposal of materials, for example, to be considered. An integrated approach is required, therefore, for a professional balanced project with the outcomes meeting the client's objectives, including compliance with both health and safety and other legislation.

Examples of where other legislation may need to be considered, depending on circumstances, could relate to roads and highways, utilities, buildings, party walls, rail and waterways. Those involved for various reasons might be the police, local authorities and others, as necessary, with activities such as appropriate liaison, compliance and obtaining of permissions, for example (see **Table 2**).

As examples, additional environment considerations where removing materials could include, where appropriate:

- waste management including waste disposal and SWMP responsibilities;

- prior agreement with the local environmental health departments in respect of environmental noise;

- application to the local authorities for licensing of crusher or screening plant;

- prior consents or other agreements with the relevant enforcer in respect of the disposal of groundwater or waste water and other fluid discharges.

Whole life cycle thinking

Whole life cycle thinking can be central to considerations for demolition and refurbishment projects in the planning and execution, including design parameters, as part of the sustainability influence (see **Figure 2**).

Understanding the high hazard to low risk imperative

Demolition activities are known to have the potential for high safety risks as mentioned elsewhere. When undertaking risk assessments, the balances of likelihood and consequences – of safety and health and other aspects – need to be considered. Often where there are demolition activities, however, the effectiveness of that balance can be critical upon the management regimes put in place during execution. Even then, consequences of an event can be very high even though the likelihood under a well managed project might and should be low. The other side of this can be seen where the risks are known to be low – but still possible – and there can be a tendency to reduce the overall resource, perhaps during planning and execution. Resulting 'events' can happen, causing problems on site which in themselves may be of 'low consequence', but are still unwelcome. When undertaking refurbishment and demolition activities some of these can have significant 'knock-on' effects and more serious consequences, however. Often, the consequence chain is not fully considered for a variety of reasons, such as:

- lack of knowledge of the facility, e.g. of what is actually there, albeit perhaps hidden;

- lack of knowledge of potential load paths resulting from intended and perhaps unintended actions, possibly where the likelihood of those unintended actions were not assessed;

- lack of knowledge of potential health hazards resulting from intended and perhaps unintended actions, possibly where the likelihood of those unintended actions was not assessed;

ICE manual of health and safety in construction © 2010 Institution of Civil Engineers

Examples of primary acts and orders in the UK where permissions are required:

a) Town and Country Planning Act 1990 (Tree Preservation and Conservation Areas Orders are covered in Section 197 et seq. of the Act)

b) Planning (Listed Buildings and Conservation Areas) Act 1990

c) Building Act (England and Wales) 1984

d) Building (Scotland) Acts 1959 and 1970

e) Highways Act 1980

f) Roads (Scotland) Act 1984

g) New Roads and Street Works Act 1991

h) Party Wall etc. Act 1996

i) Building Regulations (Northern Ireland) Order 1979 (as amended 1990 and 2009)

j) The Planning (Northern Ireland) Order 1991

k) Planning (Northern Ireland) Order 1991

l) Planning (Demolition – Description of Buildings) Direction 2006

m) The Roads (Northern Ireland) Order 1993

Note: currency and relevance should always be confirmed. See also Section 1, Chapter 1 'Legal principles'.

Table 2 List of examples of primary acts where permissions are required

■ lack of appropriate competence to varying degrees, possibly anywhere in the procurement and execution chain.

These issues patently need to be managed to positive effect so that a sensible and appropriate risk management culture prevails. An approach that could attract consideration is the approaches of the 'precautionary principle' (Europa, 2005), where the consequences of a failure may be severe, but the likelihood may be low. Caution is required when considering this approach, however, to ensure that criteria used and application of the principle is wholly appropriate for any particular application.

What are the various up-front higher level options to minimise risks?

Applying the principles of the risk hierarchy (see Chapter 1 *Legal principles* and Chapter 9 *Assessing safety issues in construction*) means that if the demolition activities can be avoided then the associated health and safety hazards and risks are managed away. To give this premise a more realistic approach, however, alternative strategies should be considered depending on the context of the proposed demolition activities.

There can be many alternatives including variations on, for example:

■ complete demolition

■ partial demolition

■ refurbishment with structural interventions

■ refurbishment without structural interventions

■ refurbishment of partially empty premises perhaps with structural interventions

■ refurbishment of occupied premises (Fawcett and Palmer, 2004) perhaps without structural interventions.

Such considerations should include many aspects including those in 'Planning, procuring and managing' below and elsewhere in this chapter and this book, including sustainability.

Additionally, safety issues such as possible structural instability (of any size) may lead to possible health hazards such as the release of fibres that may be potentially carcinogenic in form and quantity for example.

Examples of the main health hazards to refurbishment and demolition workers include those that arise from substances likely to be inhaled or ingested or likely to react with or to be adsorbed on the skin. Hence, it could be seen that the less remote the activity, the higher the likelihood of exposure to any such hazards. Also, part of the process is for contaminants on sites to be identified, which may include those present as a result of either previous uses of the site, or abandonment of materials such as containers, tanks, dust and debris including perhaps used syringes and organic deposits, e.g. bird droppings.

Planning, procuring and managing

As refurbishment and demolition are undertaken on a very wide range of types of structures, other facilities and site variations, there are many key considerations to be addressed. They should be identified as part of the process

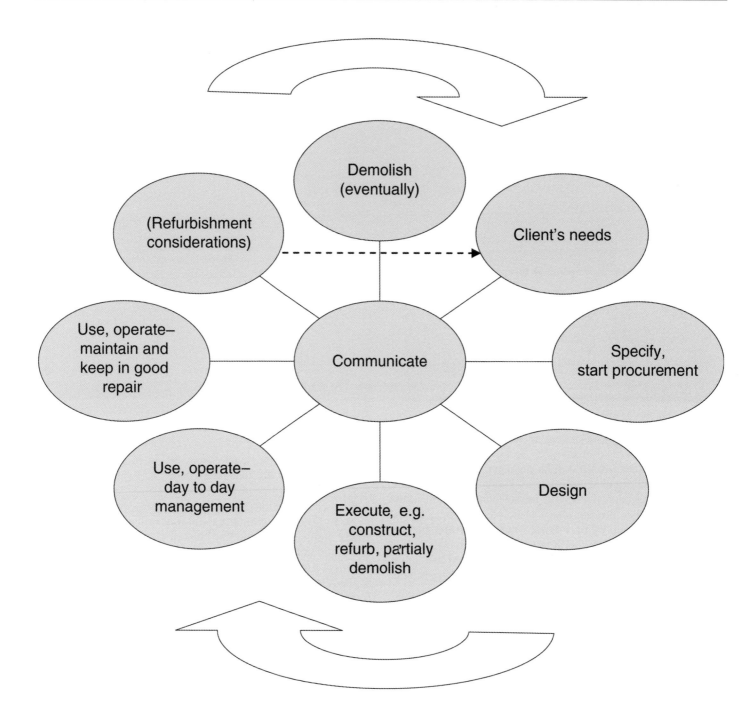

Note: considerations may include managing and designing for:
- potential future needs of the Client
- refurbishment to meet the most appropriate of Client's options
- deconstruction
- use of reused, reclaimed and/or recycled materials

Figure 2 Simplified procure/construct/maintain/refurb/demolish cycle

of looking at various options and then taken forward into the planning for the proposed works. Thus these should be considered, as appropriate, at many stages from scheme concept, through scheme development, tender development, tendering procedure and contract awarding procedures, after which they will extend to the execution processes with appropriate contract management and work on site – all to varying degrees.

Box 4 Real case history: recovery of slates from a public assembly building

A well built Victorian masonry public building with a large audi-torium was to be completely demolished. It fronted a busy pub-lic highway with public access passages to the sides and with a car park to the rear. High quality slates adorned the roof and these were to be recovered for selling on for reuse. Other materials would also be reclaimed for future use. Effective methods needed to be planned and put in place to enable safe recovery, bearing in mind the safety issues of working at height, handling and storing the materials and the health issues of working close to the slates and supporting timbers, for example.

Why was this? The further use of recovered materials for both business and environmental considerations was a major driver in determining the methods of demolition employed.

The consequences of this were that demolition by high reach machine was not employed, which would otherwise have obviated the need for people to work at height (and close to the materials).

Lessons: if remote means of demolition are precluded from use and thus keeping the workforce remote from the greatest hazards is also precluded, the application of the hierarchy of risk reduction should be applied to achieve the next most appropriate solution, bearing in mind that extra management controls and competen-cies may be required for a greater workforce which will be em-ployed in more hazardous environments.

Thus when planning refurbishment and demolition proj-ects, some considerations that should be assessed are included below, where the examples are broad and not exclusive and should be read in conjunction with each other and elsewhere in this book, as appropriate.

Project/programme management

It is prudent for the initiator of the refurbishment and/or demo-lition project to be aware of any direct responsibilities they may have, including any duty of care. The initiator of the work will usually be the owner of all or part of the facility, although there may not be sole occupancy with a number of occupiers pres-ent. The initiator will usually be referred to as the Client, who would be providing the funds. It should be noted, however, that others may be known as a client within the broader procure-ment activities and organisations further down the supply chain. Depending on their own competencies, clients may be advised to procure professional advice, as appropriate, which is also a requirement of CDM 2007.

Considerations should include many aspects, some of which, for example, are:

- the overall risk issues mentioned above;
- the site, location and extent of the works, plus
  - a detailed programme with timetable of events;
- extensive knowledge of the site;
- local requirements relating to, for example:
  - working arrangements and notices prior to commencement of work by the those undertaking demolition and refurbishment activities
  - requirements at site boundaries such as for hoardings and ar-rangements for protecting the public
  - restrictions and/or requirements for access to and egress from the site
  - the action to be taken to protect specified features of historical, or archaeological importance, for example
  - prescriptive rights of support or way, etc. and the need for, and execution of, any party wall agreements, for example;
- the specification of the works which should be clear and unam-biguous and if relevant may need to include, for example:
  - particular sequencing requirements and perhaps method of pro-posed demolition and partial demolition
  - requirements for debris management including likely types and amounts (e.g. ICE, 2008)
  - the decommissioning report for the site, the results of all rel-evant site surveys, including Asbestos type 3 surveys (see HSE, 2009, for current criteria) for example, plus any specific con-straints imposed on working methods as a result
  - a list of all documents, plans, notices and information obtained by the facility owner for incorporation in the contract, etc.
- further considerations include, as appropriate:
  - details of the machinery and other plant required, including the capacities, locations of use (BSI, 2005, 2008), etc.
  - traffic arrangements including site plan and detailing for the control of site traffic (HSE, 1998) (also see Chapter 19, *Transportation and vehicle movement*)
  - arrangements for protecting the public
  - temporary storage on-site of materials (BSI, 2005), manage-ment and removal (including reuse, recycling, recovery and disposal) where appropriate
  - contingency arrangements in case of, for example, partial col-lapses or perhaps misfires when explosives (BSI, 1998, 2000) are used (also see Chapter 20 *Fire and explosion hazards*).

Box 5 Real case history: small job + auxiliary structures

A father and son team of builders were known to work well within a local community. They decided to accept a job to work on a terraced 'two up, two down' residential property where the Client wanted the ground floor rooms knocked through into one. They had removed a significant part of the internal non-load-bearing wall between the front and back rooms where there was no evi-dence of any temporary support! At the back of the house, there was a single-storey annex. Work was being undertaken to knock through the wall that was the full height rear wall of the property into the single-storey annexe.

A catastrophic collapse occurred. There was evidence of temporary support having been installed prior to the collapse,

---

realising that they were breaking out part of a load-bearing wall. The forensic engineering investigation revealed temporary (or auxiliary) components of four adjustable screw jacks and two old rusted long I-beams (or rolled steel joists (RSJs)) with no evidence of fixity between them and with no evidence of bracing or lacing either.

Why was this? It appeared they had not appreciated the loads they were dealing with, nor the way in which they would act while they were working on the structure.

The consequence of working outside the bounds of their knowledge was that neither of them went home that day – or ever again.

Lessons: work should be planned and executed within known competencies and not extrapolated from successful experiences to other applications by unfounded assumptions.

Effective knowledge of the facility

A thorough knowledge of any site is a fundamental prerequisite to the completion of a successful project in respect of safety and health (including physical and health hazards), environmental considerations and efficiency. In this context 'the site' includes everything above, at or below ground level and within the area under consideration. The site should be assessed and surveyed sufficiently such that all relevant information is established in order to provide on a timely basis a clear understanding to those needing it. Although it may seem obvious, it can be seen that those receiving the information are expected to have the competence to understand it and act accordingly. The type and extent of information may vary at different stages of the project, but the basic premise is that most will be needed at the early stages for planning purposes.

This knowledge can be elicited in a number of ways, including a desk study with on-site information such as surveys and perhaps discussions with current and former occupiers to augment the desk study. An initial 'look around' to get a 'feel for the facility' is often appropriate, although caution is required for many reasons including possible poor conditions. Also, lone working for this activity is not usually recommended – again, for many reasons. It is also helpful to include information on subterranean, boundary and off-site features that may affect work on site.

Considerations for knowledge of the facility may include, as appropriate:

- structural form and materials including:
  - current state and condition;
- extent of decommissioning, including:
  - any residual materials;
- isolation, condition and protection of adjacent structures;
- details of the isolation and/or removal of services relevant to the work site, etc.;
- ground conditions, including:
  - previous uses
  - current state and condition, including any residues and contamination
  - ducts, tunnels, apparatus and equipment.

Surveys

The term *survey* can be misunderstood and thus result in miscommunications with the potential for significant detrimental health and safety outcomes. Hence, the purpose of any survey to be undertaken should be quite clear to both to those requiring the survey and also to those undertaking the survey. They could be, for example, for the structure, utilities, topographical features, geotechnical aspects, fibres such as asbestos, state of decommissioning and/or environmental issues. Surveys may need to be carried out around the site perimeter both inside and outside, such as for utilities.

Where 'surveys' are for a structure, the term 'structural survey' is not recommended (see CIC, 2004). This is because of the potential for misunderstandings and because the precise nature of what is required may not be achieved. A more specific term should be used, depending on the purpose which, for example, could be for:

- valuation
- structural inspection
- structural examination
- structural assessment.

Incidentally, where surveys are for valuations, again clarification should be obtained for each occasion as they can also be for different purposes.

How outputs from the on-site surveys are to be used should be considered before commissioning them in order to help with future use. For example, it is recommended that the findings from site surveys are produced in a report that details the site and buildings, the current state at that time which includes the true decommissioning status as well as any other relevant information.

Generally, full decommissioning should be achieved prior to the commencement of any refurbishment or demolition activities; however, where the site is not fully decommissioned the measures needed to achieve full decommissioning indicating any residual safety and health hazards for future attention should be included in a decommissioning action plan in a clear way.

Where time has elapsed since the above had been prepared, however, the state of the facility may have altered in a deleterious manner (Briggs et al., 1997). Hence further surveys are required to determine the (new) current state and any new residual health and safety hazards which may in turn lead to the need for reassessed risk management plans.

Principles of demolition, partial demolition and structural refurbishment

Safe and healthy approaches

As an overview approach when planning methods of work, suitable and sufficient risk assessments need to be carried out and recorded with methods, materials and equipment selected to remove or minimise risk from work activities.

Decommissioning should be undertaken before the main works start, however, and a holistic approach to consider as appropriate includes:

- planning and preparation

- devising the decommissioning procedures and activities

- creating the decommissioning action plan

- the decommissioning activities

- writing the out-of-service decommissioning report

- dissemination of the report.

Tenderers and appointed contractors that are selected are expected to look for the hazards in the job, including how and where they could be and what equipment, materials and methods of work should be used.

For demolition and refurbishment work, safety by position is a fundamental consideration. Thus safe working spaces and exclusion zones should be considered and provided for all refurbishment and demolition work with appropriate exclusion zones implemented and managed. The purpose is to protect persons so that they are not harmed as a result of any demolition activity including any handling or processing of materials. Harms to be considered include physical, biological or chemical hazards and the effects of dust, vibration and noise (see Section 4 *Health hazards*). An essential feature of safe and healthy approaches to activities is to select appropriate machinery that will help to achieve the desired and planned outcomes for a good job. Incorporated into that will be effective management of operations, including suitable competencies of drivers, for example. Many different types of machine can be used for a variety of activities, but a particular type of machine that is often referred to as a 'demolition machine' has principal component parts that are described in **Table 3**. Those parts can be identified in the diagrammatic illustration of such a machine at **Figure 3**, although there can be variations.

The less remote the activity, the higher the likelihood of exposure to any hazards considering that the main health hazards to refurbishment and demolition workers arise, as mentioned previously, from substances likely to be inhaled or ingested or possibly likely to react with or be adsorbed on the skin. Hence, contaminants on sites should be identified and taken into account. They can include those present as a result of either previous use of the site, or abandonment of materials such as containers, tanks and dust. Debris could include those mentioned previously and organic deposits, such as bird droppings, particularly if the facility had been taken out of service sometime before refurbishment or demolition are planned to start.

Component parts of exclusion zones based on BS 6187: 2000 are:

- footprint of work area

- designed drop area/zone

- predicted drop area/zone

- designed buffer area/zone.

Designed exclusion zones have outer extremities coinciding with the outer boundary of the designed buffer area/zone, although the applied location may vary (safely) beyond that for practical reasons.

It is helpful to consider the extent, shape and volume exclusion zones to be variable and dependent on the demolition activity and rate of progress. It can cover part of the site or even extend beyond the site boundary. In the latter case the consent of the adjoining owner(s) and perhaps occupiers should be obtained. Temporary traffic management (TTM) may be required; also see Chapter 19 *Transportation and vehicle movement* for further information on TTM.

Where there is restricted space on sites, e.g. within the facility or perhaps in city centre locations, the extent of the designed exclusion zones could be reassessed if there is a need to reduce the size where this could be achieved through adopting different methods of demolition or perhaps by containment, for example but only if they are adequate.

Considerations when determining safe working spaces and exclusion zones include:

- Adoption of safe working spaces and exclusion zones as part of the safety regime.

- The particular purpose of each safe working space and exclusion zone.

Attachments		Examples include: pulverisors/crushers, shears/cutters, grapples, hammers
Equipment	Dipper arm	The part to which individual attachments are connected
	Optional extension	Sometimes added between boom and dipper arm, usually to extend the reach of the machine
	Boom	Attached to base unit, supporting the optional extension and dipper arm
Base machine		Power unit with protected cab, controls and travel system

Table 3 Nomenclature of the principal parts of a demolition machine

Figure 3 Illustration of a demolition machine at full height to show the principal component parts as described in Table 3
Image courtesy of J C Bamford Excavators Limited. © J C Bamford Excavators Limited

- Exclusion zones around site vehicles and plant activities.
- The application of safe working spaces and exclusion zones:
  - designing safe working spaces and exclusion zones for each application.

On rare occasions where personnel are directly involved in the execution of refurbishment and demolition work they need to be within the exclusion zone; however, they should be located in a position of safety which relates to the stage of demolition to be extended. It should be assessed within an overall planned and effectively managed safe system of work. These personnel should not be permitted nearer the work than within the designated buffer area. Such work should only be permitted when it is inappropriate to work from outside the exclusion zone and when safe means of access and egress are implemented, for instance. Examples include:

- Demolition machine driver in a suitably protected cab environment, such as can be seen in **Figures 3** and **4**.
- Shot firers for explosive demolition within a suitably protected environment.

Approaches to structural demolition

Managing structural stability and managing structural instability could be seen as key issues. Removal or collapse of materials is a fundamental requirement of the process and the consequences of each and every part of that process should be where the actions on the facility or structure should be foreseeable and thus managed appropriately. Thus, unplanned collapses should be avoided by planning suitable methods and sequences of demolition activities, including those for structural refurbishment projects.

The nature of the structure, the interdependency of elements and other relevant criteria should be taken into account when selecting the most suitable methods. Methods of work should be clear and incorporated into documentation such as method statements which should aid competent management of the processes on site.

It can be helpful to think of unplanned collapses as the collapse of part of a facility rather an entire facility, although the latter can happen, of course. Such collapses can be thought of generally in terms, as including for example:

- premature collapse which is unintentional at a particular time because of inadequate residual structural integrity, perhaps when pre-weakened; and/or
- unintentional collapse of something not intended for demolition and occurred because of inadequate stability.

When these collapses occur, it is sometimes found that the size of the collapse can be disproportionate to the initiating event and, for example, the actions in one place were

Figure 4   Image of a 21 m high reach demolition machine working with a 2.5 tonne shear attachment
Image courtesy of Volvo Construction Equipment Division

not recognised as having the potential to cause instability elsewhere.

Thus load paths should be understood and appreciated throughout. This will help to avoid unplanned collapses – however small or large – and will aid the strategic use of inherent forces to aid efficient planned demolition of materials. The basic ways structural demolition could be considered, for example include:

- progressive demolition which is considered to be the controlled removal of parts of the structure while retaining the stability of the remainder at a particular time, by design;

- deliberate collapse mechanisms which are considered to be the removal of key structural members and perhaps weakening of others in order to precipitate or cause collapse, by design;

- deliberate removal of elements which is considered to be the removal of selected parts of the facility by deconstruction or

dismantling, for example, by design and for possible different reasons.

The above principles can generally be applied whether there is full or partial demolition, including where there are refurbishment projects with structural alterations to be carried out.

BS EN 1991-1-6: 2005 *Eurocode 1. Actions on Structures. General Actions. Actions During Execution* (BSI, 2005) gives significant relevant advice on actions to be considered during execution and specifically mentions demolition and refurbishment in its scope of application. Accidental actions are considered more fully in another code (BS EN 1991-1-7: 2006).

Retaining structural stability

Managing structural states or stability means dealing with changing circumstances and is required for avoiding unintended collapses. As well as the criteria above, provision of additional structural support which can be required to resist forces in any appropriate direction may be a beneficial option during the demolition or partial demolition processes. Such support should be designed and installed in a timely way to assist by providing, or complementing, appropriate often changing load paths. Support could be for inclusion in the final project or not intended to be part of the final job but can assist during the job. These are examples of the use of designed auxiliary support works, which are sometimes referred to as temporary works (also see Chapter 17 *Falsework* for safety hazards associated with temporary works).

These works can be in place for some time, particularly if a project is commenced but then not completed for reasons such as changes in the market and a resulting loss of funding. Examples are perhaps where a building has been demolished apart from its façades where façade retention schemes (Lazarus et al., 2003) have been constructed and left for many years, perhaps decades before work on the main project is recommenced – hence the term 'temporary' may not be seen as pertinent with 'auxiliary' seen as a more appropriate term for those supporting structures. Other auxiliary works may be very short term such as, for example, the strengthening of an element to help the reaction to an explosive charge.

The effects on the auxiliary temporary structural supports can be complex and should be assessed at suitable times by appropriately competent people to ensure that they remain adequate for any varying load patterns. These may include, for example, any induced reversal of stresses in the structure as demolition activities progress. Hence the need to understand the structure under varying load conditions is seen as important. There has been mention that a role of a 'Temporary stability coordinator' (HSE, 2004) should be considered.

Sustainable approaches

Clients and contactors should be encouraged to consider waste minimisation and recycling during the refurbishment and demolition planning stages and implement systems to prevent downstream breaches in environmental legislation such as duty of care and illegal disposal of waste. The *ICE Demolition Protocol* (ICE, 2008) can be used to demonstrate good practice in terms of environmental management at the outset, providing a transparent method for describing the materials and products available for recycling and reuse.

Additionally, the implementation of a SWMP at pre-demolition audit stage can help with identification of features such as contaminated hazards as well as potential recycling, and perhaps, reuse opportunities.

Approaches to demolition methods and techniques

There is a wide variety of demolition techniques for consideration (e.g. BSI, 2000; NFDC, 2006, 2008), with new techniques and technologies being developed, including for structural refurbishment projects. A number of methods can be used in combination or for different parts and at different times on sites. Methods of proposed work should be established after completion of suitable assessments, including consideration of various potential plant and technology options and where each will require specific precautionary measures to be established to enable the operators to execute safe activities. Safety precautions relevant to the methods of work should be clearly shown in the method statements for each stage of the works. All personnel on site should be fully aware of the requirements of the method statements, including as they develop and vary as work progresses. A preferred option for some is the incorporation of method statements on drawings related to the proposed works.

Choices should be based on minimising health and safety risks irrespective of which demolition methods are adopted. All personnel should be appropriately trained in the use of any plant and machinery for each appropriate demolition technique with which they will be required to be involved. Additionally, the choice of techniques should, in general, be considered bearing in mind the reuse and/or the recycling of materials arising from the demolition (see also the *ICE Demolition Protocol* (ICE, 2008)).

Whereas remote demolition, e.g. by machine, is generally the preferred option, hand demolition may be appropriate in some circumstances, such as for recovery of materials for reuse (see **Figures 5** and **6**). To assist with any hand demolition, however, mechanical assistance devices should be considered. These could include, for example, lifting devices for the lifting and lowering of elements once they have been released. The outcomes of risk assessments should indicate the extent to which hand demolition should be utilised, as well as any such assisting devices.

Completion of the works

The site, including various parts of the site, should be left in a safe and secure condition on completion of the refurbishment and demolition works. At the end of the job a health and safety file should be produced and include all the relevant information. Where hazards may remain and the preferred option of removal has not been implemented, they should be identified and recorded, including position. However, where they remain, they should be left in a condition such that they do not present any hazard to health or safety or to the environment, for example.

The role of the CDM 2007 Duty Holders and the health and safety file is covered elsewhere in this book (Chapter 3 *Responsibilities of the duty holders in construction design and management* and Chapter 5 *The different phases in construction*, respectively).

Training and competencies

The requirements for personnel to be competent can be assisted by involvement with the activities of various organisations and some examples are given below.

Figure 5 Image of small demolition machine working internally and handling demolished materials
Image courtesy of Bobcat. © Bobcat

The Institute of Demolition Engineers (IDE) offers individual membership to professionals involved in demolition engineering. Through seminars, regional meetings and speakers to outside organisations it encourages demolition engineers to further their continuing professional development (CPD).

It also advocates that demolition engineers join the IDE and gain accreditation by qualifying to progress through its grades: from Enrolled (basic grade) to Full Membership. A further grade, Fellowship, is possible following distinguished service to the industry and IDE.

The main parts of the cutter-crusher are shown below.

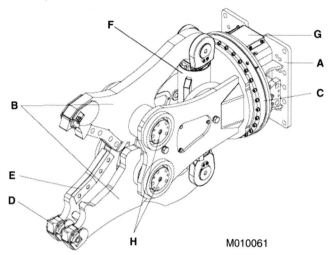

M010061

A. Mounting bracket

B. Jaw

C. Slew bearing

D. Front tooth

E. Cutting blades

F. Cylinder

G. Brake/hydraulic motor

H. Main shaft

Use crusher teeth for crushing and cutting blades for cutting only.

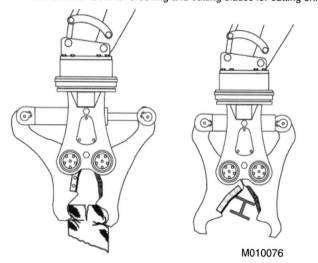

M010076

Figure 6 Example of a cutter-crusher jaw attachment and component parts (Sandvik BC2525 D-JAW, S-JAW) (images courtesy of Sandvik. © 2007 Sandvik Mining and Construction Oy)

Applicants for entry into the IDE (Enrolled grade) are expected to have at least three years of experience in the demolition industry at managerial or senior supervisory level. The knowledge gained through this experience is put to the test in the IDE's examinations which should be taken within two years of joining. The examinations consist of a multiple-choice paper which requires knowledge of current legislation and guidance as they apply to demolition activities and, on the same day, a three-hour written paper in which the candidate is expected to display knowledge of demolition related procedures and management.

Following success in the examinations, members are required to spend two years in the Associate grade. During this time they keep a diary of their role in demolition related projects and gather a portfolio of associated evidence, including training undertaken and testimonials received. The diary and portfolio, together with an updated *Curriculum Vitae*, are submitted to the Membership Committee, who, if the standard is acceptable, will invite the candidates for a professional interview before a panel of Fellows of the IDE.

The examinations and portfolio demonstrate that the aspiring Full Members have also achieved competence in the National Occupational Standards for the National Vocational Qualification (NVQ) Level 4 in Construction Site Management (Demolition). It is not necessary at the present time (2010) for this NVQ to be held by the candidate.

Upon gaining Full Membership of the IDE, Members are required to attend to their CPD and achieve IDE-CPD points which are entered on their record.

The IDE is privileged to offer the Professional membership route to the Construction Skills Certification Scheme (CSCS) Demolition Engineer Card. This is available to Full Members and Fellows providing they have satisfied the necessary CPD requirements.

Latest information is available on the IDE's website or by contacting the IDE National Secretary at info@ide.org.uk

Another organisation, the National Demolition Training Group (NDTG) was formed in 1978 with the sole objective of providing specific and relevant training opportunities to demolition workers. That objective remains at the heart of the NDTG.

Operated jointly by the National Federation of Demolition Contractors (NFDC) and Construction Industry Training Board (CITB) Construction Skills, the NDTG organises training courses across the UK that are delivered via the five NFDC regions: London and Southern Counties; Midlands and Wales; North East; North West; Scotland. The NDTG also links closely with the NFDC in the delivery and development of demolition related NVQs by way of the NFDC NVQ approved centre.

Further information on the NDTG and its courses is available from its website or the NFDC Head Office. Publications are also available.

Additionally, professional institutions such as the ICE, the Institution of Structural Engineers (IStructE) and other organisations also provide means of improving competencies by, for example, running conferences, symposia and courses. As an example, IStructE run a course through their CPD programme on demolition and refurbishment.

Summary of main points

Demolition and structural refurbishment activities are considered and part of the construction sector where decommissioning should be a fundamental early consideration.

Key considerations for planning and executing a successful project include:

- appoint the right people

- allow adequate time – at all stages

- know the facility and provide all relevant information

- make sure
  - that all involved communicate effectively and cooperate
  - proper management arrangements are in place
  - workplaces are designed correctly (as well as access and egress facilities)
  - arrangements for protection of the public are in place and maintained.

In addition, as a further summary of points to be considered where the amount of work and detail should be proportionate to the risks involved in the project may typically include:

- information about existing structures – stability, structural form, degree of deterioration, fragile or hazardous materials

- previous structural modifications, including any weakening or strengthening of the structure

- health hazards, including: asbestos and results of surveys

- security of the site, including access, egress and site hoarding

- site transport arrangements with any vehicle movement restrictions

- state of decommissioning and any contaminated land, including results of surveys

- location of existing services particularly those that may be concealed – water, electricity, gas

- ground conditions, underground structures, water courses or overhead lines where this might affect the safe use of plant, for example cranes, or the safety of ground works

- any restrictions on deliveries or waste processing, collection or storage

- emergency procedures and means of escape, including fire precautions

- traffic management and parking restrictions

- adjacent and nearby land uses, e.g. schools, hospitals, railway lines (including tunnels) and highways

- planned start and finish of the construction phase and the minimum time to be allowed between appointment of contractor (or contractors) and instruction to commence work on site.

References

Addis, W. and Schouten, J. *Design for Deconstruction. Principles of Design to Facilitate Reuse and Recycling (C607)*, 2004, London: CIRIA.

Briggs, M., Buck, S. and Smith, M. (eds). *Decommissioning, Mothballing and Revamping*, 1997, London: Institution of Chemical Engineers.

Construction Industry Council (CIC). *Definitions of Inspections and Surveys for Building*, 2004, London: CIC. Available online at: http://www.cic.org.uk/services/DefinitionsofInspectionsandSurveysofBuildings.pdf

Europa. *The Precautionary Principle*, 2005. Available online at: http://europa.eu/legislation_summaries/consumers/consumer_safety/l32042_en.htm

Fawcett, W. and Palmer, J. *Good Practice Guidance for Refurbishing Occupied Buildings (C621)*, 2004, London: CIRIA.

Health and Safety Executive (HSE). *The Safe Use of Vehicles on Construction Sites. HSG 144 – A Guide for Clients, Designers, Contractors, Managers and Workers Involved with Construction Transport*, 1998, London: HSE Books.

Health and Safety Executive (HSE). *Health and Safety in Refurbishment Involving Demolition and Structural Instability. Research Report 204 (RR204)*, 2004, London: HSE Books. Available online at: http://www.hse.gov.uk/research/rrpdf/rr204.pdf

Health and Safety Executive (HSE). *Managing Health and Safety in Construction. Construction (Design and Management) Regulations 2007. (CDM) Approved Code of Practice*, 2007, London: HSE Books. Available online at: http://www.hse.gov.uk/pubns/priced/l144.pdf

Health and Safety Executive (HSE). *Asbestos Health and Safety*, 2009. Available online at: http://www.hse.gov.uk/asbestos/index.htm

Institution of Civil Engineers (ICE). *ICE Demolition Protocol 2008*, 2008, London: ICE. Available online at: http://www.ice.org.uk/downloads//Demolition%20Protocol%202008.pdf

Lazarus, D., Bussell, M. and Ross, P. *Retention of Masonry Façades – Best Practice Guide (C579)*, 2003, London: CIRIA.

National Federation of Demolition Contractors (NFDC). *High Reach Demolition Rig Guidance Notes. For Demolition Machines of 15 Metres Working Height and Above*, 2006, Hemel Hempstead: National Federation of Demolition Contractors.

National Federation of Demolition Contractors (NFDC). *Guidance Notes on the Safe Use of Mobile Crushers in the Demolition Sector*, 2008, Hemel Hempstead: National Federation of Demolition Contractors.

Referenced standards

BS 5607: 1998. *Code of Practice for the Safe Use of Explosives in the Construction Industry*.

BS 6187: 2000. *Code of Practice for Demolition*. (Note: revised standard in preparation with widened scope, including refurbishment.)

BS EN 1991-1-6: 2005 Eurocode 1. *Actions on Structures. General Actions. Actions During Execution*.

Construction (Design and Management) Regulations 2007, 2007, London: The Stationery Office.

NA to BS EN 1991-1-6: 2005. *UK National Annex to Eurocode 1. Actions on Structures. General Actions. Actions During Execution*.

BS EN 1991-1-7: 2006 Eurocode 1. *Actions on Structures. General Actions*. Accidental actions.

Further reading

International Organisation for Standardization. *ISO/DIS 13822 Bases for Design of Structures – Assessment of Existing Structures.* London: ISO. (Note: Revision of first edition, ISO 13822: 2001.)

Websites

British Standards Institution (BSI) http://www.bsigroup.com

Construction Skills http://www.cskills.org/

Construction Skills Certification Scheme (CSCS) http://www.cscs.uk.com

Eurocodes Expert – making Eurocodes easier http://www.eurocodes.co.uk

Health and Safety Executive (HSE) http://www.hse.gov.uk

Health and Safety Executive – Abestos health and safety http://www.hse.gov.uk/asbestos/index.htm

Institute of Demolition Engineers (IDE) http://www.ide.org.uk/

Institution of Civil Engineers (ICE) http://www.ice.org.uk

Institution of Civil Engineers (ICE), Health and safety http://www.ice.org.uk/knowledge/specialist_health.asp

Institution of Structural Engineers (IStructE) http://www.istructe.org

National Demolition Training Group (NDTG) http://www.ndtg.org/

National Federation of Demolition Contractors (NFDC) http://www.demolition-nfdc.com/

Chapter 19

Transportation and vehicle movement

doi: 10:10.1680/mohs.40564.0235

David R. Bramall Department for Regional Development, Tandragee Northern Ireland, UK

CONTENTS

The terms transportation and vehicle movement relate to the movement of all vehicles and plant, and particularly transporting plant, materials and people to, from or between sites and around construction sites. The terms apply not only to the vehicles and plant directly employed by the people working on the site but also to the vehicles that deliver things to the site, such as from hire companies and waste disposal companies. They apply to the routes that everyone uses including the permanent road network and any temporary roads and haul routes to assist in these types of movement and the methods of controlling both works traffic and public traffic by the appropriate use of local site rules and temporary traffic management measures. In this chapter, we examine how to do these things without causing harm to anyone, by examining some of the main hazards involved in the various activities and some of the control measures that could be utilised to deal with these hazards. Each section finishes with a short summary of possible control measures that have already been identified. Additionally we will examine lifting operations and lifting equipment including the use of cranes and the various control measures which can be put in place to ensure the safety of the people involved in these operations.

Introduction

Driving to and from sites and the movement of plant, vehicles and people around sites might not immediately appear to be your highest priority in terms of health and safety compared with things such as confined spaces, chemicals, excavations and the like. It does, nevertheless, present us with hazards which need to be controlled just like any other hazards that we will encounter and the potential for harm is equal to any of these so called high risk areas.

The main activities and operations you are likely to experience are:

- Getting to and from the site.

- Getting in and out of the site.

- Moving around the site.

- Getting people safely past the site.

Although there are some similarities between these activities and therefore the hazards and controls will be duplicated a little, they are distinct aspects of the work and each one is worthy of consideration. Remember the hazards and controls mentioned here will apply to different personnel involved with the site, so whether you are a Contractor, Designer, Client or Coordinator, or whether you're driving a car, a delivery lorry or any other vehicle, it is worth being aware of the hazards and controls, even though it may not always be your own responsibility to directly deal with them.

Think about these four aspects of the work and consider how they could cause harm to you or to anybody else either working on or visiting the site or to anybody passing by or living near the site, and also of how the activities can be controlled and made safe.

Although the roles on site are varied, and it might not be your own responsibility to control all of the hazards that exist on the site, it will still be useful to be aware of what the hazards are and aware of some of the ways of controlling them, remembering that the same hazard can usually be controlled in different ways. For example, speed could be controlled using traffic management measures or by using slower vehicles; overturning could be controlled using more stable vehicles or by having restrictions on routes whether on site or on the road network; and the outcomes, if something does go wrong, can also be controlled maybe by vehicle design or simply using the fitted safety equipment such as seatbelts and harnesses. With this in mind, it is important to remember that hazards can be controlled in many different ways such as eliminating the hazard, using different equipment, using different methods and even just by the use of personal protective equipment (PPE), and be careful not to become blinkered into thinking your way is the only way or the right way, but that other ways could, and probably are equally valid. The important thing to remember is that the work operation, whatever it is and whatever hazards it presents, should be safe to start, no matter which control measures you select do deal with the hazards.

Getting to and from the site

So you are ready to get into the car and drive to the site. This is something you do all the time, whether it is driving to a site, driving to work or just going somewhere at the weekend. You might not automatically think of this as a hazard but remember you are enclosed inside a small metal container going quite fast and surrounded by numerous other similar containers going either in your direction or coming towards you at a similar speed, and whose actions you have no control over. So, since there is the potential to cause harm to you or to others, it is a hazard which needs to be controlled. Good driving practice is covered in the *Highway Code* (DfT/DSA, 2007) and if the guidance in that is followed then that's probably enough advice for you to be able to drive safely to and from the site. Having said that, following the *Highway Code* is nevertheless a valid control measure and should not be ignored whether you are driving your own car, a tipper lorry, a concrete delivery lorry or any other vehicle.

Vehicle checks

It is quite common to hear health and safety professionals and other people saying things such as 'you do a risk assessment every time you go out walking or every time you get in your car, or cross the road?'. Now I know you don't go through a formal process of risk assessment before you do anything and neither do I. You simply take account of the conditions and adjust your own actions to take account of these. If there's a car coming, you don't cross the road, if its icy or foggy, you slow down to a suitable speed when you're driving, you don't check your brakes, lights, fluid levels and all the other possible checks that could be done before you set off every day so why do it just because you are going to a construction site? There shouldn't be any need. You probably keep your car in good condition, you probably check your tyre pressures and oil levels occasionally and get repairs carried out when they're needed so daily checks shouldn't be necessary. You certainly aren't going to record that you have checked your tyre pressures or checked your oil or fluid levels in the car.

But, say you are driving, for example, a loaded tipper lorry, a mobile crane or a low loader instead of a private car, or that you are a supervisor or manager of the drivers of these vehicles: are these checks necessary now? With these vehicles, other hazards could be present, things like overloading or uneven loading causing overturning, poor braking, poor visibility and materials falling off of the vehicle. Maybe the vehicles are large or heavy and so the potential for injury or damage if anything does happen becomes much greater. So maybe for these types of vehicles we do need some further controls to ensure everyone's safety.

So while we don't carry out all of these checks on our own cars we do need to accept that, given the extra hazards

concerned with construction vehicles and plant, regular checks are not only worthwhile but necessary.

Imagine driving a tipper lorry with 20 tonnes of material in the tipper approaching a junction at the bottom of a hill. The vehicle braking system should be easily capable of stopping the vehicle at the junction but remember these vehicles are designed to operate safely up to their weight limit and. if this is exceeded, the braking, as well as the steering and suspension may not be working at their full capacity. You will almost certainly be able to stop with no load in the tipper but with the extra 20 tonnes on board there is a chance that you could travel straight through the junction and the consequence of doing that could be horrendous. Apart from vehicles that are overloaded, there is also the potential for poor braking if regular checks are not carried out. A simple check such as testing the brakes in the depot before entering the public road network is sufficient for a regular check, while a mechanic will carry out more detailed checks and repairs as part of the vehicle's maintenance programme.

So we have established that regular checks are necessary for construction vehicles and plant; how often should they be done and how detailed should they be? The level of checks should be based on the vehicle and the activity, but as a guide daily and weekly checks on all construction vehicles should be carried out covering such things as oil, coolants and other fluid levels, brakes, lights, cleanliness, mirrors, tyres and the like. Again it doesn't matter whether you actually drive the vehicle, but you should be aware of the hazards involved.

Loading

You probably don't carry heavy loads in your car apart from maybe a bag of coal in the boot or even worse the weekly shopping or something similar, but you often do carry very heavy or awkward loads on construction vehicles, and of course this is exactly what many of the vehicles are designed for so there shouldn't be a problem. The way you load these vehicles, however, has an effect on the performance, the handling and the safety of the vehicle. Vehicles are subject to various maximum weights. Things such as Plated Weight, Kerbweight, Gross Vehicle Weight and Maximum Towing Weight are all common terms. All commercial vehicles must carry a plate showing the vehicle and axle plated weights. Now the driver, being properly trained and licensed for the particular vehicle knows what all of these terms mean, but do you? A good guide for assessing the critical weight for a vehicle carrying loads is called the Gross Vehicle Weight (GVW) which is the total weight of the vehicle plus whatever load it is carrying. This weight is important because the design of the vehicle and so the safety of the vehicle is based on this not being exceeded. Suspension, braking system and steering are the obvious things that could be adversely affected by overloading as well as materials and items falling off the vehicle and either injuring people or damaging other vehicles or property. It is not only overloading that causes a problem but the way in which a vehicle is loaded will also affect performance of key safety features such as steering, suspension and brakes. All loads should be

ICE manual of health and safety in construction © 2010 Institution of Civil Engineers

Daily checks on vehicles			
Beacons		Wipers and washers	
Steering		Cleanliness (windows, lights, indicators)	
Fuel		Horn	
Handbrake		Seat belts	
Footbrake		Drain airbrake reservoir	
Lights and indicators		Tyre condition (flat, soft, damaged)	
Weekly checks on vehicles			
Hydraulic fluid level		Tyre pressure and condition	
Battery fluid level		Hinges and fasteners	
Coolant level		Engine oil level	
Brake fluid level		Clutch fluid level	

Figure 1 Typical vehicle and plant checklist

uniformly distributed over each axle of the vehicle so that the load and the vehicle are more stable.

Have you ever followed a tractor and trailer carrying hay bales or something similar loaded 3 or 4 m high, along a country road and watched in amazement how they manage to stay attached to the trailer? Maybe the same bales are leaning to one side or even hanging off the side of the trailer. That's uneven and unstable loading and although they are not construction related they can be compared with construction vehicles carrying much heavier and maybe much more dangerous loads.

What types of vehicle can be badly loaded? Things such as tippers, concrete delivery lorries and the like carry the load in a partly or fully enclosed body but that does not mean that they are always loaded correctly. A concrete delivery lorry, for example, might typically be a 6 cubic metre capacity vehicle and so if it carries 7 cubic metres of concrete it is overloaded,

it has the potential to spill material from the back and, because the load is essentially a liquid, it may move from side to side causing instability of the vehicle and therefore the vehicle has a greater chance of overturning. If a tipper with a GVW of 20 tonnes is weighed at 23 tonnes then it is overloaded. If it is carrying material piled high above the sides of the tipping body, spillage is a real possibility. If it is carrying different types of material and maybe equipment it could possibly be unevenly loaded and prone to overturning. Uneven loading is a distinct possibility in vehicles that are carrying loads made up of different materials, materials which can move within the body of the vehicle or even plant and equipment or where vehicles are fitted with hoists, tail or side lifts, grabs and the like which can cause some degree of uneven loading even before any additional load is added to the vehicle body.

Other vehicles, such as low loaders, are open and carry their load on basically a flat platform. Given the size and weight of some of the plant that these carry, it might be hard to believe that the load could be dislodged but it is a possibility, especially if the load isn't properly centred on the loader or if the vehicle has to negotiate rough terrain, and so all loads, regardless of their weight, shape or size, need to be fully secured on these vehicles.

Although not completely construction related, another major problem with transportation is the movement of abnormal loads. An abnormal load is simply a load which exceeds certain limits in terms of length, width, height or weight so it is certainly possible that you could be transporting abnormal loads to construction sites, things like beams or other precast units, components of fixed plant or some items of plant or

Figure 2 Overloaded vehicle

Figure 3 Unevenly loaded vehicle

equipment. If we consider the terms we can see the obvious problems with transporting these loads. If a load is very long, it will have difficulty negotiating corners and sharp bends; if it very wide it may not be able to manoeuvre on certain roads and streets; if it is high it may not be able to fit below certain bridges; and if it is very heavy it may not be able to safely travel over certain structures.

The local Road Authority will have a procedure in place for permitting the movement of abnormal loads. This will include the methods of applying for an abnormal load permit and the notice required for different weights and sizes of loads. The notice may vary from two working days for loads exceeding 38 tonnes or exceeding 5 m in width up to five working days for loads exceeding 80 tonnes, exceeding 6.1 m in width or 27.4 m in length. In addition, extra large or heavy loads exceeding 150 tonnes may require a special permit to be issued by the Roads Authority (limits may vary depending which Roads Authority controls the route).

Normally, when issuing a permit for the movement of an abnormal load, the Road Authority will specify the route to be followed and any restrictions on the timing of the movement to ensure a safe route and with minimum inconvenience caused to other road users and residents.

Traffic routes

Think about the routes that you could use to get to and from the site. Are these roads that are suitable for you to use in your own car, also suitable for large or heavy construction vehicles? If the site is on or beside a main road, the chances are that the roads leading to it are suitable for most types of vehicle. The geometry of the roads should allow for straightforward driving with no extreme bends or gradients. The structure of the road and verges should be adequate to carry any loads that are needed and any bridges over the road should have enough clearance for all types of vehicle to travel underneath. The

Roads Authorities keep bridges and structures databases and will be able to advise on the suitability of routes to and from the site depending on the types of load to be transported. This could be of help when planning larger or heavier loads in a similar manner to the movement of abnormal loads. On the other hand, if the site is a bit off the beaten track and can only be reached using lower class roads, narrow roads, roads with sharp bends and gradients, and possibly with weak verges then you need to think about the adequacy of these roads. Think about abnormally large or heavy loads using these roads. There is a strong likelihood that these could overrun the verges, expose any weaknesses and maybe lead to verge collapse and eventual collapse of the road itself; the construction traffic might make it impossible or unsafe for other road users to use that road. Anyway, there isn't much you can do about it; it's where the site is situated isn't it? Yes it is but there are still things you might be able to do to control or at least to minimise the problems. The Client might even have put some restrictions on routes which can be used for accessing the site, maybe avoiding town centres or routes with a school for example and again as a designer or as a contractor you will need to address such restrictions. Several options are available to you to deal with these problems ranging from very simple solutions to complex solutions and from low cost to extremely high cost. A simple, low cost solution may be to look at the vehicles and loads that you are proposing to use to carry out the work and ask if smaller or lighter vehicles which are more appropriate for the routes available could be used instead of the larger ones. Another solution may be to move the access to another position on the site if a higher category route exists at another part of the site, or to arrange the layout of the site to facilitate maybe a one way in, one way out method of working which may be particularly useful in bulk excavation or large scale pavement surfacing. Such a system may allow the use of major routes to and from the site to be used by loaded vehicles

ICE manual of health and safety in construction © 2010 Institution of Civil Engineers

while the minor routes can be used for unloaded vehicles hence reducing the impact of heavy loads on the structure of the road network. Another advantage of this system is that it can eliminate vehicles meeting on perhaps narrow or uneven site routes or having to negotiate the live work area to get out of the site. The more complex and so the more costly solutions might include things such as upgrading the routes either completely or at certain locations to make the route suitable for the traffic which will be using that route. These could include small local widening or verge strengthening schemes which can be carried out prior to the main construction work starting on the site. Another possible solution, and probably the most costly one, may be to construct a haul route or routes to and from the site or between different parts of the site, meaning that the large construction vehicles do not need to use the public road network. This will of course mean designing a suitable route in terms of geometry and strength and constructing this to a high standard before any work can really start at the site although it does have obvious advantages if the cost and time can be justified. Traffic can be kept well away from narrow, weak roads, even towns and villages and many of the hazards vanish. Not only the obvious, verge collapse, traffic collisions and the like but the less obvious ones such as noise, vapour and fume at nearby locations, which apart from the problems associated with health and safety, also provide ample cause for public complaints.

Box 1   Summary of controls for travelling on the public road

- Ensure drivers are properly licensed and adhere to the *Highway Code* (DfT/DSA, 2007).
- Ensure relevant vehicle checks are carried out at the required intervals and repairs take place where defects have been identified.
- Ensure routes are suitable for the types of vehicles and loads and maintain the roads and verges on the routes.
- Ensure vehicles are not overloaded, unevenly loaded and that loads are properly secured.
- Ensure trailers are properly fitted to the vehicles.
- Keep reversing on public roads to a minimum.

Getting in and out of the site

So you know how to drive and you know how to get to the site but how are you going to get in? Now this might appear obvious, especially if you are in your own car – drive through the site entrance and park in the designated parking area if one exists. Now think again about the types of vehicles that you might be using and the different types of site and how entrances and exits are constructed. Are they adequate for the traffic that has to use them?

You may have seen a similar situation to this before. At an entrance to a construction site a large delivery lorry or the like is manoeuvring into or out of the site and one of the site operatives stands in the middle of the road with a hand in the air stopping the traffic. Infuriating isn't it? It might only take a few seconds or it might take several minutes but either way if you're late for work and in a hurry, it drives you mad. Regardless of how long it takes, this should be avoided and almost certainly is easily avoidable by proper design of the entrance and by the proper design and use of temporary traffic management. Not only is it easily avoided but it is illegal to stop or obstruct traffic on a public road. There are situations where this is possible but it must be done using proper temporary traffic management techniques and signs. Whether you are a Client, a Designer or a Contractor you can do something about this. As a Client you can impose restrictions or conditions on things like traffic disruption and as a Designer you can design the scheme and indeed the accesses to accommodate the expected traffic. As a Contractor you will have included these aspects in the construction phase health and safety plan; knowing what types of vehicles and plant you will use to complete the work, and therefore what geometry you will need for entrances and exits to accommodate these vehicles, you can ensure that entrances and exits are properly constructed to suitable dimensions and gradients and maintained, and that any traffic management required is of a suitably high standard.

What type of site is it?

It could be a completely enclosed site in a city centre or town centre involving maybe demolition of existing buildings and replacing these with a new building. It could be a site within the confines of buildings where no demolition is to take place. It could be a linear site maybe on one side of a motorway, a greenfield site in a rural setting or a linear site comprising both on-line and off-line construction, for example in the construction of a new road or a road realignment. Each one is different in terms of the ease of providing accesses but all of them still require a similar approach to ensure proper accesses and each of them will involve use of the road network in getting to and from the site.

If you are going to an enclosed construction site adjacent to a public road, one or more entrances and exits will be constructed, and if these are fully designed and constructed in accordance with current standards then entering and leaving the site is no more than a normal driving manoeuvre. If the entrance is not constructed to current standards, maybe the site entrance is through an existing alleyway in a town centre or between two existing buildings and so the entrance geometry cannot be changed. All you can do is maintain the surfaces at the entrance and give consideration to using smaller vehicles and plant to do the work. If a new entrance is poorly designed, manoeuvring is difficult maybe because the radii at the entrance are too tight for larger vehicles, the gradient of the entrance too steep, the width of the entrance too narrow or the surface of the entrance not maintained properly.

The site might be on a busy road, for example on a motorway, or any other busy main road. Given the varied nature of construction work, the site could be anything ranging from a few square metres in area lasting a few minutes up to several kilometres in length lasting for several years. How do we

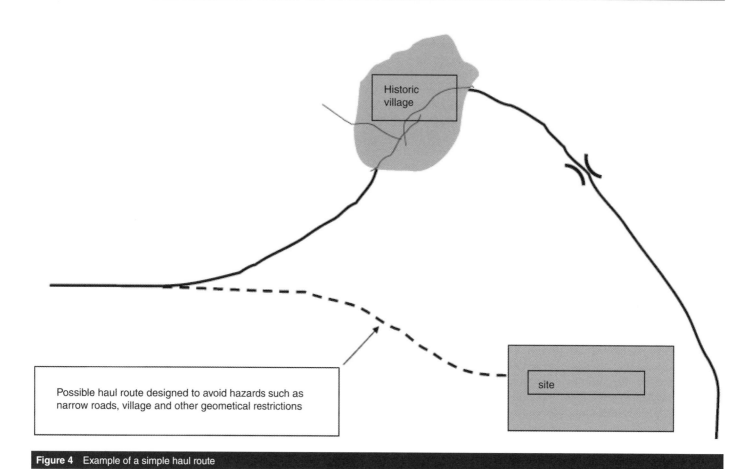

Historic village

Possible haul route designed to avoid hazards such as narrow roads, village and other geometrical restrictions

site

Figure 4 Example of a simple haul route

get in and out of these sites? Think of a long length of, say, resurfacing work on a motorway. Is one combined entrance and exit enough, how will the entrances be constructed, how can you stop other motorists following construction traffic onto the site? Given the nature of the work, more than one entrance or exit might be needed. Perhaps preparation work is continuing at the far end while final surfacing work is starting at the near end. In this case it could make sense to have separate entrances and exits for the asphalt vehicles and the other vehicles on the site and positioned at suitable locations which are conveniently situated near to the various operations.

If you are a Contractor, you will know what your method of work is going to be, the plant and equipment that you propose to use and the phasing of different parts of the work. With these things in mind you should easily be able to construct adequate entrances and exits where these are needed to allow all vehicles to safely get into and out of the site. Further advice on the standards that these should be designed, constructed and maintained to are set out in the *Design Manual for Roads and Bridges* (Highways Agency, 2009a) and the *Specification for Highway Works* (Highways Agency, 2009b) while guidance on signing requirements can be found in Chapter 8 of the *Traffic Signs Manual* (DfT 2009a, 2009b) and in the *Safety at Street Works and Road Works – A Code of Practice* (DfT, 2001).

Having established the locations of the entrances and exits, and constructed these to a high standard, how are you then going to stop other motorists following you or any of the other construction vehicles into the site? First, if temporary traffic management has been designed and constructed well, all site accesses and exits will be adequately signed and guarded and it should be very clear to motorists as well as to the site staff exactly what is going on and what action they all need to take either to get into the site or to get safely past it.

The only other thing you then need to consider is maintenance. Entrances and exits should be maintained to a high standard so that vehicles can move safely over the surfaces and without having to slow down to the extent where traffic is affected unduly by site vehicles entering and leaving the site. If you always think of the site as an extension of the public road, then the construction and maintenance requirements should become clear. In other words, any manoeuvre getting into or out of the site should be equivalent to the same manoeuvre on a public road. Potholes and other surface defects should be repaired immediately so that vehicles can move from the public road to fully within the site in one movement. Maintenance work, while necessary, should be timed so as not to cause further disruption, for example before or after peak times.

Moving around the site

Think about the types of site again and ask yourself how you can organise the site to ensure the safety of yourself, other engineers, operatives and visitors. OK, this will normally be something the Contractor will consider in establishing the site but whether you are part of the Contractor organisation or not you should be aware of the hazards and potential consequences. There is no reason why most sites can't be laid out in the same way as a permanent workplace. Think of where you work when you're not on a construction site. It might be simply an office block, or it might be a works depot or a quarry for example. There is probably a car park for staff and maybe a separate one for visitors. The car park is probably close to the office block away from maybe quarry vehicles or other works vehicles, mixing plant, loading areas and the like. Visitors are told to report to reception on arrival. Footways are provided or some sort of segregation is used to keep people and vehicles apart. There might be speed limits and one way systems in force around the premises. There might also be locations where unauthorised people or vehicles are not allowed to enter and these will be clearly signed and maybe guarded by barriers or gates.

So why not just copy the layout of your own workplace on the site? There are some problems that could arise with this; things such as lack of space to segregate traffic and people perhaps in an enclosed site, no suitable location for a single car park maybe on a long, linear site. So even though you can't always have a perfect layout to ensure safety, at least by using your own workplace as a model to work from, you should at least be able to provide a series of measures that will increase the safety of everybody on the site.

During the life of the project, work will take place at certain locations and the type of work will change as the project evolves so it will be necessary to change the layout of the site to accommodate these. During the early stages, excavations are likely to be predominant and, as work progresses, there might be a need for fixed plant to be installed, structures to be constructed and finished surfaces for pavements, floors, footpaths and walkways provided. Each stage of the work has an effect on how we can get around the site so how can we address the various hazards? Think about the typical operations on site, the plant and equipment used for these operations and what safety measures can be put in place to control the hazards. Although the site evolves through the various

operations, it is a mistake to simply change small sections of the site layout to deal with a change to one area of the site. Rather you should always look at the whole site and adapt the layout as required.

Dump trucks, delivery lorries, tipper lorries, excavators and earthmovers are all commonplace on construction sites, particularly at the early stages of the construction phase and all have particular hazards associated with them.

Vehicles such as dump trucks, tipper lorries and tractors are particularly prone to overturning due to having a high centre of gravity and this, combined with possibly poor loading and less than perfect surfaces on a construction site, means that overturning becomes a major hazard.

Vehicles such as large excavators, earthmovers and large dump trucks usually have poor visibility and certainly have blind spots where the operator or driver can see nothing, particularly at the rear of the vehicle, but also to a lesser extent to the front or sides of the vehicle, due to their construction and this can lead to collisions with other vehicles, fixed plant or structures, striking people on the site and overrunning excavations and the like.

There are numerous cases of incidents involving these vehicles either running over someone, colliding with a fixed or moving object and even driving over the edges of excavations. There are simple solutions for poor visibility which are now common, particularly in quarrying, but also on construction sites, where vehicles are either supplied with or fitted with rear facing CCTV cameras and convex mirrors to ensure all-round visibility for the operators of these machines.

The worker, who was using a mobile phone, was positioned on the road and was unaware that the vehicle was approaching.

Although the mobile phone was a major factor, the field of visibility for the lorry driver was also an issue.

This accident might have been prevented if the vehicle had been fitted with equipment such as a rear facing CCTV camera and safety mirrors to ensure all-round visibility.

Excavations

It could be a small trench 1 or 2 m deep or it could be a 10 m deep cut out of a hillside. Either way we don't want vehicles or people to fall in or to have material fall on top of them. Although excavations need their own controls for the plant and operators involved, there is also a need to protect these from other people and vehicles using the site. For vehicles that need to work in close proximity to the excavations, it might be necessary to provide properly constructed accesses and barriers capable of restraining the types of vehicle being used. For all other vehicles and people, the area can be totally segregated to avoid unauthorised access to the area. Whether the site is laid out in established routes or not, this can be achieved simply enough by providing suitable routes, not necessarily properly constructed but clearly marked and capable of carrying the loads to be imposed or simply some form of barrier which does not necessarily need to be capable of restraining vehicles but to give guidance on the route to be used. A simple system of timber posts driven into the ground and tape or rope or timber rails fixed to them should be good enough for this.

Forgetting about the drivers of the vehicles, how are you going to get people to the various parts of the site? It's unlikely that every member of the Contractor's team has a vehicle to use to follow the laid out routes to their own work area. Is it better for the operatives to walk around the site or to be transported? It depends, again on the nature of the site, but various options are possible. On a small enclosed site the best option is probably walking to the relevant work area whereas on a 10 km linear site, this is impractical and some form of transport is probably going to be needed unless the work can be planned so well that each operative can stay entirely within their own area of the site. If this is not the case, we need to have appropriate transport; something such as minibuses, seated vans, jeeps or cars. If these are available, it means that everybody can safely move around the site; however, some of the alternatives don't provide a safe means of transport. Have you ever seen anyone being transported in, for example, the bucket of an excavator or loading shovel? Hopefully not, but it does occur and whatever your position or role on the site, you should not let this happen. Some vehicles are designed for passengers; others aren't – for example a van with people sitting on the floor or on their tool boxes. Appropriate vehicles will have fixed seats fitted with seat belts. Some vehicles are designed for raising people to a height to allow them to carry out particular work such as attaching lanterns to street lights, while things like the bucket of an excavator or loading shovel are definitely not designed for this purpose or for moving people around the site.

Box 5	Summary of controls while on the site

- Competent and authorised drivers.
- Vehicle and plant checks.
- Minimise reversing of vehicles.
- Use of flood lighting if required.
- Excavation guarded by barriers or stop blocks and banksmen if required.
- No unauthorised vehicles in the vicinity of excavations.
- Surfaces of routes around the site maintained in good condition.
- Routes properly designed, constructed and maintained.
- Slopes constructed at appropriate gradients to accommodate the vehicles on the site.
- Use of rollover protection systems.
- Provide all-round visibility for site vehicles using convex mirrors and rear facing CCTV cameras.
- Properly demarcated and signed routes including one way systems and speed limits, for vehicles and operatives.
- Appropriate vehicles for transporting people around the site.
- Segregation of vehicles and people.

Getting people safely past the site

Just as you don't want unauthorised people and vehicles entering the site because of the hazards involved, neither do you want to expose them to similar hazards outside of the site. Think again about the types of site you are likely to be working on and what protection is needed.

A fully enclosed site is reasonably simple to protect. It might already be protected by surrounding buildings and a simple gate at the entrance is enough, or it might be a demolition where protection in the form of fencing, hoarding and canopies might be needed to catch falling debris. So you can protect the site easily enough, but is all of the work fully enclosed? It is probably the case that whatever type of site it is that there will be work which needs to be done on or beside the public road or it could be the case that dirt is being carried onto the road by site vehicles. It might be bringing services onto the site, constructing entrances or using part of the road for construction plant, delivery vehicles and the like. So you need protection for the public passing the site to ensure their safety and suitable temporary traffic management (TTM) arrangements are the key to achieving this no matter what type of site it is. Designing TTM is the same as any other aspect of design; it is governed by standards which have certain levels of statutory force depending on which jurisdiction you are working in. It isn't something which is an afterthought or to solve a problem when it occurs but it must be considered at the earliest design stage and incorporated into the overall design of the project.

In Great Britain, Chapter 8 is a guidance document which is considered best practice for temporary traffic management design and operations. In Northern Ireland, Chapter 8 is given statutory force for the signs and devices used under Article 31 of the Roads (NI) Order.

Depending where you are working, the TTM should be in accordance with one of two documents. The appropriate documents are Chapter 8 of the *Traffic Signs Manual* (DfT, 2009a, 2009b) and *Safety at Street Works and Road Works – A Code of Practice* (The Safety Code) (DfT, 2001). Both documents provide adequate guidance on the signing and guarding requirements which, if followed, provide safety for both the site operatives and members of the public regardless of the size and nature of the obstruction. This doesn't apply only to motorists but pedestrians using footways as well as cyclists, horse riders or anybody else that uses the roads and footways adjacent to the site.

The measure of a good TTM scheme is that there is adequate warning of the works, that the works area is well defined and that traffic can move freely through the TTM without any extra danger than the non-work situation. In other words, a motorist or pedestrian should be totally clear about what action he or she needs to take to negotiate his or her way through or past your construction site and the same standard of signing and guarding should be used for every site.

The Roads Authorities have enforcement powers under the New Roads and Street Works Act 1991 (in Northern Ireland, the Street Works (Northern Ireland) Order 1995) in terms of inadequate TTM and, since early consultation should take place at an early stage in the project, TTM requirements should be discussed at this stage and more detailed designs produced as the project progresses and are agreed with the Roads Authorities. If you need temporary orders for speed limits, no overtaking or road closures, these need early agreement and the Roads Authorities will arrange for these orders if they are satisfied that they are essential. They will not allow closures simply for the convenience to the Contractor in carrying out the work.

For this reason a systematic approach is needed in deciding on the possibilities for signing and guarding a particular work site. By examining certain criteria such as a permanent speed limit, road geometry, traffic flows, required safety clearances, the type of work and the plant and vehicles required, a range of options can be identified, any of which will be adequate for the signing and guarding appropriate to the relevant guidance – Chapter 8 (DfT 2009a, 2009b) or the Safety Code (DfT, 2001). Such an approach will assist you in justifying to the Roads Authorities that particular restrictions are necessary to safely carry out the work, including road closures, lane closures, no overtaking orders and mandatory temporary speed limits. The signing used must be as described in the relevant code, both in terms of the layout and signs used. This is vital if enforcement is to be used for exceeding speed limits, for example. If the signs used for the purpose of setting a speed limit are not the right size, not the right design, have the wrong colour of backing or even if there are insufficient repeater signs, the speed limit is not enforceable.

The signs used for temporary traffic management are included in the Traffic Signs Regulations and General Directions and any sign which does not conform to this is unlawful (the Traffic Signs Regulations in Northern Ireland).

A typical decision making tool is shown in **Figure 5**, which outlines the general processes to follow in designing TTM and if you are approaching the Roads Authority to obtain a road closure, a temporary mandatory speed limit or other type of temporary order, a copy of this with the relevant details completed, should help you to justify the reasons for any temporary orders.

Using the decision making tool does require a proper understanding of Chapter 8 (DfT, 2009a, 2009b) and the Safety Code (DfT, 2001). The purpose is to help the Designer to identify the parameters to work within; the actual temporary traffic management arrangement or arrangements used for the work will be a choice of several options identified from using the decision tool. Remember also that, often, using more than one arrangement to carry out the various stages of the work will provide the best solution in terms of safety of both the workforce and the public and in terms of problems with traffic congestion.

The use of temporary orders needs to be justified; however, the use of these is sometimes misunderstood. For example, temporary speed limits are often used within temporary traffic management arrangements, although sometimes it is questionable whether these are really necessary and questionable whether the speed limit adopted is reasonable. Chapter 8 (DfT, 2009a, 2009b) gives guidance on the use of temporary speed limits and recommends a reduction where, for example, narrow lanes are introduced, the TTM is complex or the geometry of crossovers or lane changes cannot be designed to full geometric standards.

> ## TEMPORARY SPEED LIMIT IN FORCE FOR THE PROTECTION OF THE WORKFORCE

Signs such as this one are often displayed at temporary traffic management sites. It does, however, show a lack of understanding of the TTM guidance documents within the construction industry and even among TTM specialists. Many people agree that speed limits are indeed put in place to protect the workforce. This is wrong. It gives the impression that the workforce are the only people who need to be protected. Chapter 8 (DFT, 2009a, 2009b) states that 'Temporary speed limits should not be imposed at road works sites solely for the direct purpose of protecting the workforce'. The real reasons for introducing temporary speed limits are for the safety of the general public by allowing them to safely negotiate temporary traffic management schemes, even though this has the obvious knock on effect of keeping both the workforce safe due to fewer taper strikes and the TTM crews due to fewer repairs having to be carried out on a busy, live carriageway and to protect the workforce where other traffic management techniques cannot be used.

So having designed and set up a suitable means of getting people safely past the site in terms of the temporary traffic management arrangements used, what other hazards could be present and how could we control them?

Road details and design considerations				Decisions			
1	Type and classification of road			1	Can road accommodate normal number of traffic lanes?	YES	NO
2	Road width		(m)	2	Can shuttle working accommodate traffic flows?	YES	NO
3	Permanent speed limit		(mph)	3	If NO, is a one way order necessary/appropriate	YES	NO
4	Volume of traffic			4	Is temporary widening of road required along the works?	YES	NO
5	Of which HGVs			5	Are narrow lanes, speed restrictions or no overtaking orders required?	YES	NO
6	Level of pedestrian activity			6	If NO, road to be closed. Period of closure to be minimised by planning of the works	YES	NO
7	Level of cycle activity			7	Are sight lines/forward sight distances acceptable?	YES	NO
8	Suitability of alternative/ diversion route	Yes	No	8	Can road be signed to Chapter 8/*Street Works Code of Practice*?	YES	NO
9	List any requirements, restrictions or conditions specified by the Client				If NO, what additional measures are required, or does design need to change?		
				9	Will traffic run on temporary surfaces?	YES	NO
10	Operations to be carried out				A. If YES, are these surfaces appropriate for the flows/speeds?	YES	NO
					B. Can repairs be carried out to temporary running surfaces if required?	YES	NO
11	Width of plant required		(m)	10	Timing of the work? (Date/daytime or night working)		
12	Working space required if any		(m)	**Additional comments**			
13	Extra width for delivery/storage		(m)				
14	Safety zone required (lateral clearance)		(m)				
15	Requirement for pedestrians		(m)				
16	Extra space for cones and cylinders		(m)				
17	Remaining road width		(m)				

Figure 5 Temporary traffic management decision making tool

ICE manual of health and safety in construction © 2010 Institution of Civil Engineers

Certain hazards are easily identified and controlled as long as we think logically about the whole scheme, while others might not appear as obvious. Assuming the TTM is adequately designed and set up, the main hazards are things such as slow-moving construction vehicles entering the carriageway, dirt and other debris being deposited on the carriageway by these vehicles, and the signs and equipment becoming damaged or dirty.

Box 6 Summary of controls for getting people safely past the site

- TTM properly designed, constructed and maintained to either Chapter 8 or the Safety Code.
- Use of only prescribed signs and approved equipment for TTM arrangements.
- Use of relevant temporary orders at the relevant times.
- Use of approved methods of setting up, maintaining and removing TTM.
- Operatives to stay within the designated works areas within the site; not in the safety zones.
- Adequate signs for the information of motorists.
- Use of appropriate traffic control.
- Wheel washes for vehicles leaving the site.
    - Loads well secured on vehicles leaving the site.
    - Regular sweeping and washing of the road surface.
    - Regular checks of TTM arrangements to identify dirty equipment, damaged equipment and displaced equipment and carrying out cleaning or repairs as required.

Box 7 Case study

A roadworker was killed while operating a stop/go board as part of a TTM operation to allow short term road works to be carried out. Although the TTM was designed and set up in accordance with the Safety Code and the use of stop/go was appropriate, the operative was positioned within the safety zone and an errant vehicle entered this safety zone and struck the operative.

Although the motorist was at fault, the operative should not have been positioned within the safety zone, which is designed as a zone where no people, plant, equipment or materials should be positioned.

This accident may have been prevented if the operative had positioned himself within the works area to allow a buffer zone between him and the traffic using the road.

Provision and Use of Work Equipment Regulations (PUWER)

Work equipment as well as works vehicles and plant are covered by regulations to ensure their suitability and condition for use. Work equipment is governed by the Provision and Use of Work Equipment Regulations which are designed to ensure that the hazards people face when using equipment, including mobile equipment, are assessed by the employer and eliminated or controlled. Work equipment should be:

- suitable for its intended use as regards purpose and conditions of use;
- safe for use, i.e. people's health and safety are not compromised;
- maintained in an efficient condition, in good repair and, where appropriate;
- inspected to ensure that it remains safe;
- used only by those who have received adequate information, instruction and training and have been assessed as competent;
- accompanied by suitable safety measures, e.g. protective devices, markings and warnings.

The appropriate control measures include:

- provision of guards and protection devices
- markings and warnings
- system control devices, e.g. emergency stop buttons and interlock devices
- information, instruction and training
- safe systems of work
- personal protective equipment (PPE).

The Regulations apply to any equipment used by an employee, including hammers, knives, ladders, drilling machines, power presses and, lift trucks and vehicles. Office equipment is also included.

If employees provide their own equipment and tools they are also subject to the same requirements of the regulations. Employers must ensure by inspection that personal tools comply with the regulations and prohibit the use of those that fail.

Lift trucks

A disproportionately high number of fatal accidents occur in the use of lift trucks (HSENI, 2001), 24% of all workplace transport accidents. Between 2001 and 2006 there were 21 190 injuries involving lift trucks of which 94 were fatal, 5758 were major and 15 338 were over three day lost-time injuries (LTIs). One of the most common accidents is the crushing of the driver when his vehicle has overturned. Trucks can overturn if they are turned too quickly, even at walking speed, or if they are overladen or are being driven on an uneven surface (HSE, 2009b).

The three main causes of accidents are:

- lack of driver training and poor supervision (HSE, 2009b);
- inadequate premises; and
- poor vehicle maintenance.

In addition to what has been outlined above concerning vehicle safety, there are several particular controls specific to the use of lift trucks and these include:

- Ensure that new trucks meet state of the art safety standards, are CE marked and come with a certificate of conformity.

- Obtain manufacturer's instructions for use, including maintenance and training requirements.

- Prohibit the use of lift trucks as work platforms or as a means of access to heights unless they are specifically designed for that purpose, e.g. mobile elevating work platforms (MEWPs) (see Chapter 14 *Working at height and roofwork*).

- Carry out statutory inspections every 12 months, or six months in the case of vehicles designed for use as work platforms.

- Ensure that the correct lift truck and attachment is used for each job.

- Plan all routine and non-routine jobs taking into account weight, shape and stability of the load and working conditions.

- Ensure that drivers know and follow rated capacity of the lift truck for the safe stacking height and working loads.

- Prohibit the use of the vehicles for carrying passengers.

- Develop and ensure that safe working methods are followed for suitable lift trucks and work platforms.

- Reduce noise levels where practicable and ensure that drivers have suitable means of communication.

Lifting Operations and Lifting Equipment Regulations (LOLER)

The Lifting Operations and Lifting Equipment Regulations (LOLER) require that all lifting equipment provided for use at work is:

- strong and stable enough for the particular use to which it is put; this includes the load and any attachments;

- marked to indicate the safe working load (SWL);

- positioned and installed to eliminate risk;

- used safely;

- subject to ongoing thorough examination and inspection by a competent person.

The Regulations apply to any equipment used for lifting/lowering loads, including anchoring, fixing and support attachments. Equipment includes cranes, lift trucks, MEWPs, hoists and vehicle inspection platform hoists. Accessories such as chains, eyebolts and slings are included in the regulations.

- Plan and supervise all lift operations.

- Ensure that operators are competent and are following safe systems of work.

- People lifting equipment should be marked accordingly and are safe for that purpose, thorough examinations of the equipment should be undertaken at least 6-monthly in the case of equipment used to lift people, and in respect of all other lifting equipment, annually (unless otherwise stated in schemes of examination or in the event of unusual circumstances arising that may compromise the safety of the equipment).

- Examination reports must be submitted to the employer who shall act on them accordingly.

In 2007 the Health and Safety Executive (HSE) issued a warning to construction employers to make sure that they had safe systems in place before cranes are installed on their sites. Why? Because people are being hurt as a result of poor operational practices. An HSE inspector at the time (HSE, 2007) (Press Releases from Health and Safety Commission/Executive 2007 (GNN ref 155850P)) said:

> Crane operations can present serious risks and it is therefore essential that crane installation is properly planned and implemented.

The incorrect use of cranes can have fatal consequences; for example in the USA the Department of Labor estimates that there are over 80 crane related fatalities each year. Besides the fatalities the potential for serious injuries happening, putting workers temporarily or permanently out of the industry, is always there if the crane operations are not planned and carried out with workers' safety and health in mind.

We can break crane accidents down to three critical areas where things can go wrong:

1. crane overturning
2. items being dislodged during slinging
3. people being struck during slinging operation.

So what are we to do? Is it all down to the site conditions, weather patterns or even operator error? I don't think so. It all goes back to proper planning. Remember what can go wrong and what is being done to prevent it. Well that applies here and when the HSE talks about wanting to see crane operations being planned and implemented, that is all they are getting at. Here is a question for designers. When you are sitting in front of your CAD (computer-aided design) system how much consideration do you give to how material is to be moved around the site? Remember I talked earlier about site footprints allowing space for site transportation. Where there is heavy or awkward lifting operations involving the use of cranes, can the site accommodate the size of crane needed to carry out the task? Now let's consider each of these topics in some detail.

Siting the crane

Crane Stability on Site (Lloyd, 2003) among others gives a sample checklist and documents that will help you put together your method statement/safe system of work. There is also an HSE document (HSE, 1980), *Management's Responsibilities in the Safe Operation of Mobile Cranes*, which in addition to detailing three crane accidents, for the purpose of learning from others' mistakes, gives a series of fundamental questions to ask.

The key things to remember about siting the crane to avoid the possibility of overturning are that it should be sited on level and stable ground with the outriggers fully extended and with the wheels off the ground. The outriggers should be supported on timber or steel plates (to evenly distribute the load). And also remember it isn't always possible to say whether the ground is stable just by looking at it. It may be necessary to have geotechnical surveys done to be certain.

Slinging equipment

When slinging, the weight at the centre of gravity of the load must always be known, as must the gross safe working load (SWL) for the crane. Remember, when calculating the maximum weight of the load that can be slung, you need to also take account of the weight of the chain, lifting tackle and any lifting gear. The weight of these parts of the crane form part of the gross SWL and must be deducted from it in order to determine the maximum weight of the load to be slung (i.e. the net SWL). Cranes have an indicator that alerts the operator to an overweight load or when the crane has extended its reach beyond a safe limit. Operators must be alert to the overload indicator and not make any attempt to operate the crane should the alarm sound.

Finally there is a series of hand signals (**Figure 6**) that are used to guide the crane operation and which you should be familiar with. Further guidance on these signals can be found in the HSE's *The Health and Safety (Safety Signs and Signals) Regulations 1996. Guidance on Regulations* (HSE, 2009a).

Slinging operations

Crane operation is a safety critical operation and as such should only be carried out using competent, fit and healthy operatives.

Operatives' health and fitness should be regularly reviewed. When siting the crane it should be in a position where the operator has all-round visibility, away from the edge of excavations and far from structures so that in the slewing operation no one can get trapped between the crane and the structure.

Crane operations need to be planned and executed in a very precise manner, using highly skilled personnel. BS 7121: 2006 *Code of Practice for Safe Use of Cranes* identifies three people with important roles to play in a crane operation: the crane operator, crane coordinator and the crane supervisor. Their titles are almost self explanatory but just to be sure of the differences between them:

- The crane operator is the person who is operating the crane for the purpose of positioning loads or erection of the crane.
- The crane coordinator is the person who plans and directs the sequence of operations of cranes to ensure that they do not collide with other cranes, loads and other equipment (such as concrete placing booms, telehandlers, piling rigs).
- The crane supervisor is the person who controls the lifting operation, and ensures that it is carried out in accordance with the appointed person's safe system of work.

BS 7121: 2006 provides recommendations on the management, planning, selection, erection and dismantling, inspection,

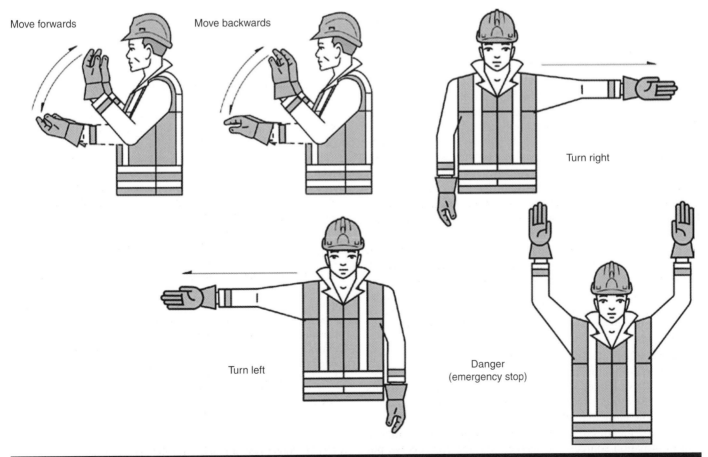

Figure 6 Signals to drivers (HSE, 2009b)

testing, examination, operation and maintenance of cranes, and the planning and management of lifting operations.

Statutory Tower Crane Register

One final point to note regarding cranes is that regulations were scheduled to come into force in April 2010, requiring tower cranes to be registered. It is expected that the details that would have to be notified to HSE are:

- the site address where the tower crane is being used;
- the name and address of the crane owners;
- details needed to identify the crane;
- the date of its thorough examination;
- details of the employer for whom the examination was made; and
- whether any defects posing a risk of serious injury were detected.

Summary of main points

At the beginning of this chapter, we said that transportation and vehicle movement and lifting equipment might not appear to be the highest priority for us in terms of the health and safety implications.

We have since examined key aspects of the construction process such as travelling on the public road, travelling within the site and ensuring the safety of the public passing by the site. As we have shown, each separate activity possesses peculiar hazards, and common hazards also run through each aspect of the process and that these need to be controlled.

We are all aware of headline stories and statistics regarding the number of motorists, cyclists and pedestrians who are killed or injured on our roads, but we do not necessarily associate these statistics with our own work, namely construction, even though the vehicles we use are much larger and heavier and many of the routes we use to move around sites are not to the same standard as the normal road network. Because construction work is one of the major causes of death and serious injury within the UK, we often associate these deaths and injuries with such things as falls from heights, manual handling, confined spaces and excavations. We have now shown that the transportation activities of construction in all of its forms contributes to these headline figures for accidents and that the potential outcome is as serious as the outcome from any other type of accident and therefore we need to put controls in place to avoid these accidents.

We have now identified many of the common hazards that are present within the field of transportation and vehicle movement and have looked at how we can deal with these hazards and how the hazards and controls will differ depending on the nature of the particular site.

We have identified some of the key legislation, guidance and information sources and learned that the controls we put in place are for the protection of, not only the workforce, but anyone else who may be affected by our activities.

Although we have identified a large number of hazards and possible controls, there will be others that may be peculiar to a particular site or a particular operation which has not been covered here; but remember, the key message is **'Every work operation should be safe to start, no matter which control choices you make'**.

Disclaimer

The views expressed in this chapter are those of the author along and not necessarily those of the Northern Ireland Department for Regional Development.

References

Department for Transport (DfT). *Safety at Street Works and Road Works. A Code of Practice*, 2001, London: The Stationery Office.

Department for Transport (DfT). *Traffic Signs Manual Chapter 8 – Part 1: Design. Traffic Safety Measures and Signs for Road Works and Temporary Situations,* 2nd edition, 2009a, London: The Stationery Office.

Department for Transport (DfT). *Traffic Signs Manual Chapter 8 – Part 2: Operations. Traffic Safety Measures and Signs for Road Works and Temporary Situations,* 2nd edition, 2009b, London: The Stationery Office.

Department for Transport and Driving Standards Agency (DfT/DSA). *The Official Highway Code 2007 Edition*, 2007, London: The Stationery Office.

Health and Safety Executive (HSE). *Management's Responsibilities in the Safe Operation of Mobile Cranes: Report on Three Crane Accidents*, 1980, London: HMSO.

Health and Safety Executive (HSE). Press Releases, 2007 (GNN ref. 155850P).

Health and Safety Executive. *The Health and Safety (Safety Signs and Signals) Regulations 1996. Guidance on Regulations*, L64, 2009a, London: HSE Books. Available online at: http://books.hse.gov.uk/hse/public/saleproduct.jsf?catalogueCode=9780717663590

Health and Safety Executive. *The Safe Use of Vehicles on Construction Sites*, HSG144, 2009b, London: HSE Books. Available online at: http://books.hse.gov.uk/hse/public/saleproduct.jsf?catalogueCode=9780717662913

Health and Safety Executive for Northern Ireland (HSENI). *Is Your Lift Truck Being Used Safety?* Workplace transport safety initiative, 2001, Belfast: HSENI. Available online at: http://www.hseni.gov.uk/index/information_and_guidance/sector_specific_guidance/construction.htm

Highways Agency (HA). *The Design Manual for Roads and Bridges*, 2009a, amendments. Available online at: http://www.standardsforhighways.co.uk/dmrb/index.htm

Highways Agency (HA). *The Manual of Contract Documents for Highway Works, Vol. 1 – Specification for Highway Works*, 2009b, London: The Stationery Office.

Lloyd, D. (ed.). *Crane Stability on Site*, 2nd edition, 2003, London: CIRIA.

Referenced legislation and standards

BS 7121: 2006 Code of Practice for Safe Use of Cranes.

Lifting Operations and Lifting Equipment Regulations 1998. Statutory Instruments 2307 1998, London: The Stationery Office.

Lifting Operations and Lifting Equipment Regulations (Northern Ireland) 1999. Statutory Rule 1999 304, London: The Stationery Office.

Provision and Use of Work Equipment Regulations 1992. Statutory Instruments 1992 2932, London: HMSO.

Provision and Use of Work Equipment Regulations (Northern Ireland) 1993. Statutory rule 1993 19, London: HMSO.

Further reading

CIRIA. *Crane Stability on Site – An Introductory Guide* (C703), 1996, Revised 2003, London: CIRIA.

CIRIA. Tower Crane Stability (C654), 2003, London, CIRIA

European Agency for Safety and Health at Work downloadable publications:

E-fact 2 – Preventing Vehicle Accidents in Construction, 2004. Available online at: http://osha.europa.eu/en/publications/e-facts/efact02/view

Factsheet 18 – Preventing Road Accidents involving Heavy Goods Vehicles, 2001. Available online at: http://osha.europa.eu/en/publications/factsheets/18/view

Factsheet 16 – Preventing Vehicle Transport Accidents at the Workplace, 2001. Available online at: http://osha.europa.eu/en/publications/factsheets/16/view

Health and Safety Executive (HSE). *Construction Site Transport Safety: Safe Use of Dumpers* (CIS 52), 2006, London: HSE. Available online at: http://www.hse.gov.uk/pubns/cis52.pdf

Health and Safety Executive for Northern Ireland (HSENI). *Rider-operated Lift Trucks: Operator Training*, 2nd edition. Approved Code of Practice and Guidance. L117, 2000, Belfast: HSENI. Available online at: http://books.hse.gov.uk/hse/public/saleproduct.jsf?catalogueCode=9780717624553

Road Vehicles (Construction and Use) Regulations 1986. Statutory Instruments 1986 1078, London: HMSO.

Websites

Crane Accidents http://www.craneaccidents.com/

Department for Transport guidance http://www.dft.gov.uk

European Agency for Health and Safety at Work (EU_OSHA) http://osha.europa.eu/en

Institution of Civil Engineers (ICE), Health and safety http://www.ice.org.uk/knowledge/specialist_health.asp

VOSA (Vehicle and Operator Services Agency) http://www.vosa.gov.uk

Overseeing Organisations (Roads Authorities) websites:

DRD Roads Service http://www.roadsni.gov.uk

Highways Agency http://www.highways.gov.uk

Transport Scotland http://www.transportscotland.gov.uk

Welsh Assembly Government http://www.wales.gov.uk/topics/transport/?lang=en

Chapter 20

Fire and explosion hazards

David W. Price GexCon UK Ltd, Skelmersdale, UK

To be able to evaluate requirements and know when to seek further assistance, an engineer needs to have sufficient knowledge of general fire and explosion hazards. This chapter provides an overview of these hazards, covering combustion, pyrolysis, diffusive fire, steady state buoyant fires, jet fire, pool fires, and flashover through to deflagration, detonation and boiling liquid expanding vapour explosions. Relevant legislation and key terms of reference, such as lower and upper explosion limits and minimum explosible concentrations, are introduced. Potential ignition sources, and classification of hazardous zone extents for gases, liquids and dusts, are identified and explained. Prevention and protection is discussed with the various risk assessment methods and the use of estimating probability and consequence to determine residual risk against acceptable risk criteria.

doi: 10:10.1680/mohs.40564.0251

CONTENTS

Introduction

HSE (2005) estimated that there are 11 construction site fires every day. There is suggested further reading at the end of this chapter, which gives good guidance on construction fire safety. During this chapter I intend to give you a basic understanding about fire and explosion without any complicated formulae or fancy words. At the end of the chapter you should take away enough information to know the basics and understand when you may need to seek further advice.

In the future you may be presented with a potential risk from fire or explosion; with the following brief notes you should be capable of applying some logic to the challenge you are faced with.

What actually is combustion?

Let us get started and go back to a very basic concept which, to be honest, not many people really understand. Organic solids don't actually burn!

Combustion is the chemical reaction of flammable gas molecules and oxygen molecules, the reaction generates heat and light and that is what we typically call fire.

The point at which the gas and oxygen (air) meet and mix together is better known as the combustion zone; complete combustion from an ideal reaction of the two sets of molecules generates a nice clean flame which for most gases produces a blue flame.

The optimum mixture of gas and oxygen within the combustion zone is called stoichiometry or a stoichiometric concentration. As mentioned, it is not my intention to baffle you with big words – however this one is quite important and hopefully it will be worth getting to know.

If the combustion is not complete within this zone, the molecules can group together without actually fully reacting and

these will form soot and it's this soot that gives off a yellow colour inside the flame. If these soot particles do not burn within the flame then they leave the combustion zone by way of the thermal currents and move away from the flame; yes, this is what we call smoke.

So now it's clear for an organic solid such as wood or other materials of construction to contribute to the combustion process they must first turn into a gas. This is achieved through a process called pyrolysis in which the solid is heated sufficiently to agitate the tightly locked together molecules and allow the solid to liquefy and vaporise.

So that about sums up combustion: we need a gas or something that can be reduced to a gas mixed with oxygen and heated with a form of ignition to complete the reaction and generate flames. Providing the fuel/gas remains available with sufficient oxygen and the heat from the combustion feeds back into the reaction at a rate which is more than the heat loss to the surrounding atmosphere, then combustion will continue.

When is a fire a fire?

I've already mentioned above that the term 'fire' is commonly used for the reaction zone of combustion where heat and light are produced; however, many different types of fires are often referred to, such as steady state buoyant fire; jet fire, pool fire and a flashover.

Diffusion flames, as mentioned above, are where the fuel and oxidiser are initially separate and combustion occurs within the zone where the gas and oxygen are mixed.

A steady state buoyant fire is a turbulent diffusion fire as a result of burning above a pyrolysing solid or liquid under conditions where the fuel has no or very little initial momentum; it's only the heat feedback mechanism that controls the rate of fuel supply and is fundamental in determining the size of fire.

A buoyant fire generally develops over time.

A jet fire is a turbulent diffused fire as a result of a fuel being continually released and propelled with a certain velocity in one or more directions.

Whereas a buoyant fire has little or no initial directional velocity and develops over time, a jet fire has almost instantaneous intensity. This also means that a jet fire can be stopped with a similar instantaneous effect when the fuel supply is isolated.

A pool fire is similar to a buoyant fire. Whereas the fuel is controlled by the feedback of heat (radiation) back to the vaporising fuel spill, the size and duration of a pool fire is fairly predictable as the spread and thickness of the liquid fuel can be determined in most cases.

A flashover is the transition between a localised fire to a situation where all the available fuel surfaces within a compartment are burning. Sometimes a flashover is described when hot flammable gases are ignited and the flame is seen to propagate through the layer of hot gases, typically formed below a ceiling within a compartment, although this can be defined as a flash fire or backdraught in which a flammable gas cloud is premixed with oxygen to create a combustible concentration and instantly ignited, similar to an explosion.

Explosions

Now this is where we start to get into some seriously interesting stuff, big 'bangs' and the like, events which have the potential to cause devastation and carnage.

With a fire, detection systems and procedures can be in place to detect the onset of fire and raise the alarm quickly enough to allow people to escape to a place of relative safety and limit the damage to property by fire containment, suppression systems, etc.

Unlike a fire, an explosion is an instantaneous event (commonly referred to as deflagration), combustion of the fuel in milliseconds and seconds rather than minutes and hours.

An explosion is a process where combustion of a premixed flammable cloud, i.e. fuel–air or fuel–oxidiser creates rapid increase of pressure. Basically pressure is generated during combustion due to the sudden rise in temperature and as you probably know due to the laws of physics temperature has a direct correlation with pressure. Depending upon the confinement and turbulence generated during an explosion, pressures can be very high and cause catastrophic damage and loss of property and life.

Yes! The same applies for a dust explosion, I can hear your mind thinking of all the nasty dusts that must be explosible – what are they? To be honest you would be better trying to list the dusts that are inert and not explosible. Your kitchen cupboards are full of dusts that are explosible such as flours, icing sugars, custard powder, etc. Most of these will generate higher pressures and be more violent than a typical natural gas explosion. Dust can lie on surfaces and be undisturbed for many years and then raised as a flammable cloud during a disturbance typically known as a primary event. If an effective igni-

tion source is present a dust explosion may occur which can be catastrophic – in nature known as the secondary event. Details of recent dust explosions and associated losses including human and business can be found on the Health and Safety Executive (HSE) and United States (US) Chemical Safety Board (CSB) websites. In particular, the CSB website has invaluable videos and information in relation to the Imperial Sugar dust explosion during 2008 which took the lives of 14 and injured 36. The HSE website has details of an explosion in a grain storage complex which killed 11 (Wright, 2000).

These dusts of certain particle size and low moisture concentration mixed with oxygen (air) react in a similar way: ignition energy causes the surface of the dust particle to pyrolyse and break down into the various elements, producing gases and radiation which ignite the surrounding dust particles creating heat and of course pressure; a chain reaction then develops throughout the dust cloud, however this time higher temperatures and turbulence are experienced. Turbulence generates a positive feedback mechanism to accelerate the mass burning rate of the combustion process.

I can here you now saying, 'Well what's the main difference between a gas explosion and a dust explosion?' Good question, although the basic dynamics of combustion are similar. With dust explosions we typically make reference to a primary event and possible secondary events. The primary event is the initial ignition of a dust cloud; from this initial event a pressure wave moves away from the point of ignition pushing unburnt fuel (dust) away, potentially raising other deposits of dust into a dust cloud and creating a turbulent atmosphere. Following the pressure wave the flame front increases the temperature of the surrounding area and propagates into and through the secondary flammable dust clouds; however, the combustion process within these secondary clouds is of increased intensity, due to the turbulence creating a positive feedback mechanism which accelerates the combustion process and so on.

An explosion will only occur if the fuel to air mixture falls within certain limits known as the lower and upper explosion limit (LEL and UEL), or minimum explosible concentration (MEC) for dust; these differ for the type of fuel in question, and the LEL/UEL/MEC are important factors that need to be established when conducting a risk assessment. This will be discussed later in the chapter.

Explosions can occur inside process equipment or pipes, in buildings or processing plants, onshore and offshore, in open process areas or confined areas. When we are talking about an explosion as an event, it is a good idea to take a look at the events leading to and after the explosion, see **Figure 1**.

Figure 1 shows what can happen if a combustible fuel or evaporating liquid is accidentally released into the atmosphere. If the cloud, formed from the release, is not within the flammability limits or if an ignition source is lacking, the flammable cloud may be diluted and disappear. Ignition may occur immediately, or may be delayed by tens of minutes, all depending on the circumstances. In the case of an immediate ignition (i.e. before mixing with air or an oxidiser) a fire will occur for the release of gas or vaporising liquid.

ICE manual of health and safety in construction © 2010 Institution of Civil Engineers

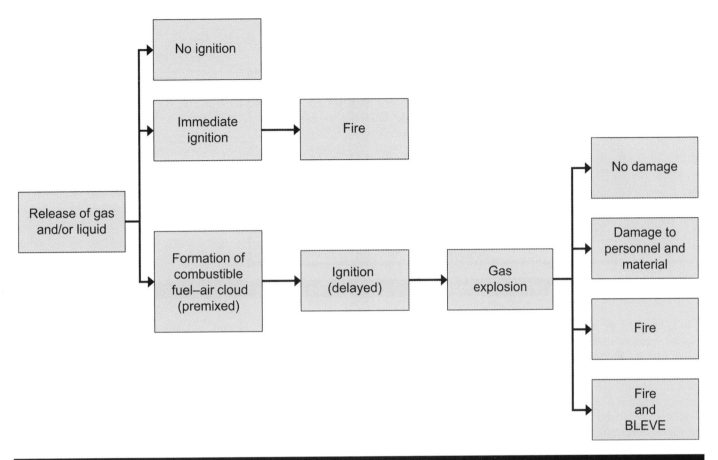

Figure 1 An event tree showing typical consequence for release of a combustible fuel or evaporating liquid into the atmosphere

The most dangerous situation will occur if a large combustible premixed fuel–air cloud is formed and ignites. The time from release start to ignition can be from a few seconds up to tens of minutes. The amount of fuel can be from a few kilograms up to several tons.

The pressure generated by the combustion wave will depend on how fast the flame propagates and how the pressure can expand away from the cloud (governed by confinement). The consequences of explosions range from no damage to total destruction. The pressure build-up due to the explosion can damage personnel and material or it can lead to accidents such as fires and BLEVEs (domino effects). BLEVE is an acronym for boiling liquid expanding vapour explosion. A BLEVE is an explosion due to the vaporisation of liquids with a high vapour pressure contained within a vessel. The failure of the vessel is often caused by an external fire. If the substance released is a fuel, the BLEVE can result in very large fire balls. Rocketing vessels are also hazards related to BLEVEs. And if you are wondering what a 'flash fire' is, this is basically an explosion that generates little pressure with thermal effects being the predominant hazard.

Fires are very common events after explosions.

When a fuel cloud is ignited the flame can propagate in two different modes through the flammable parts of the cloud.

These modes are:

i) deflagration

ii) detonation.

The deflagration mode of flame propagation is the most common. A deflagration propagates at subsonic speed relative to the unburned gas; typical flame speeds (i.e. relative to a stationary observer) are of the order of 1 to 1000 m/s. The explosion pressure may reach values of several bar (g), depending on the flame speed.

A detonation wave is a supersonic combustion wave (relative to the speed of sound in the unburned gas ahead of the wave). The shock wave and the combustion wave are in this case coupled. In a fuel–air cloud a detonation wave will propagate at a velocity of 1500–2000 m/s and the peak pressure is typically 15 to 20 bar.

In an accidental gas explosion of a hydrocarbon–air cloud (ignited by a weak source – a spark) the flame will normally start out as a slow laminar flame with a velocity of the order of 3–4 m/s. If the cloud is truly unconfined and unobstructed (i.e. no equipment or other structures are engulfed by the cloud), the flame is not likely to accelerate to velocities of more than 20–25 m/s, and the overpressure will be negligible if the cloud is not confined.

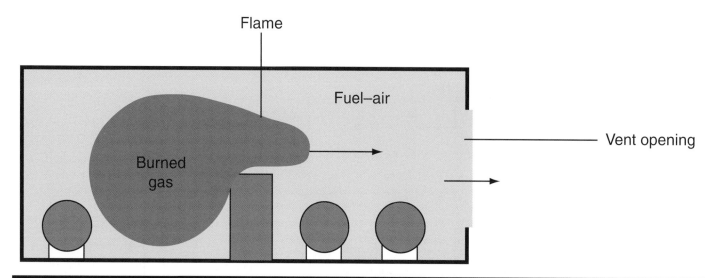

Figure 2 Gas explosion in a partly confined area with obstacles

In a building or in an offshore/petrochemical module with process equipment as shown schematically in **Figure 2**, the flame may accelerate to several hundred metres per second. When the gas is burning, the temperature will increase and the gas will expand by a factor of up to 8 or 9. The unburned gas is therefore pushed ahead of the flame and a turbulent flow field is generated. When the flame propagates into a turbulent flow field, the effective burning rate will increase and the flow velocity and turbulence ahead of the flame increase further. This strong positive feedback mechanism causes flame acceleration and high explosion pressures and in some cases transition to detonation.

In a confined situation, such as a closed vessel, a high flame velocity is not a requirement for generation of pressure. In a closed vessel there is no or very little relief (i.e. venting) of the explosion pressure and therefore even a slow combustion process will generate significant pressures.

The consequences of an explosion will depend on:

- type of fuel and oxidiser
- size and fuel concentration of the combustible cloud
- location of ignition point
- strength of ignition source
- size, location and type of explosion vent areas
- location and size of structural elements and equipment
- mitigation schemes.

Legislation

Fire and explosion legislation is similar for most developed countries. Within the UK the main drivers are: the Regulatory Reform (Fire Safety) Order 2005, statutory instrument SI 1541, commonly referred to as the RRO; and the Dangerous Substances and Explosive Atmospheres Regulations, statutory instrument SI 2776, often referred to as DSEAR, or within Europe ATEX 137 (*Atmospheres Explosible*).

In addition the Health and Safety at Work Act 1974 also drives the requirements for fire and explosion safety without being explicit.

All of the driving legislation calls upon the responsible person to conduct sufficient assessment of the risks and provide measures so far as reasonably practicable to ensure the risk from fire and or explosion is reduced to an acceptable level.

Various approved codes of practice (ACoPs) exist, with some being more prescriptive than others.

I will not attempt to list all the details of the regulations as this will most certainly send you off to sleep. All I will advise is that you only attempt to assess the risk if you consider yourself fully competent to do so and if you are not sure then seek advice from a specialist within the particular field. Be sure to verify the competence of the specialist. Further information on health and safety legislation can be found in Chapter 1 *Legal principles*.

Hazard and risk assessment

Risk assessments come in all shapes and sizes; some are basic and sufficient for the risk they are assessing, while others are more complex and try to solve the risk mathematically. In the following three subsections I will try to give you a brief description of the various methods most commonly used.

Qualitative methods

A qualitative risk assessment is an unstructured approach and typically uses a very informal technique, where the assessor will judge the risk perceived from their own knowledge or experience; this technique would typically be used to assess 'relatively low, relatively well understood and/or when there are a large number of premises to be assessed'. Providing the same assessor remained responsible for all the premises within the group being assessed, it would be reasonable to assume that the unstructured assessment of each of the premises would follow the same pattern through the various stages to identify hazards and thus the resultant evaluation of the risk would follow a similar train of thought. Thus this provides the assessor, a datum for which each premises is reviewed with resulting in a consistent philosophy across all sites, hence an iterative approach. If two different assessors conducted an assessment of the same application it would be unusual for them both to agree totally on the same level of risk.

Semi-quantitative methods

This method is typically a half-way house between a qualitative approach and a quantitative approach, where the individual risks are scored using a points scheme, lists and tables are used with defined scores for the identification of hazard and other scores for the level of safety provision; the balance of the two hence determines the residual risk remaining. This method of semi-quantitative risk assessment places some numerical value on the level of risk and therefore removes some of the subjective approach of individual assessors. This method of assessment is particularly useful for assessing the same type of risks and premises repeatedly with trained assessors who will use the same scoring system. When several of the premises have been assessed it is very quick to identify the high risk offenders and start implementing risk-reducing measures on these. first working down the list to those identified having lower risks, this can continue and constantly increase overall levels of safety as part of an overall short, medium and long term strategy.

Quantitative methods

As with the other methods discussed, a full quantitative risk assessment would normally commence with these simpler methods; however, this method would then take the risk assessment into a more complex and analytical realm. Statistical and probability data for hazard identification, frequency analysis, consequence analysis, risk acceptance and risk reduction would be calculated and compared using a very in-depth mathematical procedure, more often with a computer based model, mainly to save time when performing those interactive computations.

The following notes and tables describe how you could typically conduct a risk assessment for an explosion hazard, starting with assessing the probability of an explosion and then considering the consequence of an event.

Estimating the probability for explosions

To be able to quantify the probability for the occurrence of an explosive atmosphere, properties of the combustible material should be considered, together with how likely it is that the combustible material will be mixed with air. The probability of a specific ignition source being able to ignite the explosive atmosphere is considered based on different criteria, such as available energy of the ignition source and the ignition sensitivity of the atmosphere. For mechanically generated sparks,

collision speed, friction, contact time and physical properties of the colliding materials are included.

Whether an ignition source is capable of igniting an explosive atmosphere depends on several properties of the atmosphere, for instance the fuel concentration and the turbulence level. High probability of an ignition source does not mean that it will ignite the atmosphere.

The factors mentioned above are considered individually and form the basis for estimating how often an explosion can occur. It is not possible to give the exact frequencies for an explosion. In the following descriptions the probability for an explosive atmosphere and the probability for an ignition source are ranged from 'I' to 'V', where 'I' has the lowest probability and 'V' has the highest probability. Each 'range' (I, II, III, IV and V) describes a range in 'probability' or 'frequency'.

The probability of an explosion occurring depends on the probability of an ignition source and the probability of having an explosive atmosphere. The probability of an explosion will be the product of these two probabilities. To be conservative, the incident with the lowest probability is used as a basis for estimating the total probability. This gives a more conservative estimate than multiplying the two probabilities.

The probability for a secondary event depends on the probability for the primary event. A conservative estimate for secondary explosions would be by using the probability of the primary event.

Estimating the consequences of an explosion

The consequence for personnel (D_p) and equipment (D_e) is estimated based on the expected effect of the explosion. This is estimated based on expected damage caused by the heat, pressure or projectiles. The consequence for personnel and equipment from an explosion depends on the explosion pressure and the heat intensity from the explosion. Pressure build-up in enclosed units might cause the units to rupture resulting in heat radiation from flames, dispersion of pressure waves and flying objects.

The strength of an explosion depends on several factors, for example the initial conditions of the fuel cloud, including the fuel concentration, initial turbulence and the position of the ignition source. The properties of the combustible material are also important, including chemical composition. The properties of the explosive atmosphere will change over time, hence the time of the explosion is important for the explosion propagation.

Flames propagating out from a ruptured vessel release heat that might injure personnel or cause damage to equipment. The convective heat transfer during an explosion causes the most severe burns. Burns/damage might occur if personnel or equipment are in direct contact with the explosion flame.

Definitions

The probability or the frequency of an explosion occurring and the potential consequences is estimated from I to V, as

Probability of the formation of an explosive atmosphere		Probability of the formation of an effective ignition source	
Range, D_a	Description	Range D_i	Description
I	Very unlikely	I	Very unlikely
II	Unlikely	II	Unlikely
III	Somewhat likely	III	Somewhat likely
IV	Likely	IV	Likely
V	Very likely	V	Very likely

Probability for an explosion to occur				
Range D_e	Description	Definition	Probability for incident during 1 hour	
I	Very unlikely	< 1/10 000 per year	1.14E−8	0.000000114 % per hour
II	Unlikely	> 1/10 000 per year < 1/100 year	1.14E−6	0.0000114 % per hour
III	Somewhat likely	> 1/100 < 1/10 per year	1.14E−5	0.00014 % per hour
IV	Likely	> 1/10 year < 1 per year	1.14E−4	0.0114 % per hour
V	Very likely	> 1 per year	>1.14E−4	>0.0114 % per hour

Consequence for personnel and equipment		
Range $D_p D_e$	Description	Definition
I	Personnel	No injury
	Equipment	Marginal damage to process units. Process shut down
II	Personnel	Limited injury
	Equipment	Damage to process unit (< £10 000)
III	Personnel	Personnel injury
	Equipment	Process unit collapse and possible damage to corresponding units (> £10 000; < £100 000)
IV	Personnel	Serious personnel injury, possible loss of life
	Equipment	Significant damage to several process units (> £100 000; < £1 000 000)
V	Personnel	Loss of one or several lives
	Equipment	Plant fully damaged (> £1 000 000)

Table 1 Definitions of the probability and consequence for explosions under normal operation

described previously. The definition and description of the different values are given in **Table 1**.

Estimating the explosion risk

The explosion risk is the product of the probability of an explosion occurring and its consequences. Normally a qualitative risk evaluation is completed for each process unit. The risk level for explosions can be estimated from the matrix given in **Figure 3** based on the probability and consequence. The risk level increases from E to A.

Acceptance criteria

The risk level and the 'recommended acceptance criteria' are selected and based on the probability for human and economic

Consequence						
	V	C	B	A	A	A
	IV	D	C	B	A	A
	III	E	D	C	B	A
	II	E	E	D	C	B
	I	E	E	E	D	C
		I	II	III	IV	V
		Probability				

Figure 3 Risk matrix

	Risk level	Acceptance criteria	Recommended action
A	Very high	Unacceptable	Risk reducing measures must be implemented
B	High	Unacceptable	Risk reducing measures must be implemented
C	Medium	Medium	Risk reducing measures should be implemented
D	Low	Acceptable	Risk reducing measures can be implemented
E	Very low	Acceptable	Risk reducing measures are not required

Table 2 Risk level – definitions and recommended acceptance criteria

loss according to **Figure 3**. The selected criteria are given in **Table 2**.

An example showing the different parts of the assessment record table is shown in **Table 3**.

Prevention and protection

The starting point for any fire/explosion risk assessment is to identify potential combustible materials and/or flammable atmospheres. It is common to break down the assessment into areas: for a fire risk assessment these would typically consist of rooms, whereas for an explosion risk assessment the process would typically be broken down into various defined elements of the process.

Fire

In general, for fire risk assessments the principal factor will be to ensure that a fire can be detected and a suitable warning given providing sufficient time to allow all occupants to escape the building to a place of relative safety before conditions become untenable. In addition consideration must be given to preventing structural collapse and the provision for fighting the fire.

Buildings should be designed in such a manner as to satisfy the above statement. Various techniques can be adopted to satisfy this statement. Standard buildings within the UK would typically be designed in accordance with the Building Regulations Approved Document B (Fire Safety); this is a fairly prescriptive guide which provides sound practical advice broken down into several sections:

B1 – Means of Warning and Escape

B2 – Internal Fire Spread (Linings)

B3 – Internal Fire Spread (Structure)

B4 – External Fire Spread

B5 – Access and Facilities for the Fire Service.

Also, buildings may be designed using a bespoke assessment method based on the prediction of fire growth and smoke spread, and the influence of mitigating measures; this is typically referred to as a Fire Engineered Solution.

Both methods determine the requirement for fire detection and alarm, quantity size and location of escape routes, fire

Process unit	Probability of flammable atmosphere	Probability of ignition					Probability of explosion
		Equipment (electric and mechanical)	Hot surfaces	Electric and electrostatic sparks and discharges	Mechanical sparks	Flames	
Example	IV	II	I	I	I	I	II

EXPOSURE TO EXPLOSION

PRIMARY EXPLOSION

Probability (injury/damage)		Consequence		Risk	
Personnel	Equipment	Personnel	Equipment	Personnel	Equipment
I	II	III	III	E	D

SECONDARY INCIDENTS (inclusive explosions)

Personnel	Equipment	Personnel	Equipment	Personnel	Equipment
I	I	V	V	C	C

Comments:

EXAMPLE

Key:

Process unit	Probability	Consequence	Risk	Ignition source
the process unit to which the analysis applies	the estimated *explosion* probability. The probability of an explosion is the product of the probability for 'an explosive atmosphere' and 'ignition source'. The lowest of these probabilities is chosen. Definitions for explosion related probability (and consequences) are given in **Figure 3**.	the consequences for an event considering both personal injuries and damage to equipment. Both primary and secondary consequences are given. Definitions for explosion related probability, (and consequences) are given in **Table 1**.	the product of *probability* and *consequence*. Both primary and secondary risk is estimated.	probability for occurrence of the five most relevant ignition sources are given.

Table 3 Table summarising the different parts of the assessment record tables (example)

resisting elements of construction typically known as compartmentalisation (holding the fire within a defined compartment for a period of time, typically 30 mins, 60 mins and 180 mins) including the provision for rescue and fire fighting.

Explosion

In general, for an explosion risk assessment the availability of a flammable atmosphere will be considered in combination with an effective ignition source and an assessment of the likelihood of and severity of an event.

Various codes and Standards are available for assessing the likelihood of flammable gases, flammable liquids and flammable dusts. Within Europe the existence of these potential flammable atmospheres is classified into designated zones.

Explosive atmospheres and area classification ('ex-zones')

Explosive atmospheres will occur if the fuel concentration is higher than the LEL and lower than the UEL. However, for

ICE manual of health and safety in construction © 2010 Institution of Civil Engineers

mists these values are often different. During normal operation such atmospheres may be present at several places inside items of process plant. If parts of the equipment leak, such atmospheres may also be present outside the process units.

The process plant will typically be area classified with respect to how often a hazardous explosive atmosphere may be formed. The area classification is based upon the description of the plant and assumptions. The different zones for gases and vapours are defined as follows.

Zone extent for gases

Zone 0

A place in which an explosive atmosphere consisting of a mixture with air of flammable substances in the form of gas, vapour or mist is present continuously or for long periods or frequently.

In general these conditions, when they occur, arise inside containers, pipes and vessels, etc.

Zone 1

A place in which an explosive atmosphere consisting of a mixture with air of flammable substances in the form of gas, vapour or mist is likely to occur in normal operation occasionally.

This zone can include, among others:

- the immediate vicinity of zone 0;
- the immediate vicinity of feed openings;
- the immediate vicinity around filling and emptying openings;
- the immediate vicinity around fragile equipment, protective systems, and components made of glass, ceramics and the like;
- the immediate vicinity around inadequately sealed glands, for example on pumps and valves;
- with stuffing-boxes.

Zone 2

A place in which an explosive atmosphere consisting of a mixture with air of flammable substances in the form of gas, vapour or mist is not likely to occur in normal operation but, if it does occur, will persist for a short period only.

This zone can include, among others, places surrounding zones 0 or 1.

Zone 2 NE

Zone 2 NE indicates a theoretical zone which would be of negligible extent under normal conditions.

The size of any zone is defined as the area limited by a distance in any direction from the place where the combustible material release occurs. This means that an area, classified as zone 2, in the vicinity of an area classified as zone 0 also will include a 'transition zone 1' between zone 2 and zone 0.

Zone extent for dusts

Sources of dust release are divided into the following grades depending on the order of decreasing severity:

- Continuous
  Formation of a dust cloud: locations in which a dust cloud may exist continuously, or may be expected to continue for long periods or for short periods which occur frequently.
- Primary grade of release
  A source can be expected to occasionally release combustible dust in normal operation.
- Secondary grade of release
  A source which is not expected to release combustible dust during normal operation; if it releases, it is likely to do so only infrequently and for short periods only.

These grades of release will according to BS EN 60079-10-2 give rise to the hazardous zones, detailed in **Table 4**.

Layers, deposits and heaps of combustible dust shall be considered as any other source which can form an explosive atmosphere.

Zone 20

A place in which an explosive atmosphere, in the form of a cloud of combustible dust in air, is present continuously, or for long periods or frequently for short periods.

Zone 21

A place in which an explosive atmosphere, in the form of a cloud of combustible dust in air, is likely to occur occasionally in normal operation.

Zone 22

A place in which an explosive atmosphere, in the form of a cloud of combustible dust in air, is not likely to occur in normal operation but, if it does occur, will persist for a short period only.

Extent of zones for explosive dust atmospheres

The extent of a zone for explosive dust atmospheres is defined as the distance in any direction from the edge of a source of dust release to the point where the hazard associated with that zone is considered to no longer exist. Consideration should be given to the fact that fine dust can be carried upwards from a source of release by air movement within a building. Where the classification gives rise to small unclassified areas between classified areas, the classification should be extended to the full area, although every effort must be made not to try to blanket

Presence of combustible dust	Resulting zone classification of area of dust cloud
Continuous presence of a dust cloud	20
Primary grade of release	21
Secondary grade of release	22

Table 4 Dust zoning

an area, as the whole essence should be to reduce the presence of a flammable atmosphere and not encourage blanket acceptance.

A continuous grade of release normally leads to a zone 20, a primary grade to zone 21 and a secondary grade to zone 22. However, due to the effect of ventilation, this may be different.

Some potential points of release as described above should be considered as continuous or primary grades of release. This implies that a zone 20 or 21 will prevail around the points of release.

In the case of poor ventilation conditions, a gas zone may have to be classified as a zone 1 or even 0. Local artificial ventilation, if of sufficient volume flow, can cause the zone type to be reduced in severity, but this relies on many other factors such as congestion, reliability and availability.

According to the European Standard EN 60079-10, an *explosive atmosphere* is not necessarily considered to be a hazardous explosive atmosphere resulting in a zone.

An *explosive atmosphere* is one in which a flammable substance in the form of a dust is mixed with air and once ignited combustion spreads through the entire unburned mixture. An explosive atmosphere only exists if the MEC is exceeded and the upper explosive limit is not reached. The MEC is an example of a substance-specific characteristic, which is experimentally determined. A hazardous explosive atmosphere is an *explosive atmosphere* which may occur in such quantities as to require special precautions to protect the health and safety of the workers concerned, and is considered to be *hazardous*. An explosive atmosphere which is not expected to occur in such quantities as to require special precautions is considered as non-hazardous and does not result in a zone.

Dust layers may be ignited by hot surfaces and discharges and represent a significant fire risk.

Fittings, covers, closures, etc.

Flanges, access covers, closures, and flexible couplings unless specifically designed to be dust-tight (usually IP rated – see box below) are all considered as potential points of dust release. All these potential leakage points will have a hazardous zone associated with them; the type (usually a zone 22) and the size of the zone will depend on several factors including the grade of release.

Box 1 IP rating – International protection rating/ingress protection code

IEC 529 is the International Standard for Classification of Degrees of Protection by Enclosures.

BS EN 60529: 1992 Degrees of Protection Provided by Enclosures (IP code) is the European Standard, covering the classification of degrees of protection provided by enclosures for electrical equipment with a rated voltage not exceeding 72.5 kV. This standard can also be applicable to empty enclosures provided general requirements are met. The standard provides:

- definitions for degrees of protection with regard to: protection of persons against access to hazardous parts inside the enclosure; protection of the equipment inside the enclosure against ingress of solid foreign objects; protection of the equipment inside the enclosure against harmful effects due to the ingress of water;

- designations for these degrees of protection;

- requirements for each designation;

- tests to be performed to verify that the enclosure meets the requirements of this standard (BS EN 60529: 1992).

The rating is commonly expressed as 'IP' followed by two numbers, the first relating to protection against ingress of foreign bodies (e.g. dust) and the second to protection against the ingress of water, e.g. IP54.

The protection rating against the ingress of foreign bodies ranges from 0 (no protection) to 6 (dust-proof). The protection rating against the ingress of water ranges from 0 (no protection) to 8 (suitable for permanent immersion) (ezscreen, 2010).

Zoning in practice

With all new installations, the driving factor will be to maintain hazardous zones to an absolute minimum.

For gas installations this can be achieved by using solid piped systems and minimal flanged connections and ensuring that good ventilation is provided throughout the installation combined with regular inspection and pressure testing of lines.

For liquid installations that are under pressure, shrouds or lagging can be installed to flanged connections to minimise the extent of a flammable mist formation and to encourage rapid return to an aqueous fuel, by providing drip trays or mats beneath connections; regular inspections will identify leaks before they become a significant hazard.

For dusts the main emphasis is directed to good containment, efficient capture at source and regular general area housekeeping.

Ignition sources

As part of any fire and explosion risk assessment, a contributing factor will be the availability of an effective ignition source. Most of these are common sense, nevertheless I will detail some main ignition sources that should be considered as part of an assessment. These are, however, especially relevant for an industrial application.

International guidelines distinguish 13 different types of ignition sources. EN 1127-1 (for example) defines ignition sources into 13 groups. These are:

1. Hot surfaces.

2. Flames and hot gases (including hot particles).

3. Mechanically generated sparks.

4. Electric equipment.

5. Stray electric currents.

6. Static electricity.

7. Lightning.

8. Radio frequency.

9. Electromagnetic waves.

10. Ionising radiation.

11. Ultrasonic.

12. Adiabatic compression and shock waves.

13. Exothermic reactions (including self-ignition of dust).

Some of the main ignition sources are described below and the potential of the ignition sources for igniting flammable atmospheres are discussed.

The strength of an ignition source for igniting a specific atmosphere can be quantified through scientific laboratory tests. Based on these tests, in some cases especially for industrial applications, the maximum temperature of a surface or the energy content of an electric spark can be controlled, hence preventing ignition. However, it is difficult to evaluate if random ignition sources such as mechanically produced sparks, hot surfaces and electrostatic sparks can ignite a specific atmosphere in a given situation. The amount of energy in such ignition sources depends on a number of parameters, resulting in varying energy content of the ignition source.

Hot surfaces

Experience and statistics show that hot surfaces have caused several fires and explosions, hence the ATEX Directives and why several European standards set different requirements regarding surface temperature for equipment in contact with flammable atmospheres. Different requirements are set according to the 'equipment categories' (Category 1, 2 and 3) and these categories allow equipment to be placed in certain classified explosive zones as described earlier.

Equipment, protective systems and components for use in explosive atmospheres should comply with the following specific requirements.

For *Category 1* equipment the main rule is that the temperature of all equipment, protective systems and component surfaces which can come into contact with explosive atmospheres shall not, even in the case of rare malfunctions, exceed 80% of the minimum ignition temperature of the combustible gas or liquid in degrees Celsius.

For *Category 2* equipment the main rule is that the temperature of all equipment, protective systems and component surfaces which can come into contact with explosive atmospheres shall not exceed 80% of the minimum ignition temperature of the combustible gas or liquid in degrees Celsius during normal operation and in the case of malfunctions. However, where it cannot be excluded that the gas or vapour can be heated to the temperature of the surface, the surface temperature shall not exceed 80% of the minimum ignition temperature of the combustible gas or liquid in degrees Celsius. These values may only be exceeded in the case of rare malfunctions.

For *Category 3* equipment the main rule is that the temperature of all equipment, protective systems and component surfaces which can come into contact with explosive atmospheres shall not exceed the minimum ignition temperature of the combustible gas or liquid in degrees Celsius during normal operation.

Whether or not a hot surface is capable of igniting a flammable atmosphere, is also dependent on the size and the orientation of the hot surface and of the properties of explosive atmosphere, such as concentration, convection and turbulence level.

Flames and hot gases

Open flames, for example welding or cutting, are strong ignition sources that might ignite explosive atmospheres and combustible materials. Welding and cutting might also produce hot and burning metal fractions.

To prevent ignition from hot work (operations involving any equipment producing heat or having naked flames), management control needs to be established. These routines should include requirements for permission to work (hot work permits), no smoking policies, etc., before such work is initiated. The routines should also include requirements for removal of combustible materials and to ensure that the appropriate fire extinguishing equipment is available. Equipment and pipe systems exposed to hot work must be properly cooled before flammable materials are brought into contact with the equipment.

During hot work activities, the areas surrounding the work area should be cleaned of combustible materials and it is usual for one team member to be nominated as fire watch. They are responsible for monitoring stray embers and to conduct regular inspections after the hot work has been completed to visually inspect for signs of delayed ignition.

If a flammable vapour–air mixture comes into contact with gases which have a temperature higher than the auto-ignition temperature of the vapour–air mixture, ignition may occur.

According to ATEX, open flames are not allowed for any of the three equipment categories.

Mechanically generated heat from friction and sparks

Mechanically produced ignition sources can occur when two objects collide against each other (single or repeated impact sparks), by rubbing against each other (frictional sparks) or during grinding. The incendivity (ability to cause effective ignition) of these sparks is closely related to the type of materials that collide, the pressure with which these objects are moving against each other, the relative velocity and the properties of the flammable atmosphere involved. The incendivity of these sparks for various vapour–air mixtures can be judged on the basis of relationships between minimum ignition energy and auto-ignition temperature for each of the spark types.

If the two colliding materials are made of steel and the relative collision speed is higher than 1 m/s, ignition sources that are strong enough to ignite the flammable atmosphere may be produced.

Such sparks can be generated from maintenance activities such as hammering or, for example, by forklift truck forks hitting metal objects. It is normal for such forks to be either made of non-sparking material or coated with such a material. If they are coated, the inspection and maintenance is vital to ensure that the integrity of the coating remains unaffected by the operation. Sparks may still result if fork coatings are worn or defective.

Equipment (electrical and mechanical)

All equipment to be used in potentially explosive atmospheres (zones 0, 1 and 2 for gas and zones 20, 21 and 22 for dust) shall be approved or certified for use in such areas. All equipment shall be designed to prevent ignition in case of contact with the flammable atmosphere. Within the European Community (EC) this equipment is governed by the ATEX (Atmospheres Explosible) Directive 94/9/EC.

Three different 'equipment categories' are established for equipment used in explosive atmosphere. The three equipment categories are 'category 1', 'category 2' and 'category 3' – equipment (see **Table 5**). The categories have different requirements with respect to safety levels. **Table 5** shows an overview of the different ex-zones, the required safety level and the approved certification agencies required for the different equipment categories.

When equipment, protective systems and components are used in hazardous places, checks shall be made to see whether ignition hazards can occur, by considering potential ignition sources. If ignition hazards are possible, efforts shall be made to remove the sources of ignition from the hazardous place. If this is not possible, different measures shall be implemented with attention being paid to the following information.

The measures shall render the sources of ignition harmless or shall reduce the likelihood of occurrence of the effective ignition sources. This can be achieved by proper design and construction of equipment, protective systems and components by operational procedures, and also by means of appropriate measuring and control systems.

The extent of the protective measures depends on the likelihood of occurrence of an explosive atmosphere and the consequences of a possible incident. This is achieved by discriminating between the different categories of equipment as specified by the ATEX Directive 94/9/EC. These categories reflect the requirements of the different zones.

The criteria determining the classification into categories are the following.

Category 1 comprises equipment designed to be capable of functioning in conformity with the operational parameters established by the manufacturer and ensuring a very high level of protection. Equipment in this category is intended for use in places in which explosive atmospheres caused by mixtures of air and gases, vapours or mists or by air–dust mixtures are present *continuously, for long periods* or *frequently.*

Category 1 equipment offers a high degree of protection (redundancy) and is designed to ensure the required protection concept is still effective during a rare malfunction, and is characterised by means of protection such that:

- either, in the event of failure of one means of protection, at least an independent second means provides the requisite level of protection;
- or the requisite level of protection is assured in the event of two faults occurring independently of each other.

Category 2 comprises equipment designed to be capable of functioning in conformity with the operational parameters established by the manufacturer and of ensuring a high level of protection.

Equipment in this category is intended for use in places in which explosive atmospheres caused by gases, vapours, mists or air–dust mixtures *are likely to occur.*

The means of protection relating to equipment in this category ensures the requisite level of protection, even in the event of frequently occurring disturbances or equipment faults which normally have to be taken into account.

Category 3 comprises equipment designed to be capable of functioning in conformity with the operational parameters established by the manufacturer, and of ensuring a normal level of protection.

Equipment in this category is intended for use in places in which explosive atmospheres caused by gases, vapours, mists or air–dust mixtures *are unlikely to occur* or, if do occur, *are likely to do so only infrequently and for a short period only.*

Equipment in this category ensures the requisite level of protection during normal operation.

If dust or liquid enters the inside of an enclosure for electrical equipment, it might cause short-circuit of the equipment and a fire inside the enclosure. According to the Standard, all electrical equipment used in explosive atmospheres shall be designed in accordance with given specifications. The maximum surface temperature of the enclosure shall not exceed certain maximum temperatures.

Electrostatic discharges and sparks

Another group of potential ignition sources that might ignite explosive explosions is electrostatic discharges and electrostatic sparks caused by uncontrolled electrostatic charging. According to the statistics a significant number of explosions are initiated by electrostatic discharges and sparks.

Different electrostatic discharges and electrostatic sparks may ignite explosive atmospheres. A number of parameters affect the forming of the electrostatic discharges/sparks. The main criterion is that the generation of charge is larger than the loss to the surroundings. If so, significant electric charges may accumulate. This charge will be discharged if the voltage is sufficiently high. Different types of electrostatic discharges may occur depending on the design of the equipment. These

Ex-zone for gas (dust)	Explosive atmosphere?	Equipment category	Level of protection	Method for approval	
				Electrical equipment	Mechanical equipment
2 (22)	Not likely during normal operation	3	Normal (not cause ignition during normal operation)l	Can be declared by the manufacturer	Can be declared by the manufacturer
1 (21)	Likely but limited during normal operation	2	High (must not cause ignition even at 1 failure)	Must be approved and certified by a 'technical control agency'	Declared by the manufacturer but documentation shall be sent and stored at a 'technical control agency '
0 (20)	Continuously, often or during long periods	1	Very high (must not cause ignition even at 2 failures)	Must be approved and certified by a 'technical control agency'	Must be approved and certified by a 'technical control agency'

Table 5 Ex-zones, equipment categories and methods/agencies for approval according to the ATEX directive 1994/9/EC

include corona discharges, brush discharges, conic discharges, propagating brush discharges and electrostatic sparks. In many cases it is very difficult to determine whether or not electrostatic discharges or sparks can cause ignition.

An electrostatic spark can occur between a non-grounded and charged conducting (often metallic) object and a second grounded conductor. The non-grounded conductor is normally charged due to direct contact with charged particles such as liquid droplets or solids (dust particles). The energy in the electrostatic sparks can be up to 1 J depending on the capacitance of the charged object and the potential difference between charged and grounded object.

Exothermic reactions, including self-heating

Exothermic reactions can act as an ignition source when the rate of heat generation exceeds the rate of heat loss to the surroundings. Many chemical reactions are exothermic. Whether a reaction can reach a high temperature is dependent, among other parameters, on the volume/surface ratio of the reacting system, the ambient temperature and the residence time. The high temperatures generated can lead to ignition of flammable atmospheres and also the initiation of smouldering and/or burning of the material.

There are examples of hydrocarbons, such as oil, self-igniting or undergoing exothermic reactions at lower temperatures than the self-ignition temperature when the oil is in contact with other materials, e.g. thermal insulation materials.

Other ignition sources

Other potential ignition sources such as stray electric current, cathodic corrosion protection, lightning, radio frequency, electromagnetic waves, ionising radiation, ultrasonic, adiabatic compression and shock should also be considered when assessing risk.

Summary of main points

Hopefully now you have a clearer understanding of the basic principles behind combustion including what defines a

stoichiometric mixture and fire including diffusion fires, steady state fires, buoyant fires, jet fires, pool fires and flashover.

In addition, the phenomenan of explosions have been described including the deflagration and detonation of flammable gases, vapours and dusts, in addition to the description of BLEVEs.

The importance of congestion from obstacles within a vapour cloud explosion to generate turbulence effects and contribute to the near field and far field overpressure development has been discussed.

Information in relation to key legislation, namely the Regulatory Reform Order, the Dangerous Substances and Explosive Atmospheres Regulations and ATEX, have been addressed including the requirements for suitable and sufficient risk assessments upon the completion of a hazardous area assessment (zoning) which determines the probability and extent of a flammable atmosphere and the estimation of the consequence of a primary or secondary event taking into account the likelihood of an effective ignition source.

The main learning point is to ensure you have sufficient information available to make a thorough assessment, do not rely on assumptions and most of all know your own limitations. Don't be afraid to ask for the advice of those who specialise within the field of fire and explosion safety, after all I wouldn't attempt to act as a civil engineer, even though I have a hard hat, site boots and a high viz vest.

References

ezscreen. Understanding IP and NEMA Ratings, 2010. Available online at: http://www.ezscreen.com/ip_ratings.htm

Health and Safety Executive. *Construction Fire Safety*. Construction Information Sheet No 51, 2005, London: HSE Books.

Wright, S. *Explosion in a Grain Silo – Blaye (TD5/029) – Hazardous Installations Directorate Discipline Information Notes*, 2000, London: HSE. Available online at: http://www.hse.gov.uk/foi/internalops/hid/din/529.pdf

Referenced legislation and standards

ATEX Directive 94/9/EC (100A) The Equipment Directive.
ATEX Directive 98/92/EC (137) The User Directive.

British Gas Transco. *Procedures for Hazardous Area Classification of Natural Gas Installations*, Draft 1988. London: Institute of Petroleum.

BS EN 1127-1: 2007 *Explosive Atmospheres. Explosion Prevention and Protection. Basic Concepts and Methodology.*

BS EN 60079-10-1: 2009. *Classification of areas. Explosive gas atmospheres.*

BS EN 60529: 1992 Specification for Degrees Provided by Enclosures (IP Code).

Dangerous Substances and Explosive Atmospheres Regulations 2002. Statutory Instrument 2776 2002, London: The Stationery Office.

Health and Safety at Work Act 1974 Elizabeth II. Chapter 37, London: HMSO.

Regulatory Reform (Fire Safety) Order 2005. Statutory Instrument 1541 2005, London: BSI.

Further reading

BS EN 60079-10-2: 2009. *Classifiaction of areas. Combustible dust atmospheres.*

Building Regulations 2000, Fire Safety, Approved Document B Volumes 1 and 2, 2006 edition. London: NBS, RIBA.

BS 9999: 2008. Code of Practice for Fire Safety in the Design, Management and Use of Buildings. London: Department for Communities and Local Government.

Health and Safety Executive (HSE). *Fire Safety in Construction Work. Guidance for Clients, Designers and those Managing and Carrying Out Construction Work Involving Significant Fire Risks* (HSG 168), 1997, London: HSE Books.

Health and Safety Executive (HSE). Explosive Atmospheres – Classification of Hazardous Areas (Zoning) and Selection of Equipment, undated, London: HSE. Available online at: http://www.hse.gov.uk/fireandexplosion/zoning.pdf

Health and Safety Laboratory Document: Area Classification for Secondary Releases from Low Pressure Natural Gas Systems, MSU/2008/03.

Institute of Petroleum. *Area Classification Code for Installations Handling Flammable Fluids, Part 15 of the IP Model Code of Safe Practice in the Petroleum Industry*, 2005, London: Institute of Petroleum.

Websites

Health and Safety Executive http://www.hse.gov.uk

Health and Safety Executive, Fire and Explosions http://www.hse.gov.uk/fireandexplosion/

Institution of Civil Engineers (ICE), Health and safety http://www.ice.org.uk/knowledge/specialist_health.asp

United States Chemical Safety Board http://csb.gov

Chapter 21

Working on, in, over or near water

David N. Porter Rivers Agency, Belfast, Northern Ireland, UK

doi: 10:10.1680/mohs.40564.0265

CONTENTS

Working on construction sites which have water within their boundaries or are near watercourses or impoundments requires some prior planning to ensure the water does not become an unwanted and uncontrolled hazard. This chapter outlines the hazards which can be quantified before site work commences and suggests ways to deal with these at planning, design and construction stages. Risk assessment, training of operatives and personal protective equipment will be detailed with some examples from best practice. It is also necessary to consider the changing nature of watercourses: the banks, the bed and flow characteristics and the steps that can be taken when it all goes wrong and an emergency situation develops. The text will also discuss the selection, maintenance and use of plant including the specialist operations of pontoons, mats and workboats. Channel works, river diversion, flood banks and the constructions of sustainable urban drainage systems, cofferdams and caissons will be examined in some detail in order to identify the risks particular to these operations. The chapter finishes with a section on the main health risk, leptospirosis, posed to those working in the water environment.

Introduction

Working adjacent to water can provide some of the most beautiful working conditions in construction. However, in this environment it could be easy to overlook the unique set of hazards that is present. With some, though, the risks associated with these hazards can be easily identified and can therefore be designed out, controlled or managed in such a way as to minimise exposure. This approach will result in a safe working environment.

Water itself seems relatively benign and this can lead to a complacency when it is present on a construction site. The following sections will detail some of the risks and provide guidance on mitigation measures that can be put in place to manage the situation. In addition, consideration will also be given to the more hazardous situation of flood emergency and the enhanced risks that are present during such an event.

Planning

As with all construction, a safe site starts well before the compound is established, well before the site is handed over to the contractor, indeed, well before a contract is even let: it starts with planning. First we need to identify the hazards and to do this we must identify the watercourses or potential sources of water and we do this as part of the initial site survey. It is important not just to identify those within the site boundary: you need to look wider and in some cases deeper, as watercourses can be underground, both naturally occurring and in culverted systems. Watercourses can often look small and insufficient in low flow; they can even be dry at the time of survey if there has been very low rainfall. The natural floodplain of the watercourse may also be an issue because dry land can quickly become inundated with floodwaters in the event of heavy rain. It is also important to look at flow paths onto the site from more distant watercourses as water can flow overland, using roads, paths or railway lines as conduits. Help is normally at hand in the form of historic and modelled flood maps from local drainage authorities, Councils, roads authorities or water companies. Requests for information and flood advice should be sent in early as there is sometimes a delay in accessing this information due to the volume of enquiries. Don't forget to gather up sewer information as these are often present and offer up an additional set of challenges.

Risk assessment

Prior to starting on site, the hazards identified during the planning phase should be developed and assessed to determine the likelihood of being exposed to them and their impact upon the works programmed. There are two approaches which when used together ensure that all foreseeable hazards are adequately dealt with. First it is possible to identify generic or common hazards. For these a standard industry or company risk assessment approach can be developed. These generic risk assessments are then complemented by a site-specific risk assessment.

Generic safe systems of work

Most construction hazards can be identified and generic procedures, in the form of safe systems of work, can be drawn up. For example, there could be a well developed procedure for controlling the risks associated with overhead power lines. The risk assessment for this example would provide a step by step work plan and outline the control measures necessary. Goalposts, safe working distances and 'permit to work' documentation could all be detailed and used on each construction site where this hazard is present. This common approach should develop a familiarisation with the safe approach needed when dealing with hazards and this helps embed a safety culture into a construction organisation.

These generic risk assessments should be practical documents written in such a manner to encourage use on site. Illustrations, bullet points and plain language all help in ensuring use by site operatives.

Site-specific risk assessment

This is the single most important document in ensuring site safety when working on a site involving water. A simple method to ensure completion of this form is to draw up a checklist based on generic risk assessments, see **Figure 1** for an example of such a checklist. The items will draw operatives' attention to potential hazards and the agreed method of work. The form should also include space for detailing site-specific hazards which are not covered by the generic risk assessments. If any risks are present that are not included in the generic risk assessment these must be assessed. This assessment should be documented before the work starts and should include the construction method proposed to mitigate the risk.

During construction

Supervision

The risk assessment highlights the hazards and mitigation measures, but the practical outworking is the responsibility of the site staff and compliance is ensured through robust supervision. Check that the situation as foreseen at the planning stage is what is being encountered during construction. Ensure that the safe systems of work are being used and are not being overlooked in the interest of progress. Look out for new hazards that were not identified at the earlier stages of the design.

Training

When working in or near water it is important that trained and experienced operatives are used as the additional hazard posed by the presence of water can lead to a dangerous working environment if not treated with respect. Where operatives are not used to working near water it is important to carry out specific training prior to starting work on a site with this hazard present. In the case of companies or organisations that are constantly or commonly working near, over or in water every piece of site related training should include consideration of the hazards posed and any specific approaches required to mitigate.

It is also good practice to take every opportunity to draw attention to the hazards posed by the presence of water on the site. This can be done at site induction, at routine toolbox talks and at progress meetings. It is good practice to use the same documentation that was completed when identifying the hazards prior to construction during these safety talks. The checklist in **Figure 1** is ideal for this dual purpose and ensures that there is continuity to the health and safety management on the site. If used correctly this approach also makes these talks site specific rather than discussing construction risks in general, which often happens when site staff are asked to complete a toolbox talk without prior preparation. The generic risk assessments should also play a part in this process when the checklist identifies that the risks covered by them are present on the site. The site engineer or supervisor completing the talk should refer to the generic risk assessment to ensure that the predetermined safe system of work is adopted.

Site safety is most effective when it becomes engrained into site life and sometimes the best approaches are the simplest. A verbal reminder when leaving the tea hut about the fast moving water, the presence of a confined space or the undercut river banks may seem simple but could contribute to saving someone from danger, injury or worse.

There are some areas that do warrant bespoke and often certified training in order to minimise the likelihood of harm when working near water. These include confined space training (see Chapter 16 *Confined spaces*), use of personal buoyancy equipment and boat handling. In addition plant operators need to be familiar with the particular hazards present on riverside construction sites and understand that a more precautionary approach may be necessary than that used on a typical flat building site.

Personal protective equipment

Personal protective equipment (PPE) used when working near water is similar to that on every construction site: hard hats when there is a risk of falling material or when near plant, and high visibility jackets when moving plant or vehicles are present. Additional PPE that should be considered includes buoyancy aids, ropes and safety harnesses and rescue equipment. The level of use of each of these elements should be determined during the risk assessment by taking into consideration the site and watercourse conditions. Typically, experienced staff working adjacent to shallow, slow moving water do not need any additional protection. Conversely, fast moving, deep water or on sites with steep, slippery river banks where there is a higher risk of falling into the water, buoyancy aids should be worn. In more extreme cases operatives should wear harnesses secured to an anchor point or controlled by a banks man. At high risk sites, where work is to continue over a prolonged period of time, consideration should be given to formalised rescue arrangements including regularly inspected life belt installations, life lines and, if appropriate, boats. As with all health and safety, the mitigation must be proportionate to the risk and **Table 1** may help in determining the approach to be adopted with various levels of experience. It should be noted that it is not possible to be too prescriptive as there is huge variance in approach depending on the watercourse affecting your site; the Thames or Severn Estuary will require a different approach to a 2 m wide rural drain.

The selection of personal buoyancy equipment is dependent on the frequency and duration of use, work activity to be

Risk Assessment for DLO Works

Sheet _____ of _____

RIVERS Agency

Name of assessor _____

Date of assessment _____/ _____/ _____

Location of the works _____

_____

Description of the works _____

_____

Significant Generic Hazards	Reference in pocket book	Risk Rating H/M/L	Site Specific control measures	Residual Risk rating H/M/L
Manual Handling				
Plant / machinery				
Trips / falls / slips				
Hazardous substances				
Excavation / trenches				
O/H U/G electric cables				
Hired / plant / equipment				

Site Specific Hazards	Reference in pocket book	Risk Rating H/M/L	Site Specific control measures	Residual Risk rating H/M/L

Unforeseen hazards	Reference in pocket book	Risk Rating H/M/L	Site Specific control measures	Residual Risk rating H/M/L

If you are using equipment / machinery has the operator received the appropriate H&S training ?	Y / N
In your opinion do you consider the control measures above sufficient to allow the works to proceed ?	Y / N

Signed _____

Date _____ / _____ / _____

Noted _____

Date _____ / _____ / _____

Site staff present during toolbox talk on ____/____/____

Figure 1 Example of a risk assessment checklist
Courtesy of Rivers Agency, Health and Safety section, Belfast, reproduced with permission

	Experienced staff who routinely work near water	Staff with some previous exposure to working near water	Inexperienced staff
Shallow or slow moving clear water, narrow watercourse channel (typically less than 2 m); no debris, stable riverbed conditions	PPE as used on typical construction sites with no necessity for buoyancy aids unless risk assessment highlights the likelihood of a sudden change in conditions	Fixed land based lifebelt or personal buoyancy aids if considered necessary by risk assessment	Personal buoyancy aids, consider harness, lifelines and banksman until experience has developed
Deep or fast moving water; tidal waters; discoloured or dirty water; debris in channel; slippery or steep river banks; unstable or unknown river or seabed conditions	PPE as used on typical construction sites with the addition of personal buoyancy aids	Avoid entry unless necessary. Compulsory use of personal buoyancy aids. Lifelines and banksmen should be considered and used when available	Can be used during on the job training with very close supervision
Extreme or flood conditions; very deep water; storm conditions in tidal waters	Avoid entry unless absolutely necessary. Compulsory wearing of personal buoyancy aid. Lifelines and banksman when available. No lone working when entering flood waters	As for experienced staff but note this type of staff can be used to assist experienced staff in these conditions but should not be in lead positions	Inexperienced staff should not be used

Table 1   Indicating suitability of staff in various river and coastal situations

undertaken, proximity of assistance and the weight and size of the individual. A final consideration is the likely state of the individual during the operation of the device. Will they be capable of helping themselves? Will they be conscious? What about fatigue? What's the likelihood they will be injured?

The simplest form of buoyancy aid is somewhat bulky as it is made from permanently buoyant material but it is cost effective, robust and requires little or no maintenance. These are best suited to relatively short duration jobs as mobility is sometimes impaired and with prolonged use they can become uncomfortable. The less intrusive and more comfortable options are the inflatable lifejackets. These come in two forms: manual and automatic inflation. Manual are really only suitable where entry into water is foreseeable with sufficient time to allow full inflation. Automatic lifejackets inflate on contact with water but then require maintenance to reset the mechanism before the next use.

Where regular and extensive use is envisaged it is best to allocate each worker with their own buoyancy equipment and train them in pre-use inspection, operation and defect detection. This is particularly relevant for those using automatic inflating lifejackets as these require careful storage, handling and use. Check with the manufacturer about service intervals for self inflating equipment and note that this can normally only be carried out by approved service centres. This work may also need to be certified as evidence of maintenance to satisfy local regulations and possibly insurance company stipulations.

Changing situation

River flows

The depth and velocity of flow is never constant in a natural river system, due to the conditions that exist. Where the flow regime would influence the onsite construction it is important to ensure consideration of this potential for change during the design stage. For example, when constructing any costal works

it is necessary to consider the tidal range, low and high tide levels and meteorological effects and their impact on the site.

In river systems, establish if the water level is natural or is it manually controlled using sluices, weirs or barrages. If natural, find out if the watercourse is subject to quick changes to the flow level and/or velocity; this is known as a 'flashy catchment'. Where the water level is manually controlled or managed, it is important to establish the flow characteristics. This can generally be determined through consultation with the local drainage or water authority.

This information can in turn dictate the timetable, method or sequence of construction activities. Additional flows or discharges into the system should also be considered. These are generally not a problem on large watercourses as the volume of discharge compared with the main flow tends to be relatively small. On smaller watercourses or culverted systems, additional discharges can dramatically change the conditions and this needs to be known and taken into account at design stage. These discharges could be from water treatment plants, industry or even water level control structures.

Bed stability

Another feature of natural river systems that can change and present a health and safety risk is the bed condition. Next time you are on a sandy beach, stand in the water when the tide is going out. You will see and feel the removal of the sand from around your feet by the action of the moving water. This will cause you to sink into the soft waterlogged sand below.

A similar process occurs when you place a machine, material or temporary works into moving water. This removal of material takes place at different rates depending on the speed of flow of the water, the channel shape, hydraulic gradient of the reach and the properties of the sediment making up the bed, such as grain size, material and cohesiveness. To complicate the issue further, deposition of material from upstream can also take place. When undertaking short term works such a river maintenance, it is generally enough to be aware of this process and to keep a watching brief on the stability of the river bed and resultant health and safety issues. For longer term projects or when constructing sizable structures in the channel, it may be necessary to consider the river morphology. A morphological study will consider the potential change and impact on the river corridor and can highlight immediate and longer term issues which may affect the river bed and the surrounding land. Importantly these studies also look at environmental impacts and information gathered can help in ensuring compliance with environmental legislation.

The action of the bed has a knock-on effect on bank stability and this should be monitored during the works, particularly in granular soils. This needs to be monitored even if your works, plant or materials are not in the river as the bed can shift through natural process, influencing the bank profile and stability. Undercut banks are a common hazard and are easily mitigated, provided they are identified.

Flood emergency response

Scale of risk

During emergency flood incidents the scale of risk increases dramatically as the hazards present change as the situation develops. Floodwaters can be fast moving, deep and dirty. This combination can mask the underlying hazards, particularly potential trips, excavations and missing access chamber covers. In these situations a precautionary approach is necessary using an ongoing dynamic risk assessment approach. It is generally unfeasible for full documented risk assessments to be completed due to the changing nature of the situation so it is necessary to rely on well trained operatives using standard procedures and approaches. 'If in doubt – stay out' is good advice and if the situation deteriorates call on the fire and rescue professions for help.

Personal buoyancy aids should always be worn when responding to flood emergency situations. Staff should be advised to avoid entry into deep or fast moving water unless absolutely necessary. Where entry is unavoidable, they should move forward slowly scoping out the ground using a stick, shovel or other stiff implement to ensure that it is safe to proceed. When possible a harness, lifeline and banksman should be used. During the hours of darkness adequate lighting using vehicles, hand-held torches or flood lights should be provided and utilised. Procedures should be in place to reduce the risk or eliminate the possibility of lone working during flood situations and when entry into floodwater is anticipated. Protect from the effects of hypothermia or exhaustion by ensuring clothing does not get overly wet by wearing good quality water resistant PPE, boots, leg or chest waders; take adequate rest and food breaks and keep warm. It is good practice for those routinely involved in flood emergency duties to have spare PPE for use during protracted events and adequate cleaning and drying facilities must be provided. Hot water showers are also useful as they remove any contaminants that may have been present in the floodwaters and provide a method of quickly warning wet and cold personnel before they go home or in readiness for redeployment to assist with the emergency.

Plant and equipment

Selection and maintenance

Functionality and suitability need to be considered when selecting plant and equipment for working near water. Decisions based purely on availability generally lead to unacceptable compromises, which in turn can present unnecessary hazards. River banks and sites constrained by watercourses can often be narrow with steep slopes. These constraints minimise the space for manoeuvre, leaving little margin for error. To mitigate this risk it is good practice to define haul routes, storage areas and turning circles. Where it is necessary to place any of these features near the river bank, clearly mark the route of safe passage and if necessary place physical barriers to enforce safety zones and prohibited areas. Trained, competent and experienced plant operators are a necessity on challenging sites. Less experienced operators or those undergoing job training should not be allowed to operate in these circumstances without very close supervision.

Window guards

Window guards are typically fabricated from folded aluminium chequer plate, although recently plastic guards have been introduced. Guards are mounted to various pieces of plant and equipment to deter vandalism to the windows and to prevent forced entry to the controls. Whilst these offer excellent protection they must be removed when the machine is in use, particularly when working on, in, over or near water. This is absolutely imperative as in the event of the machine overturning they could prevent safe escape. Many experienced plant operators will also ensure that they have a hammer in the cab to assist with the removal of glass windows in the event of overturning.

Biodegradable oil

To minimise any potential environmental impact in the event of a hose or ram failure, when using mechanised plant in the vicinity of water it is good practice to ensure that the hydraulic oil is biodegradable. Modern biodegradable oils are available in a wide range of grades and offer excellent performance characteristics at a relatively modest cost increase. The use of this oil should not be restricted to construction plant as it can be included in all waterside equipment such as barrage winches, lock gates or fixed location generators. Don't forget your Control of Substances Hazardous to Health (COSHH) responsibilities by obtaining the material safety data sheet and completing a COSHH assessment for this oil.

Impact of sea water

All water will impact on plant and machinery when in use but sea water is very corrosive. Bearings, caterpillar tracks, and hydraulic rams are all at particular risk. To minimise this impact try to plan the work in such a way to prevent the machine having to enter the water unnecessarily and ensure that it is well washed down after use. Even this approach will not fully protect the machinery as spray and salt laden air cannot be avoided so some contractors keep older machines for such situations rather than exposing expensive new equipment to this harsh environment. If this approach is used it is still important to pay equal attention to the maintenance and service regimes of these older machines as to the shiny new machines.

Use of pontoons

Pontoons are extremely useful for accessing work areas in relatively still water, but extreme caution is required to ensure a safe working environment. Their use can speed up an activity by minimising the amount of temporary causeway construction which in turn can reduce the environmental impact of the works. Machines must be carefully secured to the pontoons and in turn the entire floating platform must be tethered in such a way to provide stability and manoeuvrability. Where possible the addition of a ground support should be considered, either in the form of an anchor or adjustable pile. This will help counteract the listing of the pontoons due to the movement of the machinery during a lifting or digging operation.

Use of mats

Mats are typically large wooden planks connected using angle iron straps with a steel rope lifting loop. They are used to distribute the weight of an excavator when working on very poor ground conditions. These mats are typically used in a set of three. When the machine is sitting on two mats the third can be lifted and moved ahead to allow safe passage over boggy ground. Care should be taken to ensure the machine does not slip off the mats as this will almost certainly result in it sinking into the ground and in extreme cases could result in the loss of the equipment.

Boats

This is a specialist area and this paragraph is for guidance only as a detailed interpretation of the relevant regulations will be necessary to ensure compliance. In the UK, workboats should comply with the Merchant Shipping (Small Workboats and Pilot Boats) Regulations 1998 and the associated *Safety of Small Workboats and Pilot Boats – a Code of Practice* (Marine Safety Agency, 1998). Essentially workboats must be manufactured to a suitable standard and must have 'Certificate of Compliance' issued under the code and they must be fit for purpose. Pleasure or recreational craft are not suitable as workboats unless they comply with the regulations and are certified. These codes also set out the survey regime required to ensure continued compliance.

Watercourse works

Maintenance

Open watercourse maintenance typically takes two forms: hand work and machine work. Maintenance carried out manually or by hand is generally in urban situations where access is limited. In these situations care is needed to ensure the individuals involved are not exposed to unnecessary risk from either what's in the watercourse or on the bank. Common hazards include: sewerage; oil or chemical discharge; rats or other vermin; trips and falls; manual handling and damage from hand tools, either manual or mechanised. The particular exposure to each of these risks is dependent on the site in question and should be assessed prior to starting any works, with the appropriate controls detailed in a method statement or operation safety control sheet.

In the case of maintenance carried out by machine there are an additional number of hazards, primarily: overhead or underground services; poor ground conditions; and the hazard the machine itself presents to the workforce around it. It is good practice to identify all services before site entry and where possible arrange for disconnection or realignment. As a minimum, goalposts with visible warning signage are advisable and are generally sufficient for short duration works; see HSE Guidance Note GS6 (HSE, 1997) for detailed advice. Ground conditions can also be pre-assessed but it should be remembered that during construction these may change depending on the action of the water and the type of material under foot.

When machine use is envisaged the site should be designated as a 'hard hat site' and enforced by careful supervision. The sequencing of the works is critical to minimise any conflict between workers and machine. In some cases, for example cofferdam construction using sheet piles, it is not possible to ensure separation from the plant and additional care is needed. Clear lines of communication will help make the situation safer and ensure it does not get out of control. One person should have overall control and only that person should be directing the machine operator. This person ideally should not be trying to manhandle a sheet pile into position whilst giving direction; they should be in a position to have an overview of the whole situation.

Diversion

Where a watercourse is to be diverted either as temporary works or as part of the final construction, it is necessary to consider the hydraulic capacity as well as the structural performance of the proposal. This is relatively straightforward by identifying the catchment boundary, the likely run-off characteristics and therefore the peak volume. In many areas assistance will be available from the local drainage authority. Consent or approval may also be required and in some cases planning approval. Sizing of the channel or culvert is very important even for temporary works as the volumes of water involved may be considerable and in the event of heavy rain your actions may flood the site or the wider area causing damage and presenting a danger to life and property.

Where diversion channels have been constructed you will also need to provide access over, across or through them for both plant and people. The scale of provision should be in proportion with the watercourse taking into consideration duration and frequency of use. Where public assess is to be accommodated properly constructed accesses and crossings are a must and may even be as elaborate and costly as the finished bridge or crossing. Transporting people in digger buckets or requiring them to walk the plank is not good enough and should not be tolerated.

In-channel construction

This is potentially the most hazardous operation as all the hazards outlined in this chapter can be experienced. In line with good health and safety practice, risks should be avoided where possible, so consider the construction method to see if it will avoid the risks. It may be that a different machine, such as a 'long reach' excavator, could transform the approach and therefore the exposure. It may also be possible to de-water the area by impounding the water and pumping, diverting into an alternative system, putting in place an alternative drainage system or by constructing a cofferdam. In many cases the risks cannot be avoided and they must be managed and the risk assessment process and safe systems of work must be adhered to.

It is also worth considering alternative work methods. For example, the advances in prefabricated or precast units can be used to reduce the on-site timetable and require less labour intensive work in high risk areas. There are precast retaining wall sections, arched bridges and even culvert entry and headwall units available for quick installation on site.

Cofferdams and caissons

Cofferdams and caissons are used to dewater a limited area of a site to enable construction to take place. Cofferdams are generally large open top structures, typically constructed using steel sheet piles, wales and cross-bracing. When the outside wall of the cofferdam is watertight, a pump is used to remove the water which allows construction to take place. The cofferdam is normally removed after construction. Despite the fact that cofferdams are temporary works, it is imperative that they are designed and constructed to ensure that they are safe during operation. Also think about the method of access and egress for personnel in both normal and emergency situations. Deep cofferdams may require air circulation or extraction systems when plant and machinery is being used as part of the construction process. A formal recorded inspection regime of the structural elements of a cofferdam should be put in place.

Caissons differ from cofferdams in that they often form part of the final construction. In addition they may also be designed to be totally submerged, with the working space dewatered using compressed air rather than pumping. Closed caissons are an obvious confined space and the appropriate procedures must be followed. Caissons involving compressed air offer an additional hazard due to the pressurised environment which can lead to 'caisson disease'. This is similar to 'decompression sickness' or 'the bends' which can afflict divers who rapidly return to atmospheric pressure.

Sustainable drainage systems (SUDS)

SUDS solutions are becoming more popular as designers try to mitigate the environmental impact of drainage discharge on our watercourses. This approach has its own set of hazards which need consideration. The biggest hazard presented by these drainage systems is the attraction of standing water in ponds and watercourses to children and youths and the associated drowning hazard. In highly urbanised situations the best solution may be to avoid the risk and install a traditional pipe based drainage system but when the environmental impact must be mitigated using a SUDS systems or where discharge is not possible without attenuation, access to the ponds must be carefully considered.

During new build construction the design should, as far as reasonably practicable, ensure that the system can be built dry and then made operational by the introduction of water. When retro-fitting SUDS this is not always possible and it is therefore necessary to ensure that the existing drainage system can remain operational until the new swales and ponds are constructed or design an alternative construction phase drain.

Medium and long term maintenance of the system also need careful consideration. With time, ponds silt up and clean stone in drains gets blocked with fine sediment. It is important that any maintenance does not damage or remove the impermeable liner (either clay or geo-textile) that formed the pond. From a health and safety point of view it may also be necessary to test the chemical properties of the material being removed in order to determine a safe method of removal and the disposal procedure. SUDS systems that have received run off from roads will almost certainly contain traces of rubber, oils and heavy metals. If present this material will be controlled waste and needs to be treated in line with local regulation. Inert material can normally be spread on adjacent ground but again consultation with local environmental regulations is advised.

Flood banks and flood walls

These elements of flood defence infrastructure are key when managing flood risk but it is very important to also understand their limitations. Flood banks come in two forms: clay or earth banks and solid core construction.

New earth flood banks should be constructed with a puddle clay core. This is impermeable as it is a fine, firm clay in accordance with BS EN ISO 14688. Be very careful when working with older earth or soft flood banks as the material used often originated as scrapings from the river bed, excavated material from nearby sites or substandard clays. This presents two risks on sites. First, when carrying out modifications to the flood banks or when constructing adjacent to them it is possible to impair the integrity of the existing structure and hence reduce the protection offered. Second, when in flood conditions with water being retained by the earth flood bank, it is almost impossible to determine, by visual inspection, how the bank is performing and the level of reserve capacity. Catastrophic collapse occurs often without much warning. Given the nature of the material this can result in a progressive failure with larger and larger amounts of material removal due to water passing through the breach. In flood emergency situations, earth flood banks should really only be inspected from higher ground and from a distance. The breach of the levees and the subsequent inundation of New Orleans is a dramatic example of the impact of water due to flood bank failure. This example and the associated cost in terms of human life reinforce the need for systematic inspection and maintenance of soft defences which protect dwellings and other critical infrastructure.

A solid core bank is a development of the earth bank and is an attempt to reduce the risk of collapse by the provision of a steel sheet pile core. This core stops rodent infestation which is a major cause of soft defence failure. It also adds to the structural rigidity and stability of the overall bank in flood conditions. The down side of this type of construction is that it is almost impossible to inspect the core and so material deterioration goes undetected.

From a health and safety point of view it is important to determine what type of flood bank you are dealing with, particularly when carrying out works to the structure. This knowledge will also be useful when working behind such structures as it will offer an indication of the protection offered.

The next step up in protection is flood walls. These can be constructed from most commonly available building materials with choice being determined by the level of protection required. Again, it is important to determine the limitations, as each wall is designed to protect against a certain level of flood. During the design process, the provision of some additional height known as freeboard helps to deal with the inherent inaccuracy in flood prediction techniques and the impacts of wave effect, whilst increasing the level of protection. If a flood event larger than the design flood occurs or a downstream blockage reduces channel capacity then the wall could be overtopped.

Think about your site in these conditions: where will the water go? At what point do you need to consider evacuating people, plant and machinery? Also think about the likelihood of catastrophic collapse of the wall during flood loading and ensure that the site design has considered this condition.

Tidal issues

When carrying out your works on sites affected by tidal waters or when constructing coastal defences it is necessary to consider the impact of the tide. The normal cycle of the tide is predicable and can therefore be mitigated at the design stage. The required information is contained in tide tables and is invaluable when planning, designing and constructing works subject to tidal waters.

Tide Tables

Predicted tide heights for standard and secondary ports around the world are published annually in four volumes of *Tide Tables*. These publications provide information about the timing and level of low and high tides. It is critically important to check the data as this information is generally plotted to a chart data rather than ordnance data and so a conversion is necessary, but ensure that the correct factor is used as it varies by port. Also be aware that the tables do not take into account any daylight saving time, such as that in operation within the UK, unless it is in operation throughout the year.

The conversion from chart to ordnance data is a very straightforward operation and is a matter of a simple addition or subtraction, as the following example demonstrates:

If we take Stranraer in Scotland as an example, the conversion from chart to ordnance datum is −1.40 m. The tide table for 25 July 2010 shows a predicted high tide will occur at 11:31 GMT at a height of 2.8 m. This means that to ordnance data (Newlyn) it will measure 1.4 m (2.8 m−1.4 m) and take place at 12:31 British Summer Time. (*Admiralty Tide Tables*, 2010)

Also remember that the levels given do not take into account meteorological effects such as storm surges or wind nor do they provide any information on tidal streams or currents. The information from tide tables is of great assistance when programming coastal works or working in tidal rivers because you can plan for the impact of the tide and therefore minimise any health and safety risks.

Health risks – waterborne infections

Water can contain all manner of substances harmful to the human body. Thankfully most are in a much diluted state and therefore don't present an immediate risk. Watch out for sewerage and urban drainage outfalls as concentrations are greater at these points. The greatest risk to people working near water is leptospirosis which is an infectious condition that typically manifests itself as mild flu-like symptoms. It is caused when infected urine containing the bacteria leptospires from animals such as rats, pigs, cattle, dogs and foxes contaminate a water source. This in turn enters the human body through the eyes, mouth, nose or wounds to the skin. The most serious form of this disease is known as Weil's disease, which can cause kidney failure, internal bleeding and may result in death. People working in conditions likely to be exposed to leptospirosis should advise they doctor so that in the event of an illness the symptoms can be quickly diagnosed. It is also good practice for those working in such conditions to carry a card alerting other medical staff of their working conditions and the need to consider the presence of this condition (HSE, 1999; NHS, 2009).

Conclusion and summary of main points

Water on or near a construction site can pose a hazard to the operations and personnel but with some forward planning and with the assistance of those with responsibility for managing the water environment this can be mitigated. The single most important learning point is to plan for all foreseeable situations by not assuming that the conditions will remain as they are on a calm, dry day. Throughout the planning process it is wise to take a precautionary approach as the forces involved during a flood event or a tidal storm can suddenly increase and can quickly become uncontrollable. The choice of construction materials, method and plant should be influenced by the site conditions and identified using a simple but comprehensive risk assessment. The information identified within this assessment needs to be effectively communicated to the site staff and monitored to ensure compliance. The hazards particular to this environment need to be treated with respect to ensure that they do not cause damage to plant and machinery, loss of productive time and most importantly long term impact on the health and well-being of our site staff. With all this caution, it is still good to remember that water either in ponds, watercourses or the sea provides a most beautiful backdrop for engineers to carry out their work.

References

Admiralty Tide Tables 2010, United Kingdom and Ireland Vol. 1 NP201-10, 2010, Taunton: United Kingdom Hydrographic Office.

Health and Safety Executive (HSE). *Avoidance of Danger from Overhead Electric Powerlines. Guidance Note GS 6*, 1997, London: HSE Books. Available online at: http://www.hse.gov.uk/pubns/priced/gs6.pdf

Health and Safety Executive (HSE). *Leptospirosis: Are you at risk*? Guidance Leaflet INDG84, 1999, London: HSE Books.

Marine Safety Agency. *The Safety of Small Workboats and Pilot Boats – a Code of Practice*, 1998, London: The Stationery Office.

National Health Service (NHS). *NHS Choices: Leptospirosis*, 2009. Available online at: http://www.nhs.uk/conditions/Leptospirosis/Pages/Introduction.aspx

Referenced legislation and standards

BS EN ISO 14688-1: 2002. Geotechnical Investigation and Testing. Identification and Classification of Soil. Identification and Description.

BS EN ISO 14688-2: 2004. Geotechnical Investigation and Testing. Identification and Classification of Soil. Principles for a Classification.

Control of Substances Hazardous to Health Regulations 2002. Reprinted April 2004 and March 2007. Statutory instruments 2677 2002, London: HMSO.

Merchant Shipping (Small Workboats and Pilot Boats) Regulations 1998. Statutory Instrument 1998 1609, London: The Stationery Office.

Further reading

Health and Safety Executive (HSE). *Agricultural Sheet No. 1 Personal Buoyancy Equipment on Inland and Inshore Waters*, 2005, London: The Stationery Office.

Websites

Health and Safety Executive (HSE), Control of Substances Hazardous to Health (COSHH) http://www.hse.gov.uk/coshh/

Institution of Civil Engineers (ICE), Health and safety http://www.ice.org.uk/knowledge/specialist_health.asp

Index

Page numbers in *italics* refer to illustrations separate from the corresponding text.

2012 targets, 35–37, 112–113

abandonment *see* decommissioning
absolute certainty, 45
'absolute' legal tests, 57
absorption of chemical hazards, 122, 124
acceptable risk, 56–58
acceptance criteria, 254–255, 256
acceptors, 77
access, 183
accidents, 52–53, 77–78, 79
accreditation, 87–88
ACMs *see* asbestos
ACoP *see* Approved Codes of Practice
administrative biological hazard controls, 141
adrenalin, 198–199
Advisory Roofing Committee (ARC), 173
ageing workers, 99–100
air pollutants, 156–157
ALARP *see* as low as reasonably possible
    risk reductions
alcohols, 143
alloy staircases, 170
amber category items, 57, 58
amosite asbestos forms, 127
amphibole minerals, 127
anthophylite asbestos forms, 127
anthrax, 137, 138
anthropometry, 158
appeals, 9
Approved Codes of Practice (ACoP)
    biological hazards, 135
    key Duty Holders, 29–30
    legal principles, 3, 5–6, 7–8
    procurement strategies, 114, 115–116
    project life cycles, 58–59
    safety issues assessment, 103–104
ARC (Advisory Roofing Committee), 173
Architects, 101, 102
architecture, 42, 101, 102
area classifications, 256–257
arthropods, 137
asbestos, 17, 22–23
    diseases, 127
    exposure control, 121, 122, 127
    forms, 127

occupational health, 93–95
    project life cycles, 58, 59–60
as low as reasonably possible (ALARP) risk
    reductions, 3, 5, 7, 13
asphalt, 127–128
asphyxiation, 190–191
asset information, 63–64
asset owners/occupiers, 54
Associate Institute of Demolition Engineers
    Grades, 230
asthma, 98
atmospheric monitoring, 196
auditory signal communications, 197
authorised persons, 194–196
autoclave, 143
auxiliary structures, 224, 228
    see also falsework
'avoid the hazard' health and safety steps, 52

back disorders, 23, 96
backfilling, 182
banks for water, 269–270
baseline environmental impact assessment
    studies, 145
battering control measures, 184
bed stability, 266–267
behavioural approaches, 166
benzene, 122, 128
Bilbao Declaration, 39–40, 82, 112, 114
biodegradable oil, 267
biological hazards, 23
    assessment/monitoring, 144–145
    biosafety levels, 139–140
    biowaste management, 144
    case studies, 146–147
    controlling exposure, 135–148
    disinfectants, 143
    entry routes, 139, *140*
    field kits, 142
    Good Working Practices, 142–143
    preventive measures, 140–141
    recognition, 139
    sources, 136–139
    special systems, 142

spillage management, 143–144
    stakeholder responsibilities/roles, 145–146
biosafety levels (BSL), 139–140, 147
bioterrorism hazards, 137–139
biowaste management, 144
Bird loss control models, 60
bitumen, 127–128
bloodstream hazard routes, 123–124, 146
the Board, 82–84
boats, 268
body temperature, 190
BOOT (build own operate and transfer), 116
Bragg Committee, 206
Bramall, David R., 233–247
breaches of contract, 12
breaches of duty of care, 11
breathing apparatus, 191–194, 198–200, 201
bricklayers, 157
Britain, 35–37, 40, 44, 46, 176
bronchitis, 156
Brunswick Centre, 176
BSL *see* biosafety levels
budgets, 41, 63, 91
buildability key engineering principles,
    106–107
Buildhealth NI, 46
build own operate and transfer (BOOT), 116
BuildSafe NI, 45–46
buried utility services, 182
business continuity, 85

CABA *see* compressed airline breathing
    apparatus
CABE *see* Commission for Architecture and
    the Built Environment
caissons, 269
cancer, 23, 122, 124–125, 154, 221
carbon composite cylinders, 199, 201
carbon dioxide (CO_2), 123, 198
carcinogens, 23, 122, 124, 221
carcinomas, 156
cardio-pulmonary resuscitation (CPR), 192
carpal tunnel syndrome (CTS), 20, 22, 154
Carpenter, John, 205–213

ICE manual of health and safety in construction © 2010 Institution of Civil Engineers